Cyclophanes

Volume I

This is Volume 45 of
ORGANIC CHEMISTRY
A series of monographs
Editor: HARRY H. WASSERMAN

A complete list of the books in this series appears at the end of the volume.

Cyclophanes

Edited by

PHILIP M. KEEHN
Department of Chemistry
Brandeis University
Waltham, Massachusetts

STUART M. ROSENFELD
Department of Chemistry
Smith College
Northampton, Massachusetts

I

1983

ACADEMIC PRESS
A Subsidiary of Harcourt Brace Jovanovich, Publishers
New York London
Paris San Diego San Francisco São Paulo Sydney Tokyo Toronto

ACADEMIC PRESS, INC.
111 Fifth Avenue, New York, New York 10003

United Kingdom Edition published by
ACADEMIC PRESS, INC. (LONDON) LTD.
24/28 Oval Road, London NW1 7DX

Library of Congress Cataloging in Publication Data

Main entry under title:

Cyclophanes.

(Organic chemistry, a series of monographs ; v. 45)
Includes index.
1. Cyclophanes. I. Keehn, Philip M. II. Rosenfeld, Stuart M. III. Series.
QD400.C93 1983 547'.59 82-25307
ISBN 0-12-403001-7 (v. 1)

PRINTED IN THE UNITED STATES OF AMERICA

83 84 85 86 9 8 7 6 5 4 3 2 1

Contents

Contributors to Volumes I and II

Numbers in parentheses indicate the pages on which the authors' contributions begin.

MASSIMO D. BEZOARI (359), Dow Chemical U.S.A., Plaquemine, Louisiana 70816.

K. ANN CHOE* (311), Department of Chemistry, Smith College, Northampton, Massachusetts 01063

DONALD J. CRAM (1), Department of Chemistry, University of California, Los Angeles, Los Angeles, California 90024

YUTAKA FUJISE (485), Department of Chemistry, Faculty of Science, Tohoku University, Sendai 980, Japan

YOSHIMASA FUKAZAWA (485), Department of Chemistry, Faculty of Science, Tohoku University, Sendai 980, Japan

HENNING HOPF (521), Technische Universität Braunschweig, Institute of Organic Chemistry, D-3300 Braunschweig, Federal Republic of Germany

SHÔ ITÔ (485), Department of Chemistry, Faculty of Science, Tohoku University, Sendai 980, Japan

PHILIP M. KEEHN (69), Department of Chemistry, Brandeis University, Waltham, Massachusetts 02254

KENJI KOGA (629), Faculty of Pharmaceutical Sciences, University of Tokyo, Tokyo 113, Japan

JOEL F. LIEBMAN (23), Department of Chemistry, University of Maryland Baltimore County, Catonsville, Maryland 21228

SOICHI MISUMI (573), The Institute of Scientific and Industrial Research, Osaka University, Osaka 567, Japan

*A portion of the work was done at the Department of Chemistry, Wellesley College, Wellesley, Massachusetts 02181.

REGINALD H. MITCHELL (239), Department of Chemistry, University of Victoria, Victoria, British Columbia V8W 2Y2, Canada

KAZUNORI ODASHIMA (629), Faculty of Pharmaceutical Sciences, University of Tokyo, Tokyo 113, Japan

WILLIAM W. PAUDLER (359), Department of Chemistry, Portland State University, Portland, Oregon 97207

JAMES A. REISS (443), Department of Organic Chemistry, La Trobe University, Bundoora, Victoria, 3083 Australia

STUART M. ROSENFELD (311), Department of Chemistry, Smith College, Northampton, Massachusetts 01063

IAN SUTHERLAND (679), Department of Organic Chemistry, University of Liverpool, Liverpool L69 3BX, England

Preface

These volumes are intended to provide a comprehensive review of the field of cyclophane chemistry for the period between the earlier volume in this series (*Bridged Aromatic Compounds* by B. H. Smith, 1964) and the present (generally through 1981). In reconciling the enormous growth in this field with our desire to produce a work of manageable size, certain topics have necessarily received less attention than is desirable. Within this limitation we have tried to provide a selection of topics that delineate the past and present of cyclophane chemistry and (it is hoped) point toward some of its future directions.

Bridged Aromatic Compounds provided an excellent evaluation of the field in an "adolescent" stage. Synthetic methods in general have grown in sophistication and power, making the number of potentially available cyclophanes much larger. At the time of writing of the book by Smith, a principle rationale for work in this area was to extend the understanding of molecular systems incorporating unusual structural features (e.g., strain and aryl ring proximity). These earlier results often proved useful in developing, confirming, and refining the theoretical underpinnings of the science. Cyclophanes have provided insight into the ways in which molecules distribute strain, the effects of strain on reactivity, transannular effects on chemistry and spectroscopy, and the criteria for aromatic stabilization. In the present "mature" stage of cyclophane chemistry, we are better able to intelligently define the scope of available compounds and to assess the structural and conformational accommodations made by these systems. This is a propitious time to step back and examine our progress.

Although in a field this large we cannot claim exhaustive coverage, we have tried to touch the areas of major interest by assembling a diverse (both chemically and geographically) group of authors with the shared purpose of critically reviewing our progress at this time of vibrant maturity. The ordering of chapters has been chosen to present background, theory, structure, and spectroscopy followed by a somewhat arbitrary division of cyclophanes into subgroups, roughly in order of increasing structural complexity. We are especially pleased that Professor Donald Cram's introductory description of the development of his thinking in this area reminds us of the human element in scientific research and of the joys and frustrations that we all share. The concluding two chapters point toward two of many future directions, underscoring the view that this field remains a vital and evolving area of chemistry.

Individual chapters have been written so that they may be read with little or no direct reference to other chapters. Each stands alone as a review of a particular area of cyclophane chemistry and therefore some overlap between chapters will be apparent. It has been left to each author to design an organizational approach most suited to the particular material being covered. It is hoped that the advantages of autonomy in this regard will outweigh any compromise in readability of the work as a whole. In any case, shortcomings in this area are the responsibility of the editors and not of the individual authors.

Philip M. Keehn
Stuart M. Rosenfeld

Contents of Volume II

CHAPTER 1

Cyclophanes: A Personal Account*

DONALD J. CRAM

Department of Chemistry
University of California, Los Angeles
Los Angeles, California

I. INTRODUCTION

Evolution as applied to organic chemistry can be defined informally as "one thing leading to another." A more sophisticated definition of evolu-

* We gratefully acknowledge grants from the National Science Foundation that made much of this work possible.

ISBN 0-12-403001-7

tion might be "a process of integrating scientific results against time and sieving the products against the screens of survival value." Whatever the definition, the literature of organic chemistry facilitates the identification of patterns of evolution. This chapter traces some of the resulting themes as applied to cyclophane chemistry.

Because research has always been a warm and personal matter to the author and because his first paper in the field[1] is approaching early middle age, this account occasionally is anecdotal. An attempt is made to illustrate how the tools at a given period affect the choice of research problems, how greatly this research depends on the state of evolution of the field, and how stimulating are some of the notions that ultimately turn out to be little more than notions.

II. GENESIS OF THE CYCLOPHANE CONCEPT

The evolution of our interest in cyclophanes commenced in 1945 and 1946 with the reading of Michael Dewar's ideas about the possible existence of π complexes as intermediate species in reactions such as the benzidine rearrangement.[2] At that time this rearrangement was thought to involve monoprotonated benzidine as the starting state. The resulting π complex was presumed to be a molecular sandwich held together by attraction between face-to-face, partially anionic, partially cationic π systems whose rotations and decompositions to the covalent states explained the semidine and benzidine products.[2]

In 1945, as a graduate student (without prospects), the author noted in his idea book that compounds should be studied in which two benzene rings are held face-to-face by methylene bridges substituted in their para positions. The initial idea was that π complexes could be stabilized and their properties studied by incorporating their two parts into a stable ring system. Accordingly, in the fall of 1948 (prospects had improved), Howard Steinberg, the author's first graduate student, undertook the first cyclophane thesis project. His initial objective was the synthesis of [2.2]-, [2.3]-, and [2.4]paracyclophanes (**1** : $m = n = 2$, **1** : $m = 2$, $n = 3$, and **1** : $m = 2$, $n = 4$, respectively), a task completed in 1950. By then, Hammond and Shine[3] had cast doubt on the π-complex intermediacy in the benzidine rearrangement. They observed that the cleaving species was diprotonated. More recently, Shine *et al.*[4] showed by isotope effect studies that this rearrangement is concerted. Thus, the paracyclophane concept was stim-

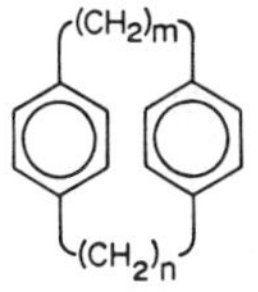

$(CH_2)_m$

$(CH_2)_n$

1

ulated by a mechanistic suggestion that proved to be little more than an attractive idea.

By the time our work was published in 1951, having benefited from the autostimulative effects of the results as they accumulated, we had a better idea of why the work was undertaken.

1. The paracyclophanes provided a unique vehicle for the study of transannular electronic effects in aromatic substitution reactions.
2. Due to the constraints of the macroring system and the shapes of the benzene rings, rotations of the benzene rings relative to one another were inhibited in the lower members of the homologous series. Thus, the system became interesting stereochemically.
3. We were interested in synthesizing intramolecular π complexes of fixed geometry by introducing electron-withdrawing groups in one ring and electron-providing groups in the other.

One reason for studying the paracyclophanes that went unmentioned in our first paper on macro rings[1] was that we anticipated that the compounds would be crystalline. At that time the main methods of purifying reaction products were fractional distillation and crystallization. Only rudimentary chromatographic methods had been developed. The synthetic organic chemist's greatest enemy before 1950 was a low yield of an oil with a molecular weight of several hundred. The conventional knowledge of the day was that compounds dominated by aromatic groups of high symmetry tended to be crystalline. We were not disappointed—[2.2]paracyclophane melts at 285 to 287°C and is only sparingly soluble in most organic solvents.

This high crystallinity worked for others as well. A polymer research group at ICI in Great Britain in the late 1940s was attempting to prepare and polymerize *p*-xylylene, the tetraene derived by high-temperature cracking of *p*-xylene. The nonpolymeric fraction from the pyrolysis was a complicated mixture of compounds from which trace amounts of [2.2]paracyclophane crystallized. Without the benefit of the usual carbon and hydrogen analyses and uv spectra, this "curiosity" was communicated in 1949 by Brown and Farthing[5] and characterized only by an unrefined crystal structure. Crystal structures at that time were rarities. When we learned of this work, Steinberg had already prepared the substance by a designed synthesis involving a high-dilution, intramolecular Wurtz reaction. The procedure involved adding $(4\text{-}BrCH_2C_6H_4CH_2)_2$ in xylene over a 60-hr period to refluxing xylene and molten sodium stirred at 7000 rpm. We hoped to achieve high dilution both in solution and at the liquid sodium–xylene interface and thus obtain a high yield. The yield was 2.1%. This sequence of events initiated us into the nonexclusive club of those who have been "scooped." The experience stimulated a reexamination of the literature, which turned up the fact that even the concept of the face-to-face paracyclophanes was not ours. Reichstein and Oppenauer[6] had attempted to prepare [*m.n*]paracyclophanes **(1)** in 1933. Furthermore, Baker *et al.*[7] in 1950 reported the synthesis of [2.2]metacyclophane (**2**: $m = n = 2$).

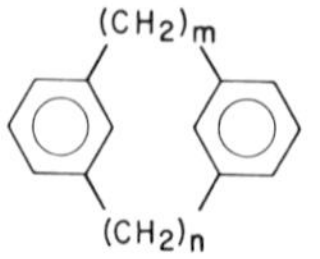

$(CH_2)_m$ $(CH_2)_n$ **2**

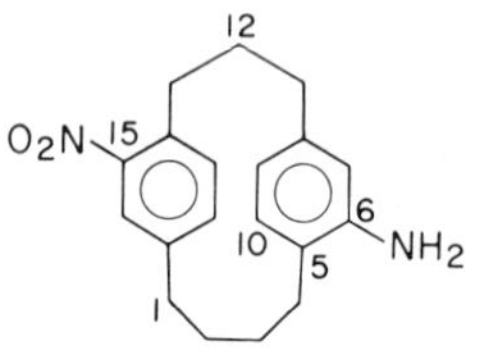

O_2N 15 12 10 6 5 1 NH_2 **3**

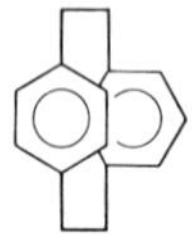

4

At an American Chemical Society meeting in New York in 1951, the author described his "new" class of compounds to Cheves Walling and asked his opinion of the name *paracyclophane* for the compound class. Pausing thoughtfully, Walling replied, "The memory is already too burdened with trivial names for organic compounds." In spite of this reasonable answer, we introduced the class name in our first paper. Emboldened by the nomenclature systematization suggestions of Schubert *et al.*[8] in 1954 and pressed by the need to name our growing number of substituted cyclophanes, we adopted in 1955,[9] and used throughout our subsequent papers, the system illustrated by **3**[9] and **4.**[10]

III. THE PAST

A. Conversion of [2.2]Paracyclophane to Other Cyclophanes

The author has watched many fine investigators draw with reverence the structures of organic compounds they had synthesized for the first time (or even the fifth). After all, their syntheses were the products of their imaginations and much hard work. This attachment was particularly evident in the case of researchers who prepared compounds with intriguing symmetry properties or chiral elements. Investigators of this kind might be termed *chirophiles*. It seems that the selection of certain research problems is dominated by a subliminal aesthetic judgment. Certainly, the author would not have selected cyclophanes as a field for investigation had they not possessed beautiful symmetry properties. The compounds and problems that we have most enjoyed have involved the manipulation of symmetry properties, a point that is illustrated in the following account.[11–13]

The Marsh discovery[14] that [2.2]paracyclophane **(5)** could be made in good yield by heating p-$CH_3C_6H_4CH_2N(CH_3)_3OH$ made the compound a convenient starting material for other cyclophanes whose initial preparations[1] caused my graduate students considerable anguish. When free of the radical-chain inhibitors that were added during the preparation of [2.2]paracyclophane **(5)** via the p-xylylene route, **5** upon treatment with NBS underwent bromination in its bridge to give substitution products that could be readily separated. These were hydrolyzed to provide ketones **6, 7,** and **8.**[15] Roger Helgeson found that when these ketones were

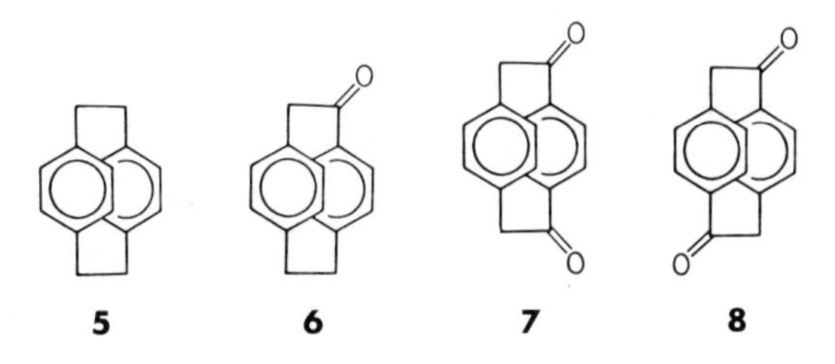

treated with diazomethane they readily ring-expanded to give homologous ketones, which in turn further ring-expanded to provide (after reduction) paracyclophanes **9–13** in practical yields.[15]

$(CH_2)_m$
$(CH_2)_n$

9 $m = 2$, $n = 3$
10 $m = 2$, $n = 4$
11 $m = n = 3$
12 $m = 3$, $n = 4$
13 $m = n = 4$

Treatment of **5** with $HAlCl_4$ provided Larry Singer and Roger Helgeson with the first of several syntheses of [2.2]metaparacyclophane **(4)**[10] in 44% yield. Irradiation of this cycle (Robert Gilman and Mary Delton) caused it to rearrange to [2.2]metacyclophane **(14)** in 43% yield.[16] Thus, [2.2]paracyclophane, which is now commercially available, is a potential starting material for a variety of other cyclophanes.

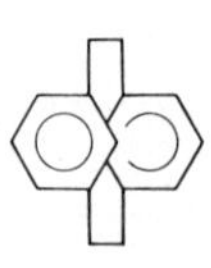

14

B. Torsional Chirality in the Cyclophanes

Appropriately, the optimism of a thesis supervisor is tempered by the realism of his graduate students. When the author first suggested in about 1953 to Normal Allinger, the second graduate student to work on cyclophanes, that he study the symmetry properties of monosubstituted [2.2]paracyclophanes, Allinger expressed grave doubts about the viability of such a project. Perhaps he had in mind our 2% yield in the Wurtz synthesis of [2.2]paracyclophane and the molten sodium, stirred at 7000 rpm in a baffled Morton stirring flask for 60 hr. His skepticism changed to enthusiasm, however, when he found a 25-g sample of [2.2]paracyclophane on his desk one morning, the gift of an industrial patron. Accordingly, he prepared a series of derivatives. Catalytic reduction of **5** gave the fully saturated derivative, only one stereoisomer being produced out of the theoretically possible six. The cause of the stereospecificity and the configuration of the product was probably never more simply explained. The principle that two objects cannot occupy the same space at the same time would have been seriously violated had any isomer been produced other than that of the cis,cis,anti,anti configuration **(15)**.[17]

The examination of molecular models indicated that in the series **16–19**, rotation of the benzene rings with respect to each other should be possible with **19**, nearly possible with **18**, and greatly inhibited with **16** and **17**. Monoacetylation of **5**,[17] **11**,[18] **12**,[19] and **13**[20] provided the methyl ketones, which were oxidized to the acids **16**,[17] **17**,[18] **18**,[19] and **19**,[20] respectively. The smaller the ring system, the greater the rate of acetylation.[19] One acetyl group in one ring deactivated both rings toward further acetylations. These transannular effects greatly aided the syntheses. Racemates **16–18** were readily resolved into their enantiomers, whereas **19** could not

$(CH_2)_m$ CO_2H HO_2C $(CH_2)_n$

15

16 $m = n = 2$
17 $m = n = 3$
18 $m = 3$, $n = 4$
19 $m = n = 4$

20

be resolved. Thus, at ordinary temperatures the aromatic rings of **16** to **18** could not rotate with respect to one another on the graduate student time scale, but **19** rotated very rapidly.[17–20] This work predated ^{1}H-nmr spectroscopy.

In later work the monocarboxylic acid of [2.2]metaparacyclophane **(20)** was prepared and resolved into its enantiomers. With a combination of polarimetric and ^{1}H-nmr methods, Daniel Hefelfinger demonstrated that the meta ring rotated with respect to the para ring at available temperatures, but not the para with respect to the meta.[21]

C. Orbitally Unconjugated, Unsaturated [*m.n.*]Paracyclophanes

Classically conjugated but orbitally unconjugated cyclophanes **21–24** were prepared by Kenneth Dewhirst, a graduate student of 1958 vintage.[22,23] Ultraviolet spectral comparisons of **5** and of **21** to **24** with open-chain models showed both how unconjugated were the systems and how badly bent were the benzene rings of **5, 21,** and **22.** The multiple bonds of **23** and **24** were only partially conjugated. The absence of ^{1}H-nmr spectroscopy in 1958 prevented us from observing the probable equilibration of **23** between enantiomeric forms **23a** and **23b.** With a blend of optimism and innocence, we tried with acid or light at 25°C to convert **24** to **25,** the latter of which would at that time have been the first [18]annulene. The "god of research results" did not smile favorably on us in this venture,

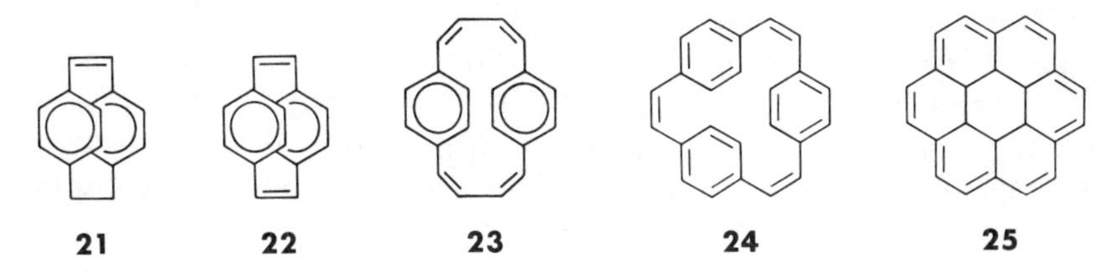

21 22 23 24 25

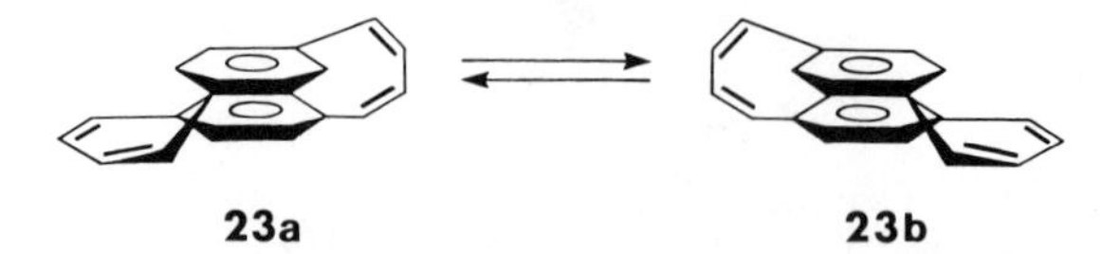

because **25** was not obtained. Almost 20 years later the light-induced conversion did yield to the time-trained hands of Boekelheide *et al.*[24] Interestingly, compound **25** reacted with oxygen at room temperature to give coronene.[24]

D. Bent Benzene Rings versus Transannular Electronic Effects

In our earliest paper[1] we suggested that both the bent benzene rings and the enforced proximity of two π systems were responsible for the abnormal uv absorption spectrum of [2.2]paracyclophane.[1] In the [*m.n*]paracyclophane series, the uv spectra become more like those of open-chain model compounds in passing from [2.2]- to [2.3]- to [3.3]- to [3.4]- to [4.4]paracyclophane.[25] Only the last cycle possesses a normal spectrum. To identify the cause of these abnormal spectral effects, we prepared the homologous series of [*m*]paracyclophanes. We hoped to make the methylene belt short enough so that its benzene could be pulled into a tub. Although compounds **26–28** were readily prepared through the acyloin

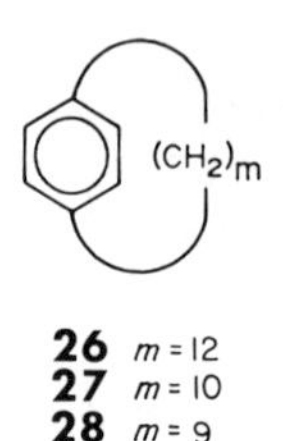

26 $m = 12$
27 $m = 10$
28 $m = 9$
29 $m = 8$

ring-closing reaction, no cycle was obtained when the attempt was made to make **29.**[25] This strained compound was prepared through the 1,6- to 1,6-cycloaddition reaction formulated.[26]

In the uv spectra of the members of this homologous series, as the methylene belt became shorter, the bands in the 260- to 270-nm range moved to longer wavelengths and lost their fine structure. In [2.2]paracyclophane the bridgehead carbons of the benzene rings are bent out of the

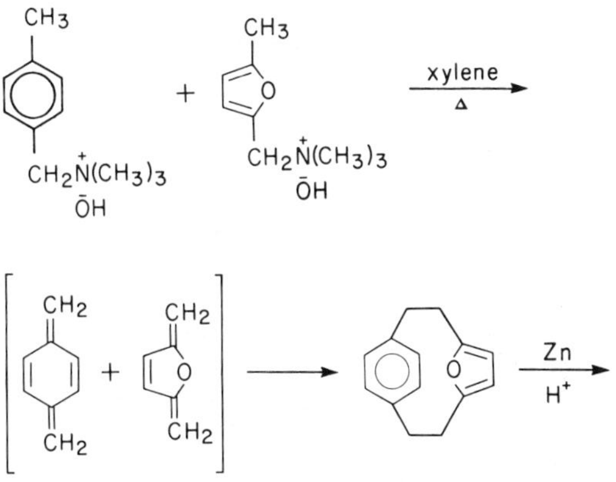

CH_3 CH_3 $CH_2\overset{+}{N}(CH_3)_3$ $\bar{O}H$ + $CH_2\overset{+}{N}(CH_3)_3$ $\bar{O}H$ $\xrightarrow[\Delta]{\text{xylene}}$

$[CH_2 + CH_2] \longrightarrow \xrightarrow[H^+]{Zn}$ **29**

planes of the other four by 13°.[27] Allinger calculated this kind of deformation in **29** to be about 20°.[28] The spectral abnormalities in **5** are much greater than those in **29,** which bears out the original suggestion that transannular electronic perturbations seriously affect the uv spectrum of **5.**[1,29]

Transannular influences are also apparent in the π–π complexes between the homologous [*m.n*]paracyclophanes and the π-acid tetracyanoethylene. The π-base strengths decrease in the order [3.3] > [4.3] > [2.2] > [4.4] > [6.6] ~ $(p\text{-}CH_3C_6H_4CH_2)_2$. Except for the position of [2.2]paracyclophane in this series, the order correlates with the distance of the two benzene rings from one another. The closer the two rings, the greater becomes the π-base strength, the nonbound benzene ring releasing electrons to the bound ring, as in **30.**[30] As expected, electron-with-

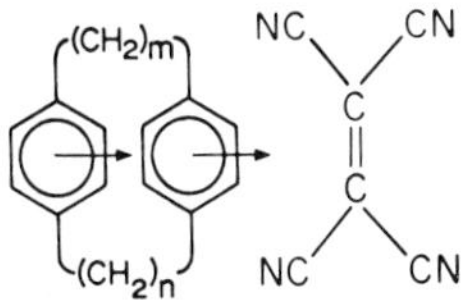

$(CH_2)_m$ $(CH_2)_n$ NC CN C=C NC CN

30 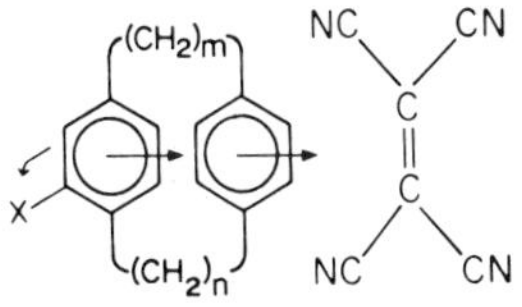 **31**

drawing groups in the noncomplexed ring decreased the π basicity of the complexed ring, as suggested in formula **31.** The homologous series of π complexes provided a rainbowlike series of deep and beautiful colors, ranging from yellow to deep purple. This series provides a visible example of how homologous compounds can differ widely in their electronic properties.[30]

E. Transannular Directive Influences in Substitution Reactions of [2.2]Paracyclophanes

Although posed in our first paper on cyclophanes,[1] the problem of transannular directive influences of one substituent on the point of electrophilic substitution in the second ring of monosubstituted [2.2]paracyclophanes could not be studied until the structures of disubstituted [2.2]paracyclophanes were determined. In about 1967 ^{1}H-nmr spectra provided the tools and Hans Reich provided the skills for an attack on the problem.[31] We had anticipated that transannular resonance effects would dictate the points of entry of incoming electrophiles. Instead, the generalization emerged that predominant substitution occurred pseudo-gem to the most basic positions or substituents in the already substituted ring. For example, bromination of **32** gave only pseudo-gem product **34,** whereas bromination of **33** gave 16% of pseudo-*ortho*-**35,** 26% of pseudo-*para*-**36,** and 6% of pseudo-*meta*-**37** product. Furthermore, in the bromina-

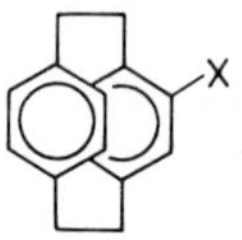

32 X = $COCH_3$
33 X = Br

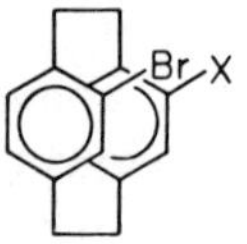

34
(pseudo-gem)

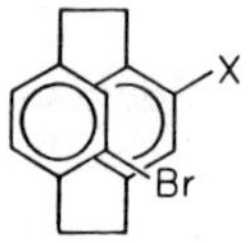

35
(pseudo-ortho)

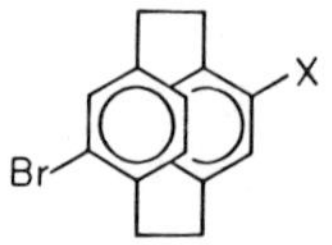

36
(pseudo-para)

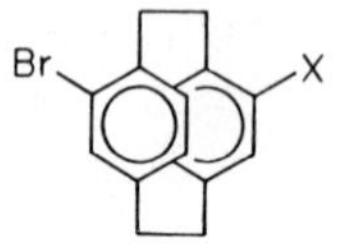

37
(pseudo-meta)

tion of **38,** deuterium was transferred from ring to ring to give **39,** and the reaction gave a $k_H/k_D = 3.7$ isotope effect. These results were interpreted in terms of the mechanistic scheme formulated.[32]

Graduate students are sometimes better off if they do not consult their thesis advisers before they try something new. In 1970 Elizabeth Truesdale decided to put the above explanation to the test by chloromethylating acetyl[2.2]paracyclophane **(32),** a compound extremely insoluble in water. She added finely divided **32** to a solution of paraformaldehyde in concentrated hydrochloric acid to produce a 55% yield of the desired **40.**

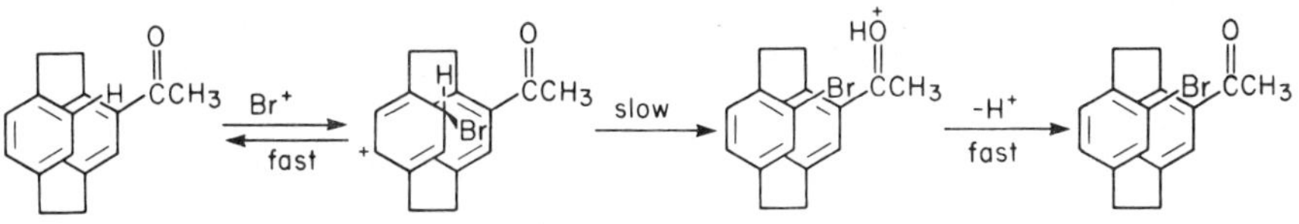

32

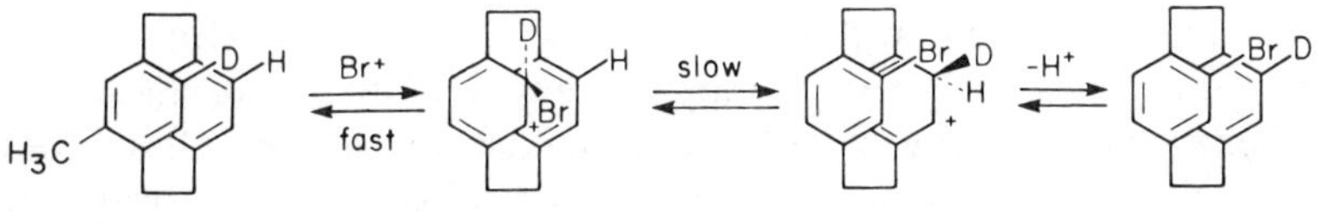

38 39

This reaction represents an insoluble-solid to insoluble-solid transformation. Had she consulted her thesis supervisor, he would given her poor odds that the reaction would work. She went on to convert **40** to **41** and **42.** The fact that the methylenes are crushed together in **41** (CPK molecular models, with one face of each of the two benzene rings shaved by about 15%) made them prone to transannular hydride migration reactions. When **41** was treated with Ag_2O, **43** was produced.[33] Our plan to use this approach to introduce additional bridges into **41** was terminated by the appearance of the beautiful work of others bent on the same objective. This effort culminated in the elegant synthesis of superphane.[34]

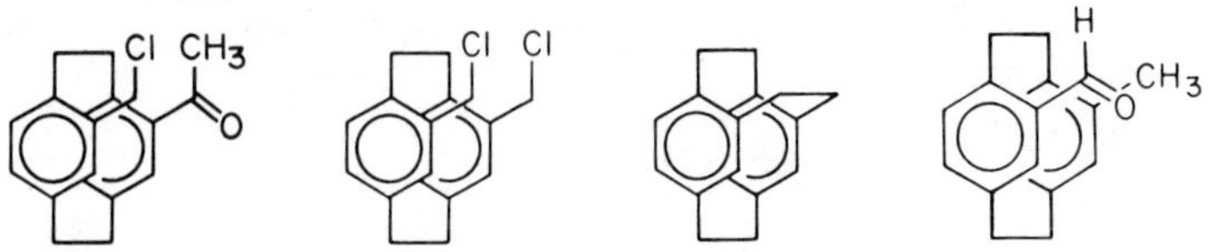

40 41 42 43

F. Ring Openings, Closings, and Expansions of Substituted [2.2]Paracyclophanes

A chemical consequence of the 31 kcal/mol of strain energy in [2.2]paracyclophane[35] is the capacity of this compound to form benzyl–benzyl biradicals reversibly at 200°C. This reaction not only provided Reich and Delton with interesting thesis material, but also exercised the author's fascination with the use of stereochemistry to solve mechanistic problems. When optically pure **44a** was heated to 200°C, it racemized.[36] A

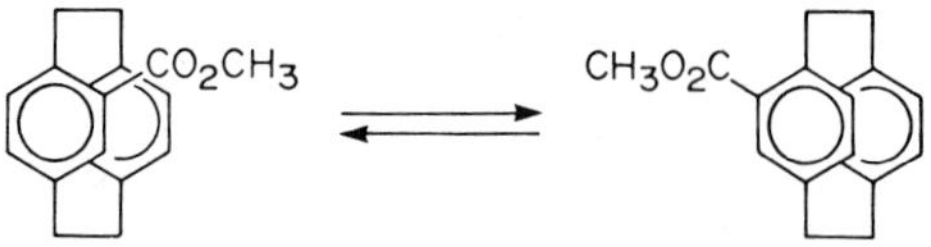

44a 44b

glance at the shaved CPK molecular models of [2.2]paracyclophane convinces even skeptics that the benzene rings cannot rotate with respect to one another without bond-breaking processes. Only one of the two benzyl–benzyl bonds cleaves. This was demonstrated by the absence of *p*-xylylenes as reaction intermediates. When disubstituted diastereomers **45** or **46** as racemates were heated to 200°C, the same equilibrium mixture of the two was formed. Similarly, **47** and **48** equilibrated. Neither **45** nor **46** gave either **47** or **48,** nor did **47** or **48** give either **45** or **46.** None of the four starting materials produced either dibromide or diester. Clearly, only one benzyl–benzyl bond broke during these interconversions.[36]

To answer the question of whether one or both rings rotate with respect to one another, optically pure **45** was partially converted to **46.** Recovered **45** was partially racemized, and **46** was totally racemic. Thus, the open-chain intermediate lasted long enough to "forget" its configurational origins.[36]

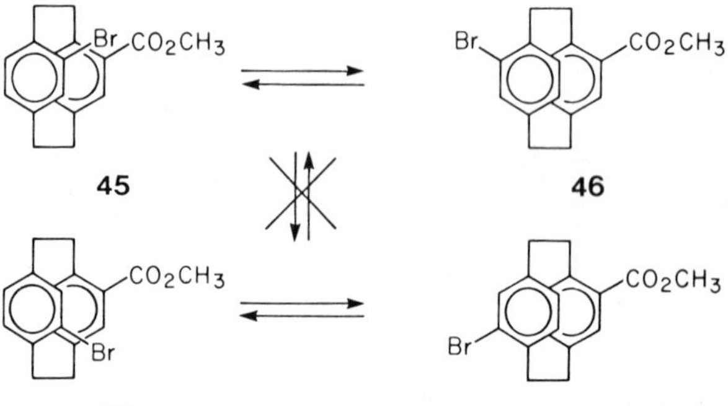

45 46

47 48

Dimethyl maleate or fumarate copolymerizes with styrene by a radical mechanism to give ABAB copolymers. This fact suggested that the diradical formed from [2.2]paracyclophane might add to these alkenes to give ring-expanded product. In the event, the hoped for 1,2- to 1,12-cycloaddition reaction was observed, and **49** and **50** were produced.[36]

5 + $CHCO_2CH_3$ ‖ $CHCO_2CH_3$ —200°C→ 49 + 50

Another interesting use of this cleavage–recombination reaction was made by the conversion of **51** to **52**[37] in 90% yield at 160°C.

51 —160°C→ 52

Photolytic racemization of optically pure **53** occurred readily in methanol with 254-nm light. Similarly, **55** racemized in acetone with >270-nm light. Although other explanations exist, the most attractive one for the nonsensitized reaction involves **54** as intermediate. For the sensitized reaction, intermediates **56** were suggested.[38]

53 ⇌ (hν; CH_3OH) 54

55 ⇌ (hν; CH_3COCH_3) 56

IV. THE PRESENT

Synthetic organic chemists fall into two groups: those who prepare old, naturally occurring compounds and those who prepare new compounds. The synthetic targets of the former group are provided by the evolutionary chemistry of nature. The synthetic targets of the latter group are designed by the investigator. Because there is a vast ocean of possible new compounds, the selection is guided by any of a wide number of objectives.

Many investigators (the writer included) have long dreamed of designing and synthesizing compounds that would imitate the desirable properties of enzyme systems. The catalytic properties and specificities of enzymes depend on their capacity to complex and orient substrate (and a coenzyme) in such a way that several functional groups in the complex can act cooperatively in causing a chemical reaction. The high reaction rates and specificities depend on the catalytic groups being organized before complexation to converge on a binding cavity close to the seat of reaction. Long protein chains that are organized conformationally by ring systems, by steric effects, by hydrogen bonds, and by solvent interactions provide the semirigid support structures for the active sites in enzymes.

The most challenging aspect of designing enzyme imitators is providing for the convergence of catalytic groups on an organized binding site in compounds of minimal molecular weight. We recognized in the 1950s that the rigidity and double-decked structure of [2.2]paracyclophane made this unit **(57)** possibly useful in the construction of binding sites on which

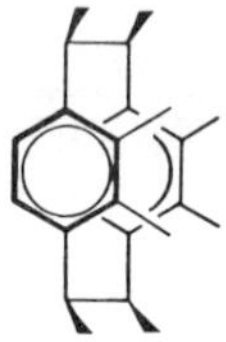

57

functional groups converged. This unit contains 16 substitutable sites, eight of whose bonds generally point in one direction, with the other eight pointing in the opposite direction.

Much later, hosts **58–61** were prepared and the complexing capacities they exhibit toward metal salts, alkylammonium, and several alkylenediammonium salts were examined by Helgeson *et al.*[39] Structural information concerning the complexes in solution was provided by ^{1}H-nmr spectra. These studies made use of the capacity of the ring currents of the benzene rings to produce chemical shifts in neighboring protons of complexed guests.

The spherands represent an extreme example of the use of the cyclophane concept to design an empty cavity suitable for encapsu-

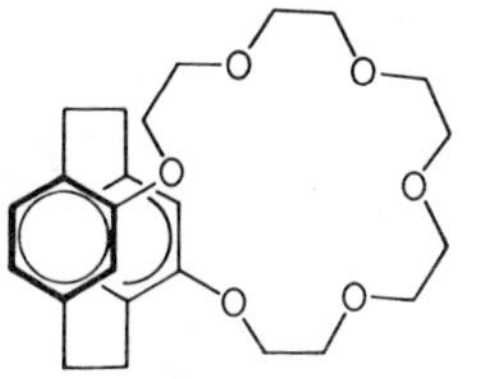

58

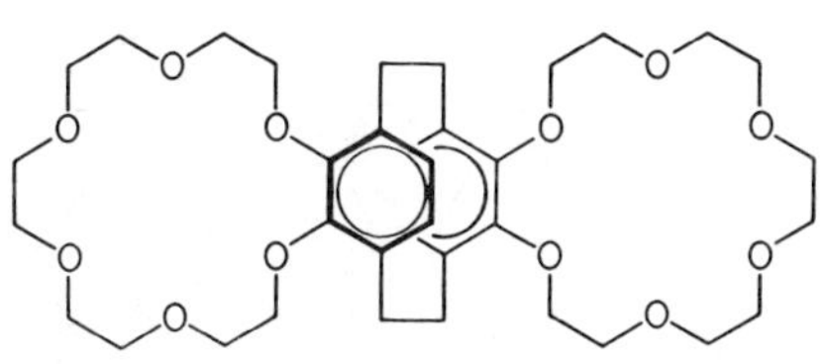

59

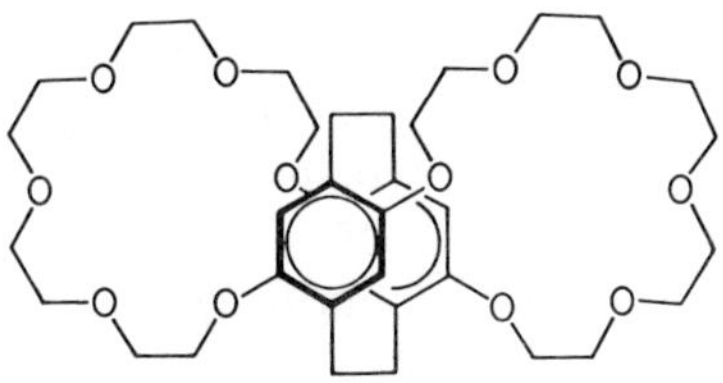

60

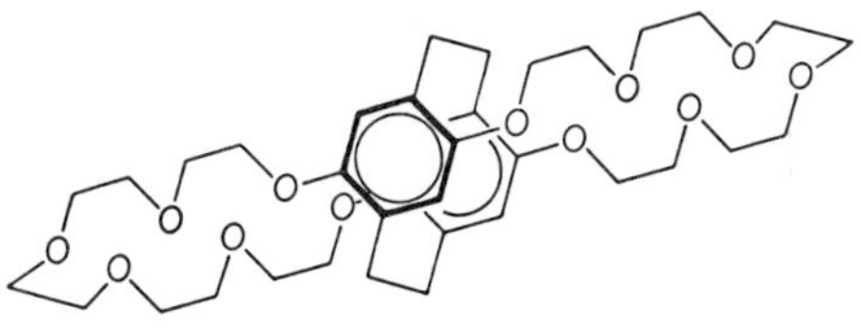

61

lating specific guests. For example, 2,8,14,20,26,32-hexamethoxy-5,11,17,23,28,33-hexamethyl[0.0.0.0.0.0]metacyclophane **(63)** was synthesized by Takahiro Kaneda *et al.*[40–42] in five steps from *p*-cresol in 8% overall yield. The decomplexation of the lithiospherium chloride **62** could be accomplished only by phase transfer, the driving force being crystallization of the free host **(63)** from the medium. The crystal structure of **62** is shown in **64** and shows the Li^+ completely encapsulated by octahedrally arranged oxygens. The complex in projection resembles a snowflake in shape. The crystal structure of **63** is shown in **65,** and its hole lined with 24 electrons is clearly visible.[41] Unlike the crowns and cryptands whose cavities are organized only during complexation of ions (in the free hosts, the cavities are filled with methylene groups), the cavities of the spherands are rigidly organized during synthesis rather than during complexation.[42]

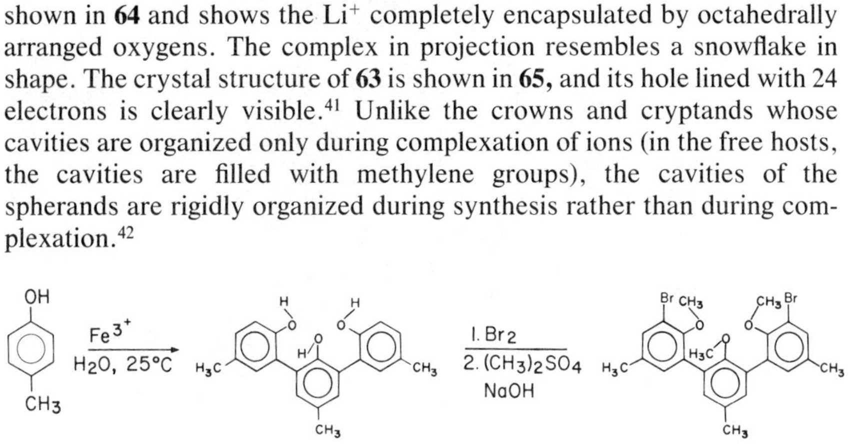

OH, CH_3 — Fe^{3+} / H_2O, 25°C — 1. Br_2 / 2. $(CH_3)_2SO_4$, NaOH

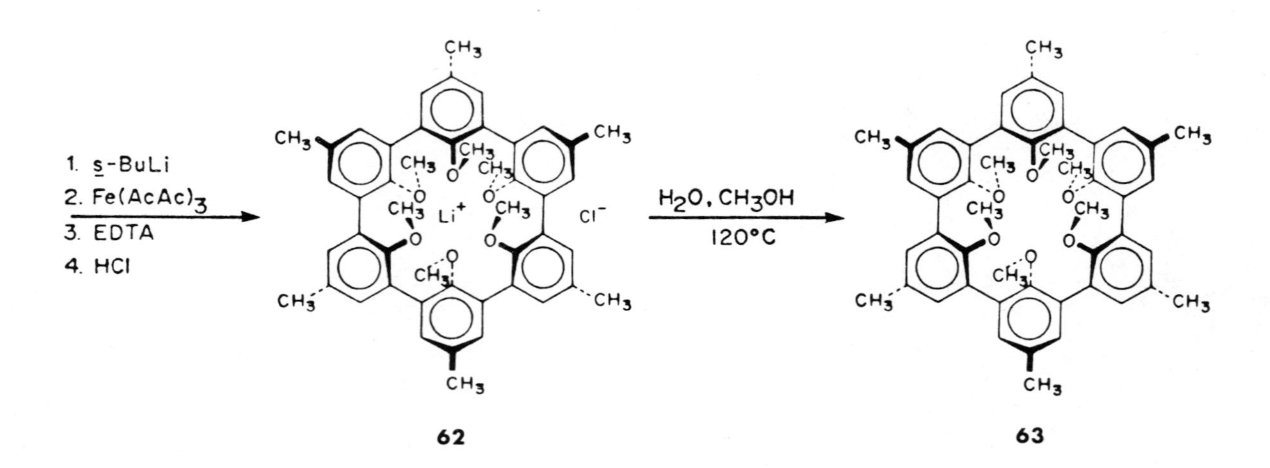
1. s-BuLi
2. Fe(AcAc)3
3. EDTA
4. HCl
Li+
Cl−
62
H2O, CH3OH
120°C
63

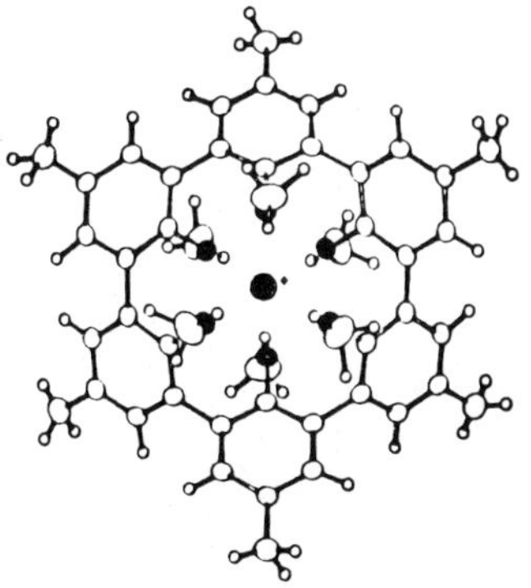

64 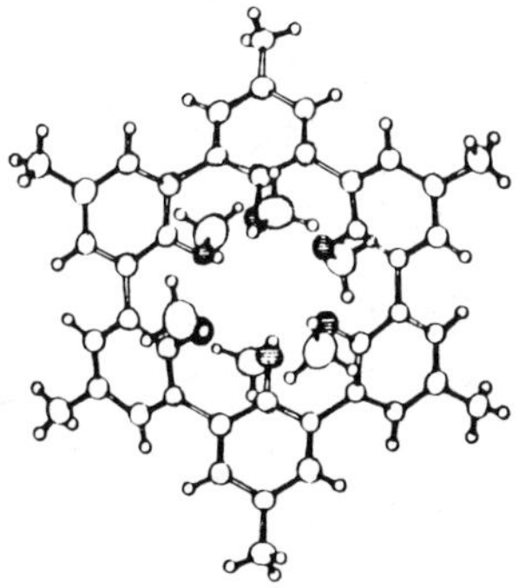 **65**

66

Spherand **63** specifically complexes only Li^+ and Na^+ and rejects K^+, Rb^+, Cs^+, Mg^{2+}, and Ca^{2+}. The binding free energy of **63** in chloroform saturated with water at 25°C toward lithium picrate is >23 kcal mol^{-1}, and that toward sodium picrate is 19 kcal mol^{-1}. These values exceed those of the open-chain model compound **66** by >17 and >13 kcal/mol, respectively. This comparison dramatically illustrates the importance of rigid preorganization to specificity and high binding energies. The binding free energies of **63** are the highest yet observed for Li^+ and Na^+.[43]

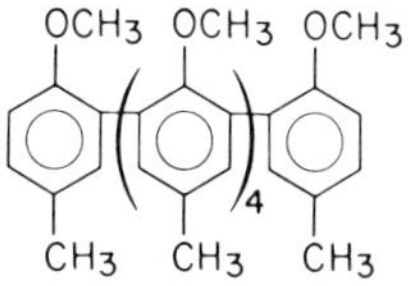

V. THE FUTURE

The enforced cavity of **63** points to the unsolved problem of synthesizing closed-surface compounds with cavities large enough to imprison ordinary organic and inorganic compounds. A variety of interesting questions arise about the properties (in all phases) of encapsulated guests that cannot escape from the enveloping hosts. Although the enforced cavities of **67**,[42] **68**,[44] and **69**[44] provide a start on the synthesis of such hosts (e.g.,

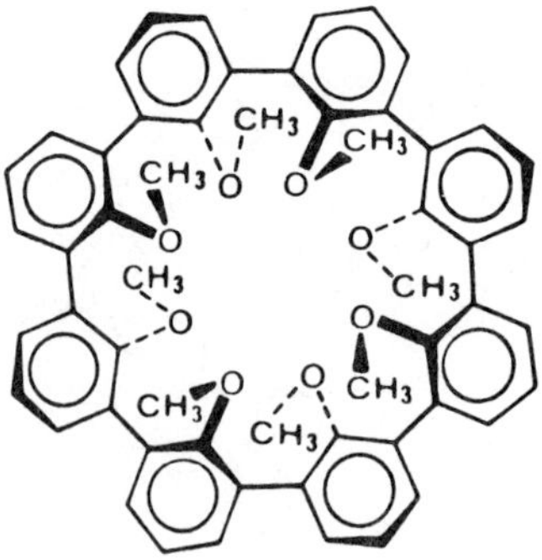

67 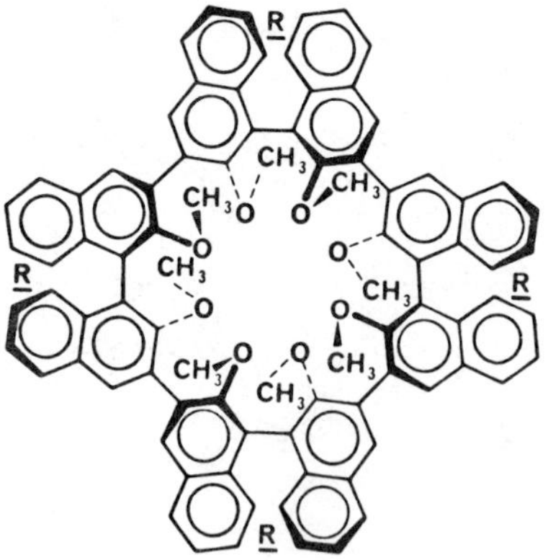 **68**

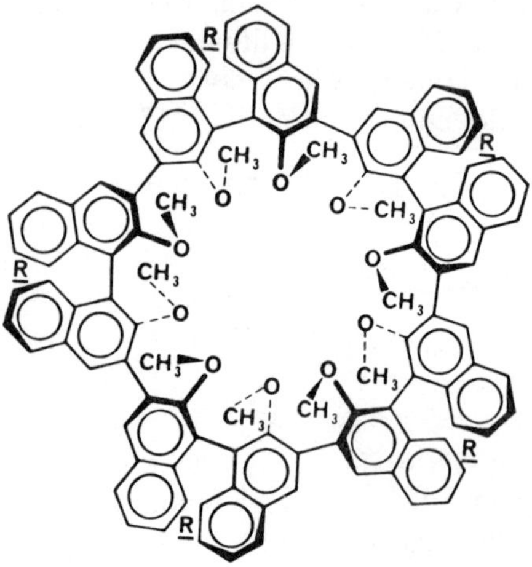

69

the cavity of **69** is large enough in CPK molecular models to embrace ferrocene), the holes in their surfaces are large enough to allow guests to enter and depart from their cavities. We suggest the term *cavitand* to describe that class of compounds which contain enforced cavities large enough to embrace chemical compounds or ions.

With the use of CPK molecular models, we designed cavitand **70,** in which A groups can be CH_2, O, or other short-spacing units. The model with A = CH_2 is spherical and contains a cavity large enough to imprison 12 "model" molecules of water. We resort to Mercator projection for-

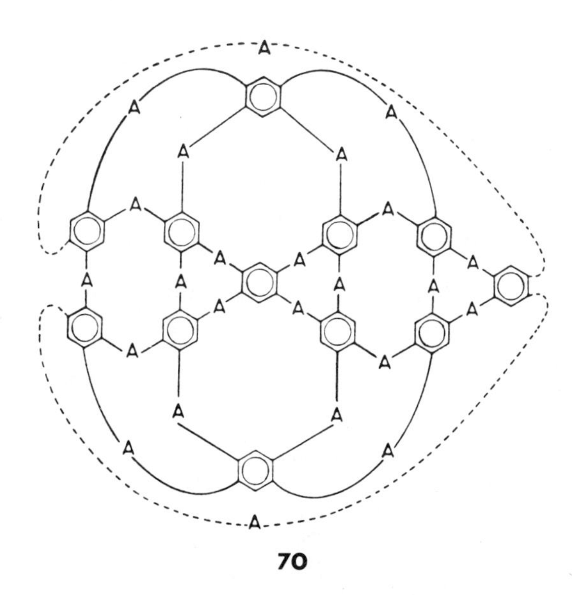

70

mula **70** to represent the compound. Gerhardus Mercator was a Flemish geographer, who in 1594 invented the Mercator projection maps of the world for navigational purposes. The spirit of our use of the drawing is that it might be useful for synthetic navigation. Notice that the "skin" of **70** is composed of 12 benzene rings linked together in their 1, 2, 4, and 5 positions by A groups. When the A groups are CH_2, **70** can be regarded as composed of four [1.1.1]orthocyclophane units attached to one another through 12 A units. Alternatively, **70** is visualized as containing six [1.1.1.1]metacyclophane units whose member benzene rings are common to other similar units.[45] In formula **71** the viewer looks into the cavity through a greatly expanded [1.1.1.1]metacyclophane hole. We are attempting to synthesize **70** and simplifications of it.

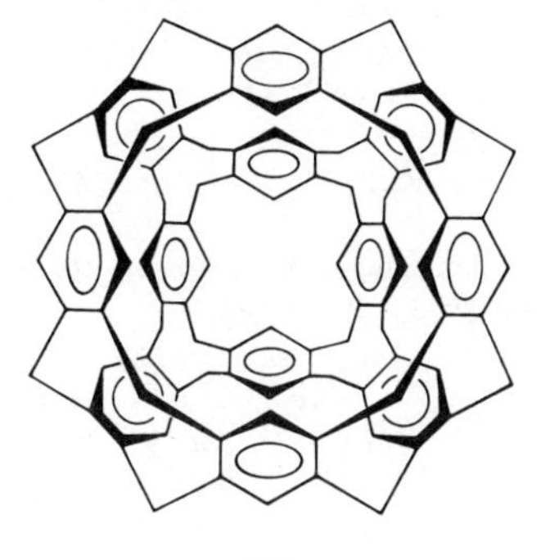

71

REFERENCES

1. D. J. Cram and H. Steinberg, *J. Am. Chem. Soc.* **73,** 5691–5704 (1951).
2. M. J. S. Dewar, *Nature (London)* **156,** 784–784 (1945); *J. Chem. Soc.* pp. 406–408 (1946).
3. G. S. Hammond and H. J. Shine, *J. Am. Chem. Soc.* **72,** 220–221 (1950).
4. H. J. Shine, H. Zmuda, K. H. Park, H. Kwart, and A. G. Horgan, *J. Am. Chem. Soc.* **103,** 955–956 (1981).
5. C. J. Brown and A. C. Farthing, *Nature (London)* **164,** 915–916 (1949).
6. T. Reichstein and R. Oppenauer, *Helv. Chim. Acta* **16,** 1373–1380 (1933).
7. W. Baker, J. F. W. McOmie, and J. M. Norman, *Chem. Ind. (London)* p. 77 (1950).
8. W. M. Schubert, W. A. Sweeney, and H. K. Latourette, *J. Am. Chem. Soc.* **76,** 5462–5466 (1954).
9. D. J. Cram and J. Abell, *J. Am. Chem. Soc.* **77,** 1179–1186 (1955).
10. D. J. Cram, R. C. Helgeson, D. Lock, and L. A. Singer, *J. Am. Chem. Soc.* **88,** 1324–1325 (1966).
11. D. J. Cram, *Rec. Chem. Prog.* **20,** 71–93 (1959).
12. D. J. Cram and J. M. Cram, *Acc. Chem. Res.* **4,** 204–213 (1970).
13. D. J. Cram, R. B. Hornby, E. A. Truesdale, H. J. Reich, M. H. Delton, and J. M. Cram, *Tetrahedron* **30,** 1757–1768 (1974).
14. F. D. Marsh, U.S. Patent 2,757,146 (1956); H. E. Winberg, F. S. Fawcett, W. E. Mochel, and C. W. Theobald, *J. Am. Chem. Soc.* **82,** 1428–1435 (1960).
15. D. J. Cram and R. C. Helgeson, *J. Am. Chem. Soc.* **88,** 3516–3521 (1966).
16. R. E. Gilman, M. Delton, and D. J. Cram, *J. Am. Chem. Soc.* **94,** 2478–2482 (1972).
17. D. J. Cram and N. L. Allinger, *J. Am. Chem. Soc.* **77,** 6289–6294 (1955).
18. M. Sheehan and D. J. Cram, *J. Am. Chem. Soc.* **91,** 3544–3552 (1969).
19. D. J. Cram, W. J. Wechter, and R. W. Kierstead, *J. Am. Chem. Soc.* **80,** 3126–3133 (1958).
20. D. J. Cram and R. W. Kierstead, *J. Am. Chem. Soc.* **77,** 1186–1190 (1955).
21. D. T. Hefelfinger and D. J. Cram, *J. Am. Chem. Soc.* **92,** 1073–1074 (1970); **93,** 4754–4772 (1971).
22. K. C. Dewhirst and D. J. Cram, *J. Am. Chem. Soc.* **80,** 3115–3125 (1958).
23. K. C. Dewhirst and D. J. Cram, *J. Am. Chem. Soc.* **81,** 5963–5971 (1959).
24. T. Otsubo, R. Gray, and V. Boekelheide, *J. Am. Chem. Soc.* **100,** 2449–2456 (1978).
25. D. J. Cram, N. L. Allinger, and H. Steinberg, *J. Am. Chem. Soc.* **76,** 6132–6141 (1954).
26. D. J. Cram and G. R. Knox, *J. Am. Chem. Soc.* **83,** 2204 (1961); D. J. Cram, C. S. Montgomery, and G. R. Knox, *ibid.* **88,** 515–525 (1966).
27. H. Hope, J. Bernstein, and K. N. Trueblood, *Acta Crystallogr., Sect. B* **B28,** 1733–1743 (1972).
28. N. L. Allinger, L. A. Freiberg, R. B. Hermann, and M. A. Miller, *J. Am. Chem. Soc.* **85,** 1171–1176 (1963).
29. D. J. Cram, R. H. Bauer, N. L. Allinger, R. A. Reeves, W. J. Wechter, and E. Heilbronner, *J. Am. Chem. Soc.* **81,** 5977–5983 (1959).
30. D. J. Cram and R. H. Bauer, *J. Am. Chem. Soc.* **81,** 5971–5977 (1959).
31. H. J. Reich and D. J. Cram, *J. Am. Chem. Soc.* **91,** 3534–3543 (1969).
32. H. J. Reich and D. J. Cram, *J. Am. Chem. Soc.* **91,** 3505–3516 (1969).
33. E. A. Truesdale and D. J. Cram, *J. Am. Chem. Soc.* **95,** 5825–5827 (1973); *J. Org. Chem.* **45,** 3974–3981 (1980).
34. Y. Sekine and V. Boekelheide, *J. Am. Chem. Soc.* **103,** 1777–1785 (1981).

35. C. Shieh, D. C. McNally, and R. H. Boyd, *Tetrahedron* **25,** 3653–3665 (1969).
36. H. J. Reich and D. J. Cram, *J. Am. Chem. Soc.* **91,** 3517–3526 (1969).
37. M. H. Delton and D. J. Cram, *J. Am. Chem. Soc.* **94,** 1669–1675 (1972).
38. M. H. Delton and D. J. Cram, *J. Am. Chem. Soc.* **94,** 2471–2478 (1972).
39. R. C. Helgeson, T. L. Tarnowski, J. M. Timko, and D. J. Cram, *J. Am. Chem. Soc.* **99,** 6411–6418 (1977).
40. D. J. Cram, T. Kaneda, R. C. Helgeson, and G. M. Lein, *J. Am. Chem. Soc.* **101,** 6752–6754 (1979).
41. K. N. Trueblood, C. B. Knobler, E. Maverick, R. C. Helgeson, S. B. Brown, and D. J. Cram, *J. Am. Chem. Soc.* **103,** 5594–5596 (1981).
42. D. J. Cram and K. N. Trueblood, *Top. Curr. Chem.* **98,** 43–106 (1981).
43. G. M. Lein and D. J. Cram, *J. Chem. Soc., Chem. Commun.* pp. 301–304 (1982).
44. R. C. Helgeson, J.-P. Mazaleyrat, and D. J. Cram, *J. Am. Chem. Soc.* **103,** 3929–3931 (1981).
45. D. J. Cram, *Science* **219,** 1177–1183 (1983).

CHAPTER 2

The Conceptual Chemistry of Cyclophanes

JOEL F. LIEBMAN

Department of Chemistry
University of Maryland Baltimore County
Catonsville, Maryland

ISBN 0-12-403001-7

I. INTRODUCTION

A. Experiment and Theory

This chapter deals with the interrelationships between structure and energetics in the class of compounds generically described as cyclophanes. We first discuss the results of experimental and theoretical studies of these species. It was deemed desirable not to separate these methodologies because, to be at all meaningful, two of the major "theme songs" of cyclophane chemistry, aromaticity and strain, generally require both experimental determinations and theoretical constructs. The word *experimental* as used herein includes quantum chemical and molecular mechanical calculations, which can be referred to as "numerical experiments," as well as more obvious "experimental" experiments.

B. Aromaticity and Strain

Both aromatic character and strain energy are defined in terms of reference states, biases, or just "feelings" as to how molecules should behave. However, because both aromaticity and strain have been thoroughly reviewed,[1,2] the precise nature of these concepts need not be cited other than to assert simplistically that *aromatic* means that the species is more stable than the reference states would suggest, whereas *strained* means that the species is less stable. In general we refer the reader to Greenberg and Liebman[2] for more information about the interrelations of structure and energetics of organic molecules. (In particular, the reader should note pp. 153–177 and 235–237, on which cyclophanes are most explicitly discussed.) In addition, aspects of both subjects are reviewed in the *Annual Specialist Reports* of the Chemical Society on aromatic and alicyclic hydrocarbons, respectively.

C. Molecular Energetics: Data, Estimations, and Assumptions

1. Are There Enough Data?

It will be soon seen that there are both too many and too few data. There is considerable information on the molecular ions of the cyclophanes, from which a coherent picture of the orbitals of these species has evolved. However, there is a marked paucity of direct, calorimetrically determined information on the heats of formation of neutral cyclophanes. Quantum chemical and molecular mechanical calculations have filled in some of the important gaps, but experimental confirmation is almost totally lacking. Both ion and neutral energetics data, along with whatever other data or concepts are needed to obtain these numbers, are presented here. In particular, estimations of thermochemical quantities are also made.

2. Energetics of Isomers and Other Species That Have the Same Number of Carbons

It will be seen that the relative energetics of a set of isomers is easier to deduce than that for an arbitrary set of organic compounds. For example, consider the acid-catalyzed rearrangement[3] of [2.2]paracyclophane **(1)** to [2.2]metaparacyclophane **(2)**. At least for thermal isomerization reactions, it is convenient (and often imperative) to make the assumptions that

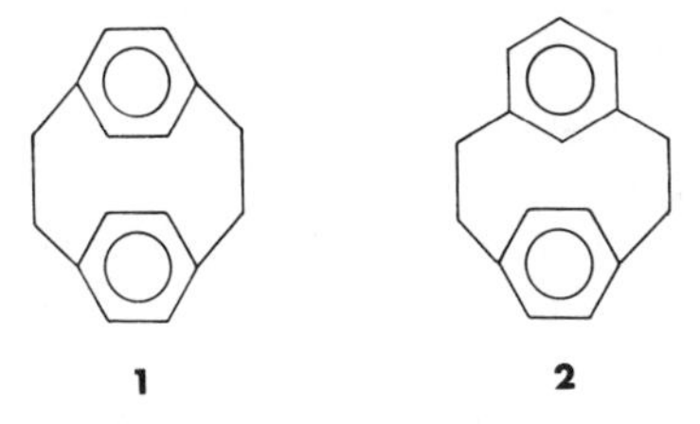

if species A rearranges to species B, then B is more stable than A. More precisely, we assume that Gibbs free energy changes (or, equivalently, equilibrium constant data) are to be generally paralleled by changes in heat of formation[4] and that condensed-phase data (i.e., liquid, solid, and solution) are equivalent to those derived from isolated gaseous molecules. ⟨For pure liquid hydrocarbons the latter assumption is usually accurate to 1.5 kcal/mol. An even simpler assumption is that the heat of vaporization of an arbitrary hydrocarbon depends solely on the number of carbons[5] n_c (Eq. 1):

$$\Delta H_v(\text{hydrocarbon, liquid}) = (1.11 \pm 0.03)n_c + (0.70 \pm 0.19)\ \text{kcal/mol} \qquad (1)$$

A related expression has been suggested[6] for cata-condensed aromatic hydrocarbons.

$$\Delta H_s = 1.98n_c \tag{2}$$

Although not general, it suggests that considerations of the energetics of isomeric species parallel those in solution, the pure liquid, or the ideal gas.⟩*

Returning to our assumptions and the isomerization reaction of the cyclophanes, we thus deduce that [2.2]metaparacyclophane is more stable, that is, has a lower heat of formation, than its [2.2]paracyclophane isomer. Direct calorimetric measurements[7] of the heats of formation and of vaporization on both cyclophanes confirm this deduction.† For reference and calibration of our assumptions, solid [2.2]metaparacyclophane is 3.3 (±0.3) kcal/mol more stable than solid [2.2]paracyclophane, whereas the difference is increased to 5.4 (±0.8) kcal/mol for the corresponding gaseous species. It is important to recall our assumption that, if A rearranges to B, then B is anticipated to be more stable than A. It was not stated that if A does not rearrange to B, A is more stable than B. In particular, the [2.2]metaparacyclophane formed by the above-cited rearrangement of [2.2]paracyclophane does not immediately rearrange to the [2.2]metacyclophane, even though the last species is 10.1 (±1.5) kcal/mol more stable than the metapara isomer in the solid phase and 11.4 (±1.6) kcal/mol for the corresponding gases. More precisely, the yield of [2.2]metacyclophane is significantly less than that of the metapara isomer, a most surprising finding given the relative stabilities of the isomeric [2.2]cyclophanes.‡ This can be explained by drawing the parallel between [2.2]paracyclophane and two *p*-xylene molecules and of [2.2]metaparacyclophane and one each of *m*- and *p*-xylene. We ignore for now all transannular or interring effects that additionally distinguish the cyclophanes

* These brackets ⟨ ⟩ are used throughout the chapter to denote particularly theoretical statements that may be omitted by a more experimentally oriented reader with no loss of continuity.

† All thermochemical citations in this chapter, unless otherwise explicitly cited, refer implicitly to Pedley and Rylance's compendium.[7] ⟨All thermochemical quantities and citations in this chapter, unless otherwise stated, refer to the ideal gas state at 298K at 1 atm pressure. Although perhaps not the customary state for considering organic compounds by the typical experimental organic chemist, it has the virtue of being a thermodynamically well-defined state and is the usual state to be employed when universality is desired.⟩

‡ Some doggerel may be useful to keep this point in mind:

There once was a [2.2]cyclophane
That had some 30 kcal of strain
Acid added, got better
One para link went meta
But why did one para remain?

from each other and from the xylenes. The *ipso*- and 2-carbons of *p*-xylene are of comparable basicity[8]; that is, the proton affinities[8–10] of both sites are essentially equal. Protonation of the *ipso*-carbon of [2.2]paracyclophane to form species **3a** is precisely that which is needed to allow for the rearrangement in acid to lose one para ring to form the metapara isomer. However, *m*-xylene is more basic than its *p* isomer, whether it be protonated at the 2- or 4-carbon. Analogous protonation of the [2.2]metaparacyclophane results in "inert" carbonium ions (e.g., **3b**) that cannot rearrange to [2.2]metacyclophane.

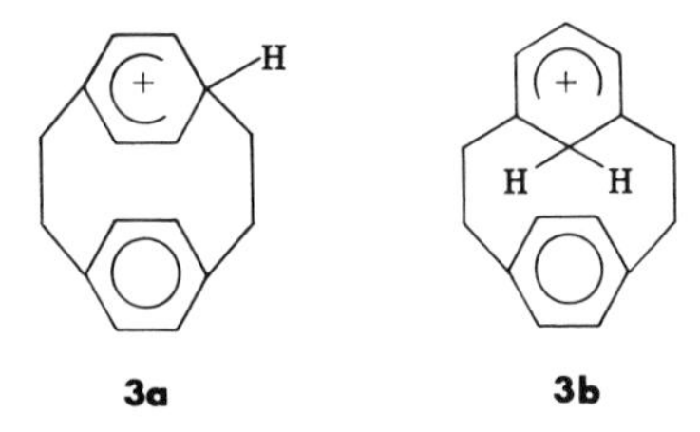

We note in addition that [2.2]metaparacyclophane can be photolyzed to the [2.2]metacyclophane[11] in satisfactory yield. This finding, however, cannot be used to prove that the metacyclophane is the more stable, because energy is being put into the system in the form of light.

D. Introduction to Interrelationships between the Chemistry of Cyclophanes and That of Other Species

The second section of this chapter explicitly considers the interrelationships of cyclophane chemistry and that of other chemical species. This choice of topic arose from the author's optimism that cyclophane chemistry, however strange or spectacular, should have its parallels, if not roots, in more conventional species. Simultaneously, cyclophane chemists should be able to provide insights into other areas of chemistry if by no other means than having a different set of footnotes to cite in their articles.

1. A Preliminary: The Benzidine Rearrangement and Cyclophanes

Indeed, Cram's pioneering research in this area was motivated in part by attempts to better understand the rather ancient acid-catalyzed benzidine rearrangement of hydrazobenzene[12,13] (see Chapter 1). Paralleling this is the observation that one reason[14] for the high reactivity of [2.2]paracyclophane in reactions involving electrophilic attack is that it entails a stabilized intermediate **(4a)** analogous to that invoked in the benzidine rearrangement.[15] An unbridged noncyclophane analog of this complex **(4b),** the protonated benzene–neutral benzene cluster, has been studied,[16] and for this analog a binding energy of 11 kcal/mol has been obtained.

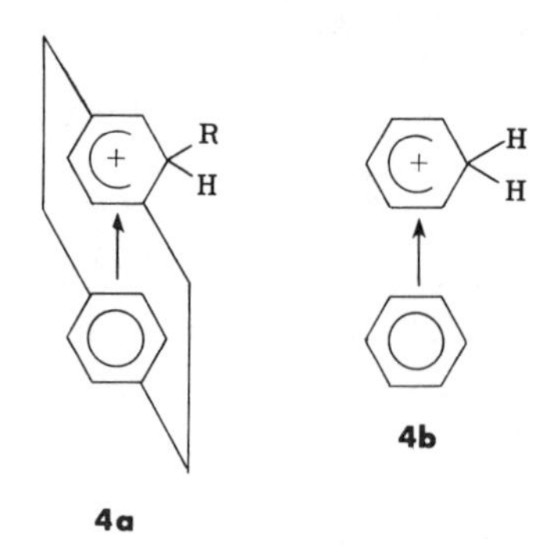

4a **4b**

2. A Preliminary: Hydrogen-Bonded Dimers and Cyclophanes

A second example may also be useful. Over 20 years ago, Anet[17] deduced the presence of doubly hydrogen-bonded dimers of *m*-nitrophenol **(5a)** and *m*-nitrobenzoate esters **(5b)** from analysis of their nmr spectra taken in CCl_4, which is a low dielectric constant, nonpolar, and poor hydrogen-bonding solvent. In a discussion (1981) with him about this, I found myself quite skeptical of these species and the structures presented until we discussed the parallels of these dimers with (1) the Watson–Crick hydrogen-bonding arrangement of nucleic acid bases in DNA[18] and (2) [4.4]metacyclophanes if one considers the N and O of the nitro group (or

O
N—O··· H—O
O—H··· O—N
O

5a

CH_3O
C=O···H—O
$CH(CH_3)_2$
$(CH_3)_2CH$
O—H··· O=C
OCH_3

5b

C and O of the ester) and the O and H of the hydroxyl group to be constituent atoms of the bridges. It is interesting that Anet had done his experimental study at Ottawa, moved to UCLA (Cram's institution), and only years later became aware of this parallel of these dimers with cyclophanes.[19] It is also noteworthy how few cyclophanes lacking carbons in the bridges are known. Interestingly, most of these species have polysulfide bridges, for example, the isomeric hexathia[3.3]ortho-, hexathia[3.3]meta-, and hexathia[3.3]paracyclophanes (**6a, 6b,** and **6c**).[20]

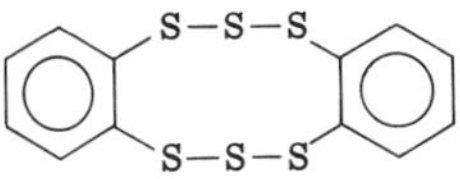

S—S—S S—S—S S—S—S

S—S—S S—S—S S—S—S

6a 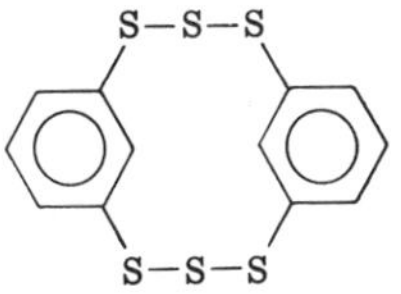 **6b** 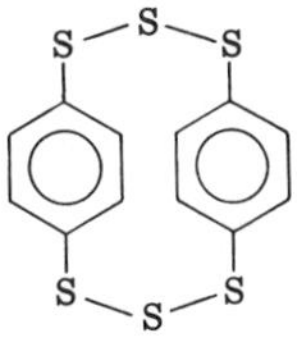 **6c**

In a chapter such as this it is the author's choice, if not obligation, to present salient examples of these interrelationships between the chemistry of cyclophanes and more conventional chemistry. I have chosen examples almost exclusively from the 1980, 1981, and early 1982 chemical literature.* The intent of this chapter is to encourage further research involving cyclophanes as well as to chronicle the results already obtained.

II. ENERGETICS CONSIDERATIONS RELATED TO CYCLOPHANES

A. Neutral Cyclophanes

1. Strain Energies of [*n*]Paracyclophanes

The [*n*]paracyclophanes are the conceptually simplest class of cyclophanes, but the heat of formation remains unmeasured by calorimetry for any of these species. Although theoretical chemical calculations have been reported, the results seem contradictory and are most assuredly

* In particular, I have surveyed the *Journal of the American Chemical Society, Journal of Organic Chemistry, Tetrahedron* and *Tetrahedron Letters, Angewandte Chemie (International Edition), Chemische Berichte, Journal of the Chemical Society Perkin(s)* and *Chemical Communications, Bulletin of the Chemical Society of Japan* and *Chemistry Letters,* and *Helvetica Chimica Acta.*

TABLE I

Strain Energies (SE) and Bending Angle ($\Delta\theta$) of [*n*]Paracyclophanes[a]

n	SE (mol mech)	$\Delta\theta$ (mm, deg)[21]	SE (MNDO)	$\Delta\theta$ (qc, deg)[22]
7	21	18.2	32	19.9
6	29	22.4	45	25.3
5	39	26.5	63	31.7

[a] Strain energies expressed as kcal/mol.

incomplete. There are but three species for which both molecular mechanical[21] and quantum chemical MNDO[22] calculations have been reported: $n = 5$, 6, and 7. Although the geometries are comparable for the two calculational approaches (i.e., $\Delta\theta$ or angle of bending of the benzene ring), there is a significant difference in predicted strain energies (Table I). In the absence of experimental data, we know of no unequivocal way of deciding which set of numbers to use. However, there are other approaches that may be employed to provide alternative estimates. The results are displayed in Table II, and the flaws discussed afterward.

The first approach asks a simple yes–no (y/n) question: Is the [*n*]paracyclophane **7a** more stable than its corresponding Dewar analog **7b**? The experimental crossover point from no to yes is between $n = 5$ and 6.[23,24]

$(CH_2)_n$ $(CH_2)_n$

7a 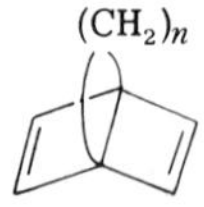 **7b**

Because the energy difference[25,26] between the "normal" and Dewar benzene is ~60 kcal/mol, a "no" answer to the question suggests that the [*n*]paracyclophane is strained by more than this 60 kcal/mol.

In the next approach, labeled br, [*n*]paracyclophanes are considered to be bridgehead or anti-Bredt olefins **(8a)**.[27,28] As such, we can estimate the strain energy by taking twice that found for the suitable bicyclo[*n*.2.2]alkene **8b** and subtracting that of the bicyclo[*n*.2.2]alkane **8c.**

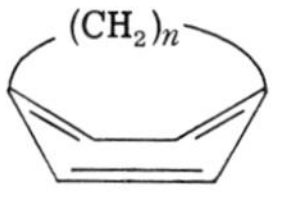

$(CH_2)_n$ $(CH_2)_n$ H H $(CH_2)_n$ H

8a 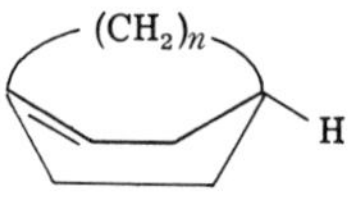 **8b** 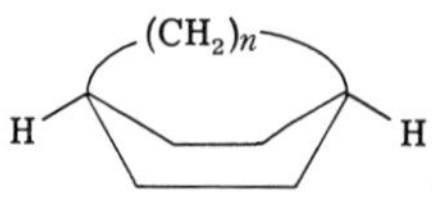 **8c**

TABLE II
Strain Energies of [*n*]Paracyclophanes[a]

[*n*]	mol mech[b]	MNDO[c]	y/n[d]	br[e]	qcm[f]	qcq[g]
10	15	—	<60	—	—	—
9	13	—	<60	—	—	—
8	17	—	<60	—	4	—
7	21	32	<60	—	9	11
6	29	45	<60	—	14	18
5	39	63	>60	—	19	27
4	—	—	>60	36	—	—
3	—	—	>60	56	—	—

[a] Strain energies expressed as kcal/mol.
[b] Literature molecular mechanical calculations.[21]
[c] Literature MNDO calculations.[22]
[d] A yes–no question: Is the [*n*]paracyclophane more stable than its Dewar isomer?
[e] Setting the paracyclophane strain energy equal to 2 × SE(anti-Bredt bicyclo[*n*.2.2]alkene) − SE(bicyclo[*n*.2.2]alkane).
[f] A harmonic oscillator model using the distortion angle $\Delta\theta$ from literature molecular mechanical calculations.[21]
[g] Same as described in footnote *f* but taking $\Delta\theta$ from literature MNDO calculations.[22]

(All bicycloalkene and bicycloalkane strain energies were taken from Maier and Schleyer.[28])

The third approach (qc) utilizes quantum chemical ⟨small basis, *ab initio*⟩ calculations on the amount of energy needed to distort a benzene ring.[29] Ideally, the energy of distortion ΔE equals $\frac{1}{2}k(\Delta\theta)^2$. ⟨This corresponds to the idealization of a harmonic oscillator model. To verify the validity of this, we optimized the fit of $\ln(\Delta E) = m \ln(\Delta\theta) + \ln(b)$, in which m is the best exponent fit in $\Delta E = b(\Delta\theta)^m$. The result was that $m = 1.981$, $b = -3.567$, and the correlation coefficient $r = 0.99998$! Alternatively, one may assume a quadratic fit and evaluate $\ln(\Delta E) - 2 \ln(\Delta\theta) = k$ and then average k. This gives $k = -3.599$ (± 0.029).⟩ This approach clearly neglects any effects due to the polymethylene bridge, and in addition the need to decide which $\Delta\theta$ to use creates ambiguity. Fixing this bending angle from molecular and quantum chemical calculations, respectively, generates the qcm and qcq numbers.

Table II shows numerous ambiguities as well as omissions. Moreover, each of the approaches just described is flawed in some way. For example, in molecular mechanics it is not obvious whether the available force fields and parameters are appropriate for distortions as marked as those found in the smaller paracyclophanes. An opportunity to check this was lost when Allinger *et al.*[21] did not also calculate the strain energies for any [*m.n*]paracyclophanes for which there exist calorimetric data (see following discussion). The MNDO quantum chemical calculation possibly has analogous flaws because it is not obvious whether semiempirical calculations are appropriate here. ⟨Indeed, some of the literature is "skirmish" about the general validity of semiempirical versus *ab initio* calculations.[30–32]⟩

The final quantum chemical (qc) approach, as earlier admitted, also suffers because it does not take into account the polymethylene chain. Furthermore, the simple quadratic dependence of bending energy on angle eventually must fail for large enough distortions. It must also be noted that *ab initio* chemical theory as normally practiced may well be inappropriate in these cases. ⟨Not only may *d* orbitals be quantitatively, and qualitatively, essential here, as with strained small rings,[33] but we expect configuration interaction[34] to be important. In particular, we note that one of the degenerate, highest occupied molecular orbitals becomes 1,4 antibonding and so is destabilized on bending,[22,35] whereas one of the degenerate, lowest unoccupied molecular orbitals is 1,4 bonding and is likewise stabilized on bending.[35] This suggests that the "doubles excitation,"

$$\phi^2_{\mathrm{HOMO}}\phi^0_{\mathrm{LUMO}} \rightarrow \phi^0_{\mathrm{HOMO}}\phi^2_{\mathrm{LUMO}}$$

will be increasingly important as the benzene ring bends out of planarity and so will preferentially stabilize the [*n*]paracyclophanes.⟩

Nor are the other approaches without flaws. The yes–no approach is clearly simplistic because the value of 60 kcal/mol neglects any effect of the polymethylene chain, whereas the experiment implicitly includes it. How great an effect this chain has is difficult to estimate. For example, in the paracyclophane the chain is clearly strained, but the Dewar analog is an [*n*.2.2]propelladiene and so is also strained.[36] The second, br approach, when used to compare [*n*]paracyclophanes with [*n*.2.2]bicycloalkenes, implicitly neglects the fact that twisting an olefin is conceptually distinct from bending a benzene ring because of the presence of the aromaticity in the latter.

2. [*n*]Metacyclophanes

Quantitative data for [*n*]metacyclophanes are sparser still. It is experimentally known that conditions that led to successful syntheses of [5]- and [6]metacyclophanes **(9a)**[37,38] resulted in the Dewar analog for $n = 4$ **(9b)**.[39] It is well established[28] that bridgehead bicyclo[*n*.3.1]alkenes are considerably more stable with the double bond on the "3" bridge than on the "1" bridge. Using the earlier br logic and one each of the strain energies for the "3"- and "1"-bridge olefins **9c** and **9d,** respectively, we

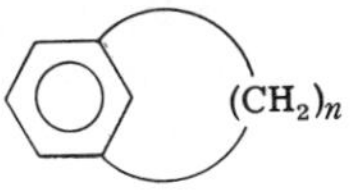

(CH$_2$)$_n$ (CH$_2$)$_n$ (CH$_2$)$_n$ (CH$_2$)$_n$

9a 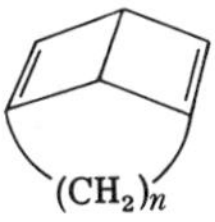 9b 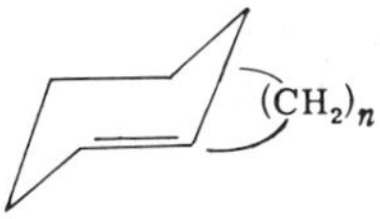 9c 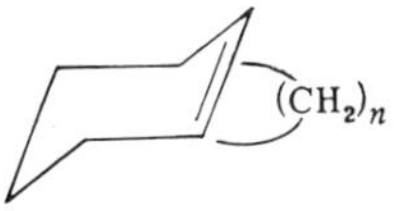 9d

deduce a strain energy of 40 kcal/mol for [4]metacyclophane. Finally, since a [5]metacyclophane precursor fails to aromatize, whereas the higher analogs readily did so (see Section III,D,4) this suggests that the strain energy of [5]metacyclophane exceeds the resonance energy of benzene, ~36 kcal/mol.

Turning to benzenoid analogs of [*n*]metacyclophanes, we consider the [*n*](1,3)naphthalenophanes **10.** It is interesting that species (where X = Cl,

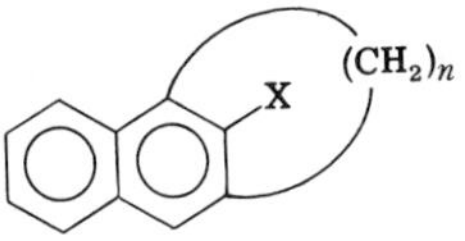

X (CH$_2$)$_n$

10

Br) with $n = 6$ and 8 have been isolated, whereas that with $n = 5$ has been suggested to be a reaction intermediate[40] but appears not to be isolable. This might seem discordant with the seeming stability of [5]metacyclophane, especially because naphthalene is more flexible than benzene.[29] Two explanations are offered now. The first argues that the greater flexibility of the less strained species allows it to rearrange into even less strained species such as the isolated "orthonaphthalenophane" (2,3-cycloheptenonaphthalene, **11**). The second explanation notes that the 2-halo substituent undoubtedly introduces additional strain. This suggests carrying out the earlier experiments with either H or F instead of Cl or Br.

3. Nonbenzenoid [*n*]Cyclophanes

[*n*](2,6)Tropylidenophanes **12a** are known to interconvert with [*n*.4.1]propelladienes **12b.** For $n = 4$, the equilibrium lies to the left,

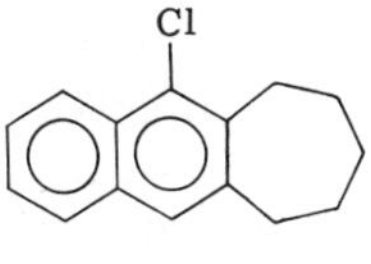

Cl (CH$_2$)$_n$ CH$_2$ (CH$_2$)$_n$

11 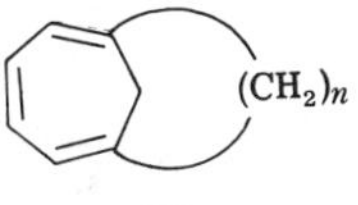 12a 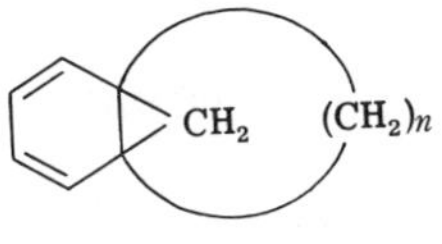 12b

whereas for $n = 3$ it lies to the right.[41] Some of the related $[n](2,6)$troponophanes **13a** have also been synthesized.[42] The $n = 6$ species could not be protonated, suggesting that the hydroxy substituent is too big in **13b.** However, the same authors also chronicled a lack of success in

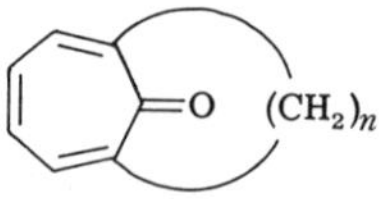

13a 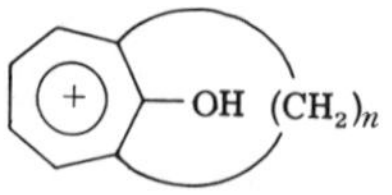 **13b**

forming a [6]tropyliophane without the "offending" OH group. This strongly suggests that the inhibition of solvation may be an important factor in these failures. Gas-phase proton affinity measurements[9,10] would be helpful in this regard. Because the unbridged tropone is thermodynamically unstable relative to benzene + CO by approximately 3.9 (±0.8) kcal/mol, it is not obvious whether small troponophanes will decompose to the corresponding $[n]$orthocyclophane and CO. Turning to larger values of n, we note that [9]troponophane appears to be normal.

In contrast to the lack of data for metacyclophanes or (2,6)troponophanes, experimental thermochemical data are available for the related $[n](2,6)$benzotroponophanes **14a,** although only for $n = 5$ and 12 and for the "acyclic" 2,6-dimethylbenzotropone **14b.** One index of the strain energy of these species is the difference in the heat of formation of the 2-R, 6-R′ benzotropone derivative **14c** and that of RCH_2COCH_2R'. For $R = R' = CH_3$, $R = R' = —(CH_2)_5—$, and $R = R' = —(CH_2)_{12}—$ the derived

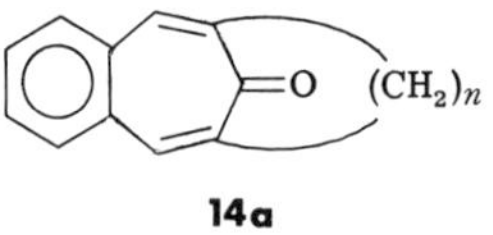

14a 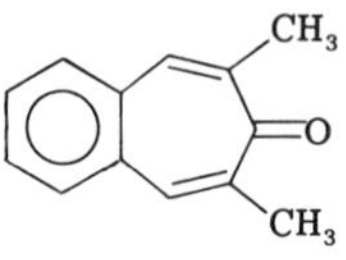 **14b** 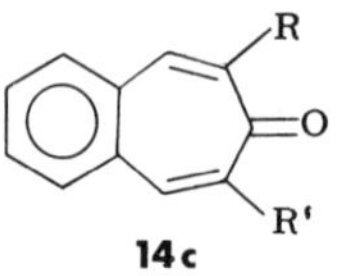 **14c**

differences are 61.2 (±2.7), 97.7 (±3.2), and 60.9 (±4.3) kcal/mol. The equality, within error bars, for $n = 12$ and the unbridged species suggests that [12](2,6)benzotroponophane is unstrained. In contrast, the [5] analog is strained by about 36 kcal/mol, essentially equal to the estimate given for [5]metacyclophane from the aromatization logic discussed previously.

4. $[m.n]$Cyclophanes

Although quantum chemical calculations are absent here, there are some molecular mechanical and experimental data. Table III presents the experimental heats of formation and derived strain energies. Strain en-

TABLE III

Strain Energies of Various [*m.n*]Cyclophanes[a]

Compound	ΔH_f	SE_1[b]	SE_2[c]	SE_3[c]	SE_4[d]	SE_5[d]
[2.2]Para	57.6 (±0.7)	32[e]	28	32	32	29
[2.2]Metapara	52.2 (±0.4)	24	22	27	27	24
[2.2]Meta	40.8 (±1.6)	12	15	17	15	12
[3.3]Para	31.0 (±0.5)	12	—	—	15	12
[6.6]Para	−18.6 (±2.3)	−7	—	—	−5	−8
[1.8]Para	6.9 (±1.8)	2	—	—	6	3

[a] Strain energies (SE) and heats of formation (ΔH_f) are expressed as kcal/mol. All the strain energies in this table would change if the more recent value of the heat of formation of [2.2]paracyclophane (58.9 (±0.8) kcal/mol) by Nishiyama *et al.*[45a] were used.

[b] See molecular mechanical calculations of Shieh *et al.*[43]

[c] Molecular mechanical calculation of Lindner.[44]

[d] See text.

[e] The average of the molecular mechanical calculations of Shieh *et al.*[43] and Boyd.[45]

ergy approaches SE_1–SE_3 are defined in the articles cited. We commence with the following simple test of the accuracy of the molecular mechanical calculations. Consider the energetics of Eq. (3), in which the discrepancy is to be identified with the change in strain energies:

$$\text{[2.2]Paracyclophane} + m\text{-}CH_3C_6H_4CH_3 + m\text{- (or } p\text{-) } CH_3C_6H_4CH_3 = \text{[2.2]meta- (or metapara-) cyclophane)} + 2p\text{-}CH_3C_6H_4CH_3 \quad (3)$$

Because the heats of formation of the isomeric xylenes are equal within experimental error and so may be canceled out, the difference in the strain energies of the three isomeric [2.2]cyclophanes "merely" equals the difference in the heats of formation. Simple arithmetic shows that SE_3 is apparently more accurate than either SE_1 or SE_2. Because adding a constant to both sides of Eq. (3) cannot affect anything, we may equally accurately use

$$\text{[2.2]Paracyclophane} + 2\ CH_3CH_2(CH_2)_2CH_2CH_3 = \text{[2.2]meta- (or metapara-) cyclophane} + 2\ CH_3CH_2(CH_2)_2CH_2CH_3 \quad (4)$$

Generalizing this to more arbitrary cyclophanes, we may consider and so define SE_4 by the discrepancy of the relationship

$$\text{[2.2]Paracyclophane} + CH_3CH_2(CH_2)_mCH_2CH_3 + CH_3CH_2(CH_2)_nCH_2CH_3 = [m.n]\text{cyclophane} + 2\ CH_3CH_2(CH_2)_2CH_2CH_3 \quad (5)$$

It is surprising that the difference between SE_4 and SE_1 is nearly a constant 3 (±1) kcal/mol except for [2.2]paracyclophane itself. Because this cyclophane is probably the most studied and best understood, and as such a potentially good candidate as a paradigm, this discrepancy is also disconcerting.

To define SE_5 we deduce the strain energy of [2.2]paracyclophane by the reaction

$$\text{[2.2]Paracyclophane} + 4\ CH_3CH_2CH_3 = 2\ CH_3C_6H_4CH_3 + 2\ CH_3CH_2(CH_2)_2CH_2CH_3 \quad (6)$$

Using this reaction (very closely related to that used elsewhere[43]) we find a strain energy of 29.0 (±0.8) kcal/mol for [2.2]paracyclophane. Accepting this new value for future studies, we use it in Eq. (5) and name the resultant discrepancy SE_5, our final and most recommended strain energy approach for [*m.n*]cyclophanes.

5. [2.2]Orthocyclophane

For completeness and applications later in this section, we now consider [2.2]orthocyclophane. No thermochemical data exist for this species. However, viewing it as a dibenzocyclooctadiene suggests a simple way of estimating its heat of formation. Namely, Table IV shows that benzoannelation of four- to six-membered rings is accompanied by an increase in heat of formation of ~7 kcal/mol. [An acyclic example of benzoannelation, *o*-xylene versus (*Z*)-2-butene, gives 6.2 (±0.2) kcal/mol.] This suggests that the heat of formation of [2.2]orthocyclophane is $\sim 14 + 2 \times 7 = 28$ kcal/mol, considerably less than that of its meta, para, and metapara isomers.

TABLE IV

Energetics of Benzoannelation, Where $\Delta^2H_f = \Delta H_f(\text{Benzocycloalkene}) - \Delta H_f(\text{Cycloalkene})$

Cycloalkene	ΔH_f (kcal/mol)	ΔH_f(benz) (kcal/mol)	$\Delta^2 H_f$ (kcal/mol)
Cyclobutene	37.5 (±0.4)	47.7 (±0.2)[a]	10.2 (±0.5)
Cyclopentene	7.8 (±0.4)	14.5 (±0.4)	6.7 (±0.6)
Cyclohexene	−1.1 (±0.1)	5.8 (±0.5)	6.9 (±0.5)
1,4-Cyclohexadiene	26.3[b]	31.6[c]	5.3
		38.2[d]	2(6.0)[d]

[a] See Roth *et al.*[46]
[b] See Benson and O'Neal.[47]
[c] The experimental data are for the liquid; we used the generally more accurate, two-parameter Eq. (2) of Chickos *et al.*[5]
[d] This is for dibenzoannelation.

6. Multibridged Cyclophanes

Strain energy calculations[44] showed that the $[2_3](1,2,4)$cyclophane is more stable than the $[2_3](1,3,5)$ isomer, whereas experimental isomerization studies on the former showed it to be less stable than the $[2_3](1,2,4)(1,2,5)$ isomer.[48] There is no direct thermochemical information on the $[2_3](1,2,3)$ isomer. However, indirect evidence may be given. Heating of the $[2_4](1,2,3,5)$cyclophane in the presence of hydrogen atom donors resulted in a dimethyl derivative of the $[2_3](1,2,3)$cyclophane.[48] Let us assume that the stability of the three intermediate benzyl radical pairs that can be formed by simple C—C bond homolysis and the stability of the dimethyl derivatives that are formed by hydrogen atom abstraction mimic that of the parent triply bridged cyclophanes. From this we deduce that this most recently isolated[49] triply bridged species, the $[2_3](1,2,3)$cyclophane, is the most stable. These results on $[2_3]$cyclophanes are *a posteriori* reasonable because the (1,2,3) isomer incorporates two [2.2]orthocyclophane and one [2.2]metacyclophane substructures, whereas the $[2_3](1,2,4)$ isomer incorporates one each of the ortho-, meta-, and para-, the $[2_3](1,2,4)(1,3,5)$ isomer incorporates one ortho- and two metapara-, and the $[2_3](1,3,5)$ isomer incorporates three [2.2]metacyclophane substructures. ⟨This substructure analysis of relative strain energies is not, however, quantitatively justified.[50]⟩ This result is corroborated if we recall that the analogous reaction of $[2_3](1,2,4)$cyclophane resulted in a derivative of [2.2]orthocyclophane, the most stable doubly bridged cyclophane.

We can estimate lower bounds of the heat of formation of numerous bridged cyclophanes by employing our earlier assumptions relating ΔG and ΔH for isomerization reactions. For example, we begin a simple estimate of the heat of formation of $[2_3](1,2,3)$cyclophane by recalling the thermal isomerization reaction of **15** to **16.** Although no data exist on the

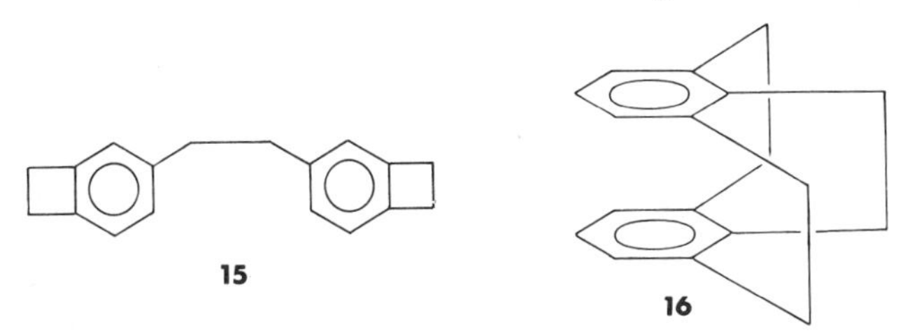

heat of formation of the benzocyclobutene **15,** it can simply and reliably[26] be estimated from Eq. (7):

$$2\ \Delta H_f(\text{benzocyclobutene}) + \Delta H_f(C_6H_5CH_2CH_2C_6H_5) - 2\ \Delta H_f(C_6H_6) = 2(47.7) + 34.2 - 2(19.8) = 90.0\ \text{kcal/mol} \quad (7)$$

As such, the heat of formation of the cyclophane may safely be assumed to be less than 90.0 kcal/mol. Analogously, the heats of formation of $[2_3]$(1,2,4)cyclophane and $[2_5]$(1,2,3,4,5)cyclophane are less than 90 and 146 kcal/mol, respectively. Equivalently, because the strain energies of cyclobutene and benzocyclobutene are comparable, ~31 kcal/mol, we may deduce that the strain energy in **15** is about 2×31 kcal/mol and so in **16** is less than 62 kcal/mol. Analogously, the strain energy of the $[2_5]$(1,2,3,4,5)cyclophane is less than $4 \times 31 = 124$ kcal/mol.

Likewise, that benzo(1,2;4,5)bicyclobutene dimerizes to $[2_4]$(1,2,4,5) cyclophane[51] suggests that the strain energy in this cyclophane is less than 4×31 kcal/mol. However, we can do better than that. An intermediate in this reaction is no doubt a cyclooctadi(benzocyclobutene)[bis(benzocyclobuteno)cyclooctadiene] **(17),** because such a five-ring compound (with

17

$X = CH_3$) is isolated from the pyrolysis of the 3,6-dimethylbenzo[1,2;4,5] bicyclobutene.[51] Assuming that the middle three rings are strainless, we deduce the strain energy of this cyclophane to be less than that of two benzocyclobutenes, or ~62 kcal/mol.

We conclude by noting that no experimental thermochemistry exists on any of the multibridged cyclophanes. Strain energy calculations[44] give either 36 or 49 kcal/mol for the $[2_4]$(1,2,4,5)cyclophane, consistent with the value given in the previous discussion, and either 60 or 79 kcal/mol for $[2_6]$(1,2,3,4,5,6)cyclophane, or "superphane" **(18).**[52–54]

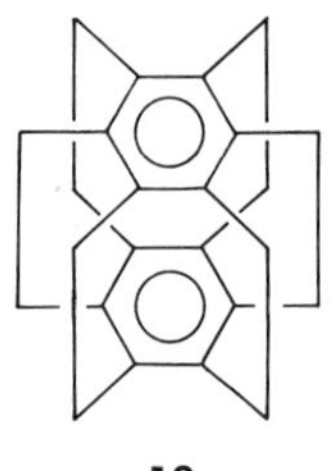

18

Molecular and quantum chemical calculations on the remaining cyclophanes and calorimetric experiments on all of the cyclophanes are sorely needed.

B. Photochemistry and Excited States

1. Isomerization, Cycloaddition Reactions, and Syntheses

Photochemistry has long been part of cyclophane chemistry, both in isomerization studies, for example, [6]paracyclophane ⇌ its Dewar isomer (**7b:** $n = 6$)[55] and [2.2]metaparacyclophane → [2.2]metacyclophane,[11] and in cycloaddition reactions, for example, of the [2.2](1,4)naphthalenophanes.[56,57] Various cyclophane syntheses have utilized photochemical reactions, for example, some sulfur extrusion steps in multibridged cyclophane syntheses,[53] a one-step synthesis[58] of triscyclobutano[2_3](1,3,5)cyclophane **(19a)** from tristyrylbenzene **(19b)**, and cyclization of an

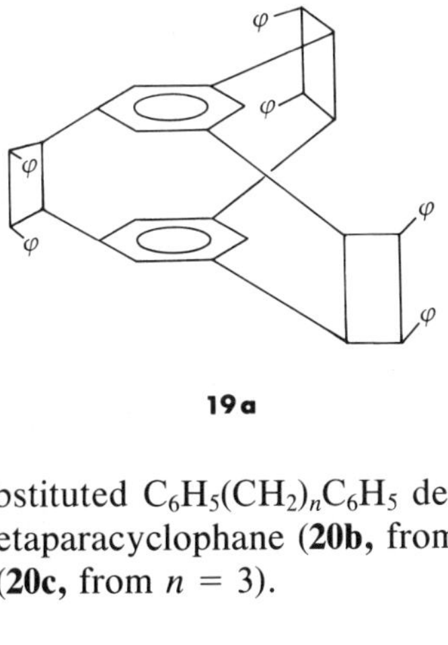

19a

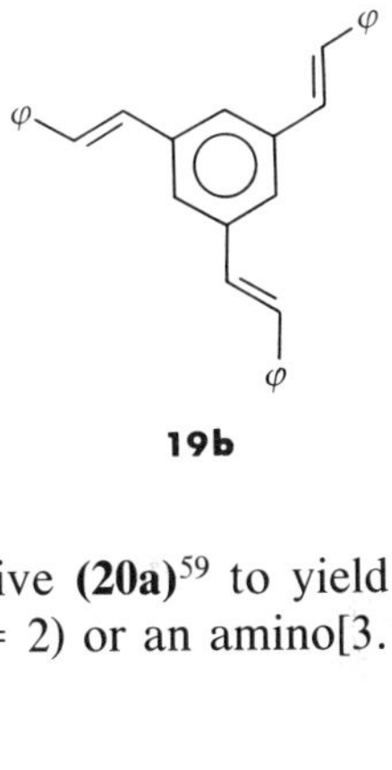

19b

m,p′-disubstituted $C_6H_5(CH_2)_nC_6H_5$ derivative **(20a)**[59] to yield either an aza[2.3]metaparacyclophane (**20b,** from $n = 2$) or an amino[3.1]metacyclophane (**20c,** from $n = 3$).

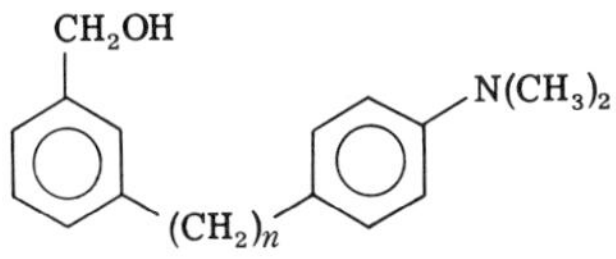

20a 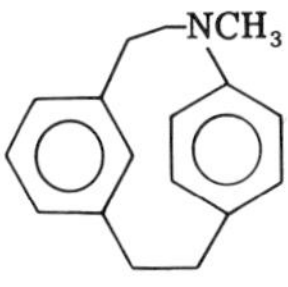 **20b** 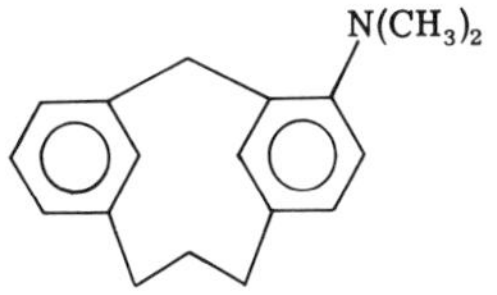 **20c**

Given these successes, it is surprising that such studies appear to be quite rare. We wonder if this seeming lack of interest is due to the presumed high strain energy of the cyclophanes, the intended high energy of the precursors, and thus the lack of need for photochemical excitation. For example, *p*-xylylene is high enough in energy that even "forbidden" reactions such as its dimerization to [2.2]paracyclophane proceed readily.

Using data on *p*-xylylene[60] and on [2.2]paracyclophane,[7] we deduce that this reaction is exothermic by ~42 kcal/mol (see also Section III,B,2).

2. Excited States and Spectral Surprises

With regard to excited states, ever since cyclophanes were first systematically studied,[61,62] it has been apparent that spectroscopic "abnormalities [may be] attributed to interstitial [interring] resonance effects and to distortion of the benzene rings from planarity."[63] Skipping over about 30 years of results and insights, we note some relevant recent results. The first is for benzene clusters bound by van der Waals interactions and presumably without any ring distortions.[64,65] These ground state benzene oligomers are apparently composed of perpendicular rings,[64,65] whereas the excited states have parallel rings,[66,67] for example, $(C_6H_6)_2$ in **21a** and $[(C_6H_6)_2]^*$ in **21b.** In addition to the change in geometry upon excitation,

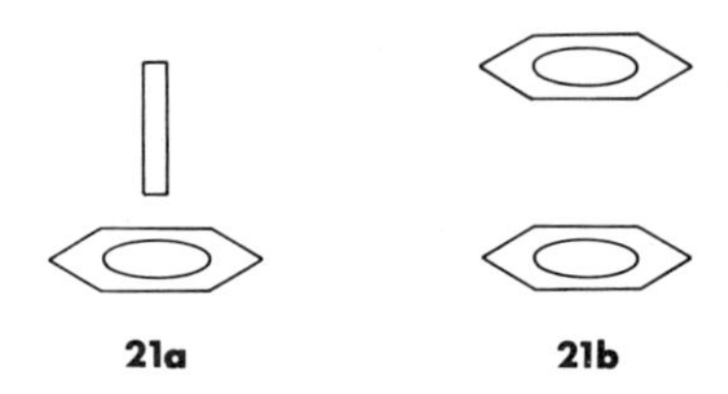

these species also show interesting changes in λ_{max} and excited state lifetimes as a function of aggregation.[66,67] Another study, of [2.*n*](9,10)anthracenophanes,[68] can be compared with the earlier studies of [2.*n*]paracyclophanes.[61,62] Because the study of cyclophanes was carried out in room temperature cyclohexane solution, whereas that for the anthracenophanes was performed either in a methylcyclohexane–decalin glass at 77K or with the anthracenophanes as "guest molecules in the photoisomer crystals [p. 199],"[68] new investigations under a common set of conditions would have to be performed before unambiguous comparisons could be made. Furthermore, in that [1.*n*](9,10)anthracenophanes have been suggested to be relevant,[68] we note that [1.*n*]paracyclophanes are well characterized[69,70] for facile comparison.

A third study contrasts three sets of cyclophanes in which, for each set, there are two species with the same structural formula but with different arrangements of the rings. These alternate species are generally referred to as diastereomers (if chiral) or as syn–anti isomers. For each of the three sets, the [3.3](2,6)-,[71] [3.3](1,5)(2,6)-,[71] and [3.2](1,4)naphthalenophanes,[72] significant spectral differences (both λ_{max} and ε_{max}) within each pair were observed.

3. A Caveat

We close this section with a warning. Photosensitized oxidation[73] of benzo[2.2]paracyclophane **(22a)** in $CH_3OH–C_6H_6$ solution resulted in the formation of 4-methoxy[2.2](1,3)naphthalenoparacyclophane **(22b).** The

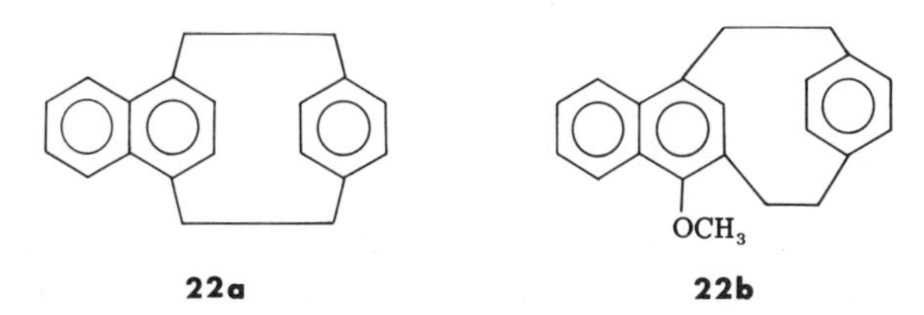

formation of the latter is not a photochemical rearrangement of a para- to a metaparacyclophane. Instead, the sole photochemistry component of this reaction appears to be the formation of singlet oxygen. Subsequent addition to form the endoperoxide, methanolysis and accompanying hydroperoxide formation, and loss of HOO^- results in a carbonium ion that rearranges in the same manner as that responsible for the acid-catalyzed para- to metaparacyclophane rearrangement noted earlier.[3] Loss of HOO^- is a surprising result but may also be invoked for the oxidation of dihydrosuperphane in air to the parent superphane.[54] We conclude by emphasizing that cyclophane rearrangements may be assisted by heat, light, and acids and that the nature of the reaction mechanism is rarely *a priori* obvious.

C. Charge-Transfer Complexes, Radical-Cations, and Ionization Potentials

1. Tetracyanoethylene and [*m.n*]Paracyclophanes

The first study of complexes of cyclophanes was described[74] soon after the synthesis[75] and powerful π-accepting properties of tetracyanoethylene (TCNE) were reported[76]. In particular, by the use of absorption transition energies as a criterion, the following order of cyclophane π basicity was derived: [3.3] > [3.4] > [2.2] > [3.6] > [2.3] > [4.6] > [6.6] > [5.6] ≈ [5.5] ≈ [4.5] > [2.4] > [4.4] > *p*-xylene. Various factors were deduced to be important contributors to this nonobvious order: (*a*) the degree of participation of each ring in the delocalization of charge modulated by interring distance and by ring nonplanarity and (*b*) the degree of hyperconjugation, modulated by σ–π inseparability and distortion of the —CH_2— groups. However complicated, the variation in absorption transition energies is nonnegligible; whereas [4.4]paracyclophane and *p*-xylene are equal within 2 kcal/mol, there is a 15 kcal/mol difference between the [3.3]- and [4.4] paracyclophanes. Strong solvent effects were also seen; for the TCNE complex of [3.4]paracyclophane, for example, there was an 88-nm shift on changing from CH_2Cl_2 to *tert*-BuOH. There is also an additional, essentially inseparable solvent dependence on the particular cyclophane. For example, in CH_2Cl_2, the λ_{max} of the TCNE complex of [3.4]paracyclophane is greater than those for [2.2]-, [4.4]-, and [6.6]paracyclophanes, whereas the inequality is reversed in *tert*-BuOH.

In principle, clarification might be achieved by chemical or electrochemical oxidation to form the radical-cation and directly yield an energy of ionization or oxidation potential. It is surprising that few cyclophane radical-cations have been observed in solution: 4,5,7,8-tetramethyl[2.2] paracyclophane **(23a)**[77] [which more readily forms the four-ring assemblage $(PC)_2^{\dagger}$ **(23b)**[78]], [2.2](9,10)anthracenophane **(23c)**,[79] and various 5,13-disubstituted [2.2]metacyclophanes **(23d)**.[80] Likewise, the thermochemistry of solid salts ⟨via Madelung-type reasoning⟩ would also be useful. Although such salts are precedented by $(naphthalene)_2^{\dagger}PF_6^-$ and $(perylene)_2^{\dagger}BF_4^-$,[81] we know of no analogous cyclophane salt.

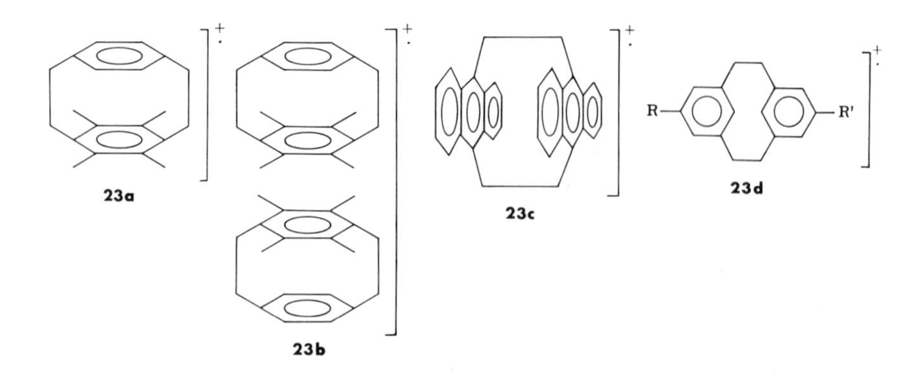

2. Multibridged Cyclophanes and Ionization Potentials

In this subsection, considerations analogous to those delineated in the previous discussion are immediately apparent. However, solution complexation studies with TCNE or other powerful π-acids have generally been thwarted because of Diels–Alder adduct formation.[82] For the latter cyclophanes, unlike most of the [*m.n*]cyclophanes mentioned earlier, there are many data and corresponding interpretations[83–87] for the related gas-phase quantity, the molecular ionization potentials.[88,89] This set of papers[83–87] presents a qualitatively and quantitatively unified orbital picture of these cyclophanes in which the orbitals of the the two rings are combined (see Section III,A,3 and hexaprismane in particular) and modified by the bridges via through-bond and through-space interactions. These papers also suggest that "the benzene moieties in [cyclophanes] . . . do not differ significantly in their electronic buildup from benzene itself [p. 2000]."[90] Despite a strong personal interest in ions, the author is somewhat brief here because (*a*) these papers represent a coherent, already packaged picture, (*b*) a thorough quantum chemical description would be too lengthy to include, and (*c*) it is almost impossible to compare the gas-phase and solution results because there are few comparable data on neutral species and ions. There are almost no cyclophanes for which solution complexation and gas-phase ionization potential data both exist. One fortunate example is for the [2.2]- and [3.3]paracyclophanes. The energy difference for the solution study, taken as the difference in transition energies,[74] is 54.9 − 47.7 = 7.2 kcal/mol. In contrast, the difference in ionization potentials[83] is 8.1 − 7.7 = 0.4 eV ≅ 9.2 kcal/mol. The agreement is encouraging.

3. Ionization Potentials of [*n*]Paracyclophanes

Data on the ionization potentials of [*n*]paracyclophanes are sparse. Measurements seemingly have been reported only for the parent [6], [7], and [8] species, 8.00,[22] 8.21,[22] and 8.17[91] eV, respectively, and for one

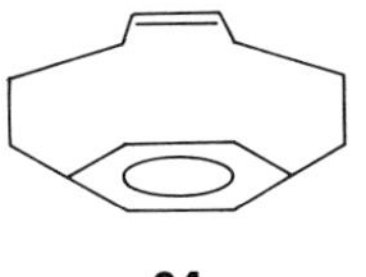

24

olefinic derivative **(24)**, for which a value of 8.37 eV was given.[91] It is surprising that there is no simple trend for the ionization potentials of the

[*n*]paracyclophanes and that a seemingly distant olefinic linkage so raises the ionization potential.

D. Radical-Anions of the Cyclophanes

1. [*m.n*]Paracyclophanes

In contrast to the rich and increasingly systematic chemistry of cyclophane radical-cations, considerably less is known about the corresponding radical-anions. The pioneering study, performed almost 25 years ago,[92] showed that the [1.8]- and [2.2]paracyclophanes are much more readily reduced than the [3.4], [4.4], and [6.6] species. In addition, these preliminary studies demonstrated that the degree of delocalization of the extra electron, that is, over one ring or two, depended on the number of bridging methylene carbons.

Somewhat over a decade later, it was shown[93] that the radical-anions of [*m.n*]paracyclophanes with m and $n = 2$ or 3 have the extra electrons delocalized, whereas for m and $n = 4$ (and presumably larger values) the extra electron is seemingly localized on one ring. In contrast, these authors noted that a two-carbon chain was sufficient to localize the extra electron in the radical-anions of open-chain diphenylalkanes. An interesting complication in these studies is the degree of association or pairing of the radical-anion with the gegen-cation. In particular, not only is there the obvious question of whether an association exists, but there is also the problem of the location of the cation, for example, by the polymethylene chain or on the quasi-sixfold axis of the benzene ring.[94] Codeposition of the cyclophane or diphenylalkane with an alkali metal and subsequent analysis of their vibrational and electronic spectra should prove highly instructive in determining the radical-anion structure, as it has for the simplest "acyclic" [*n*]cyclophane, benzene itself.[95] Indeed, both the benzene$^{\overline{\cdot}}$ K^+ or Rb^+ ion pair[92] and the cyclophane anions[88] discussed elsewhere were described as green. Such observations, although hardly precise, are nonetheless suggestive of similarities between the anions. Furthermore, it is well established that the parent benzene radical-anion, as well as its simple alkyl derivatives, is unbound, that is, unstable with respect to the hydrocarbon and a free electron, in the absence of solvent or ion pairing.[96] The anion of a formally related derivative of the [*m.n*]cyclophanes, two anthracene molecules plus an electron, has been theoretically studied.[97] It was shown that there is significant geometric and electronic reorganization in this model. It would thus be interesting to investigate the various [*n*]- and [*m.n*]paracyclophanes, as well as other cyclophanes, to ascertain whether they form bound anions in the gas phase.

2. Other Cyclophanes

A plethora of cyclophane radical-cations are described earlier in this chapter, and, with the number and diversity of cyclophanes described in this and other chapters, one might surmise that radical-anions of numerous cyclophanes other than the simple [*m.n*]paracyclophanes and their benzo analogs have been reported. It would also be reasonable to expect numerous comparisons between radical-cations and radical-anions. Surprisingly, neither expectation is realized. There are but two species known to the author for which both the radical-cation and radical-anion appear to be stable in solution: 4,5,7,8-tetramethyl[2.2]paracyclophane[78] and [2.2](9,10)-anthracenophane.[79] Synthesis of numerous radical-anions has apparently been thwarted by subsequent reaction. For example, reduction of the simple [2.2]metacyclophane resulted in the formation of the tetrahydropyrene radical-anion, and reduction of the [2.2]metacyclophanediene likewise results in the formation of pyrene radical-anion.[98]

It is interesting that, whereas [2.2]paracyclophane has a rich chemistry with generally no skeletal rearrangement, the thermodynamically more stable [2.2]metacyclophane often reacts by rearrangement and subsequent loss of hydrogen such as in the analogous radical-cation reaction.[99,100] The triply bridged $[2_3]$(1,3,5)cyclophane forms a dimethyltetrahydropyrene derivative, and, again, the related triene forms the pyrene radical-anion.[98] The seemingly stable radical-anion of $[2_4]$(1,2,4,5)cyclophane, a species incorporating both meta- and paracyclophane structural features, has been described and enjoys complications of ion pairing analogous to that of the more simple, doubly bridged paracyclophanes.[101] Additional studies of other multibridged cyclophanes are promised,[101,102] and their results are eagerly awaited.

We note in addition that numerous cyclophanes undergo Birch reduction, and so there are indirect data on the radical-anions of these species. However, the motivation for these studies has been to obtain some hydrogenated product and not to compare relative stabilities of the cyclophanes, their radical-anions, or any cyclophane versus its derived anion. The absence of quantitative data, other than product yields, has discouraged this author from chronicling these results.

Finally, we note the electrochemical reduction of [2.2.2.2]paracyclophanetetraene and related four-carbon-bridged analogs to form radical-anions, dianions, and even one tetraanion. Although a discussion of these neutral and ionic species is deferred to Section III,A,5, this does suggest that the electrochemistry of cyclophanes will prove to be both interesting and instructive.

III. INTERRELATIONSHIPS AND INSIGHTS REGARDING OTHER CLASSES OF COMPOUNDS

A. Insights about Alicyclic Chemistry

1. Cycloalkane and Bicycloalkane Derivatives, Including Cycloalkynes

We shall begin with a thought (*gedanken*) experiment. Imagine a medium cycloalkane or bicycloalkane derivative with a few lengthened single bonds and enlarged carbonyl groups and/or heteroatoms (e.g., **25**).

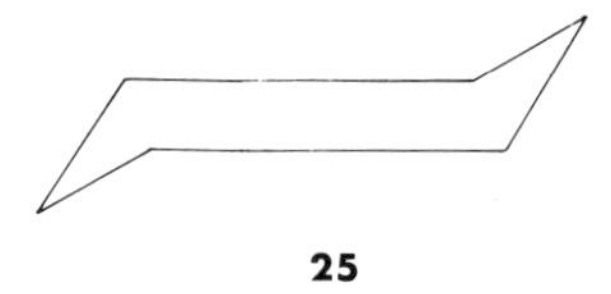

25

The analysis of such a modified compound could prove useful in attempts to understand the more conventional ring systems with normal bonds and atoms, because these purportedly conventional or normal compounds have long been known to be far more complicated than drawings of their structural formulas would suggest.[103,104] Such species and the corresponding nmr analysis of the conformational processes have been realized using cyclophanes[105] and "bicyclophanes"[106] wherein a para-substituted benzene serves as a mimic of an extended C—C bond and a meta-substituted benzene is the mimic of a carbonyl group or more general 2-coordinated heteroatom. Likewise, a 1,3,5-trisubstituted benzene mimics the tertiary >CH— group. Indeed, this type of analysis has also been used to generate analogs of triptycene in efforts to better understand the effects of nonplanarity on the stability of triphenylmethyl ions and radicals[107] (see **26a** and **26b**).

Earlier the analogy of C—C≡C—C and an extended C—C single bond had been made.[108] This suggests that a "modulation" and modification of this concept can be offered by using species containing various numbers of —CH_2—CH_2— ($r_{\text{end-to-end}} \cong 1.5$ Å), —CH_2—C≡C—CH_2— ($r \cong 4.1$

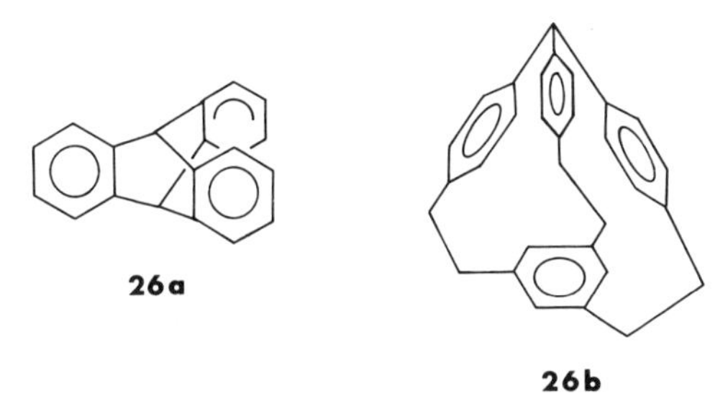

Å), and $—CH_2—p\text{-}C_6H_4—CH_2—$($r \cong 7.2$ Å) units. That this may be interesting is apparent from the strain energies of [n]paracyclophanes[21] and of cycloalkynes with n-methylenes,[109] for example, [8]paracyclophane versus cyclodecyne, and the derived difference in strain energies ΔSE (see Table V). For sufficiently large n, the strain energy of both classes of species must equal zero because the benzene, the acetylene, and the bridging polymethylene chain can adopt strainless geometries. How ΔSE varies with n until it settles down at the limiting value of zero should prove to be instructive.

If care is exercised in using this analogy, the results can be highly educational. For example, one may compare [2.2]- and [3.3]paracyclophanes with cyclobutane and cyclohexane, respectively, by replacing two $—CH_2—p\text{-}C_6H_4—CH_2—$ groups in the cyclophanes by $—CH_2—CH_2—$ groups in the cycloalkanes. The strain energies are 32,[45] 12,[43] 27,[110] and 1[110] kcal/mol, respectively. These numbers seem quite logical. Indeed, if we subtract twice the eclipsing energy of ethane (found in the 1,2 and 3,4 bonds of cyclobutane but not in the [2.2]cyclophane), the strain energy of cyclobutane is reduced to ~20 kcal/mol. The difference in strain energy between the cyclophanes and the corresponding cycloalkanes is then nearly identical for both cases.

TABLE V

Strain Energies of Paracyclophanes and Cycloalkynes and Their Difference (ΔSE)[a]

n^b	Paracyclophane	Cycloalkyne	ΔSE
5	39	31	8
6	29	21	8
7	21	16	5
8	17	10	7

[a] Strain energies expressed as kcal/mol.
[b] Number of bridging methylenes in the paracyclophanes and number of nonacetylenic carbons in the cycloalkynes.

However, the formally related comparison between [6.6]paracyclophane and cyclododecane, with associated strain energies of -5[43] and 7[110] kcal/mol, respectively, appears to be quite enigmatic. These results can be understood[111] in terms of destabilizing "overlap" or transannular H---H repulsions in an otherwise stabilizing "diamond-lattice" type of structure in the latter,[112] whereas the analogous transannular interactions in the former may be stabilizing. ⟨Boyd *et al.*[43] suggested some long-range van der Waals interactions without stipulating their nature. We recall the experimental finding[64,65] that two benzene molecules will stick to each other in a T-shaped, perpendicular or prolate asymmetric top arrangement.⟩ Analogously, one may also compare cyclic diynes and [*m.n*]paracyclophanes, for example, 1,5-cyclooctadiyne[113] (strain energy, 28.9 kcal/mol) with [2.2]paracyclophane (strain energy, ~32 kcal/mol). It is regrettable that the only other cyclic diyne for which we have thermochemical data is 1,8-cyclotetradecadiyne,[43] which corresponds to the essentially totally ignored [5.5]paracyclophane.[114] Dare we compare this tetradecadiyne with [6.6]paracyclophane since both contain "macrorings"? It is the author's opinion that this comparison is fraught with major conceptual difficulties. The first difficulty is that, whereas [2.2]paracyclophane is significantly strained, and [3.3]paracyclophane moderately so, [6.6]paracyclophane seemingly has a negative strain energy. Do we dare interpolate to the [5.5]paracyclophane? The second difficulty is that, whereas "diamond-lattice" structures are found for both cyclotetradecane and cyclohexadecane,[115] they have significantly different strain energies, 15.0 and 5.5 kcal/mol, respectively.[110]

2. Dodecahedrane

A second example of the interrelationships between alicyclic and cyclophane chemistry returns us to Woodward's suggestion[116] of dimerizing triquinacene **27** to dodecahedrane **28.** Although interest in this approach

27 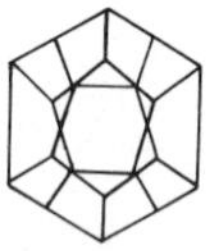 **28**

may have waned because of Paquette's synthesis of dimethyldodecahedrane,[117,118] the author finds it interesting to chronicle this dimerization approach and note modifications. A review[119] documents the unsuccessful attempts to dimerize the parent triquinacene. Likewise, Paquette's earlier approaches of joining two triquinacene units (in various degrees of saturation or substitution)[120] have not come to fruition. However, another approach remains untried: If one were to use two triquinacenes joined by two or more bridges, the entropy loss upon formation of the dodecahedrane would be minimized. This general idea and the synthesis of a suitable *anti*-dithia[3.3](2,6)triquinacenophane have been reported.[121]

3. Prismanes

A third example is Eaton's success at synthesizing pentaprismane **(29a).**[122] Recalling the earlier successes at synthesizing triprismane (prismane) **(29b)**[123] and tetraprismane (cubane) **(29c),**[124] one may inquire about the eventual synthesis of the novel benzene dimer, $(CH)_{12}$, hexaprismane **(29d).** From the fact that Eaton has been responsible for synthesizing both

tetra- and pentaprismanes, one may immediately, and correctly, surmise that he is trying to make hexaprismane.[125] This exotic substance may appear to be quite distinct from any cyclophane. However, it is a rather logical end point of the process of scrunching two benzene molecules together more and more closely. An intuitively appealing correlation diagram may be drawn in which it is seen that cyclophanes with their ~3-Å interring separation distance are but an intermediate stage in this process. ⟨This is formally analogous to the united and separated atom limits of diatomic molecules as the orbital levels change with internuclear separation. From the generally useful equivalence of benzene and other annulenes with two-dimensional atoms,[126] one may write for the electron configuration of the π orbitals of benzene, s^2p^4. For cyclophanes with two rings we have the "repulsive" $\sigma^2\sigma^{*2}\pi^4\pi^{*4}$, and for hexaprismane we have the closed-shell $\sigma^2\pi^4\delta^4\phi^2$ and an accompanying "sextuple bond." Furthermore, from the theoretical equivalence of cyclopropane and benzene and other cycloalkanes with [*n*]annulenes via the concept of "σ aromaticity,"[127] we deduce that prismane may be related[128] to "[0_3]cyclophane" and cubane to "[0_4]octatetraenophane." (For a related but more calculational study, see Minkin and Minyaeev.[128a])⟩

4. Long Hydrocarbon and Polymer Chains

Finally, we note the relevance of cyclophane chemistry to the unraveling of the conformations of long-chain hydrocarbons and more general polymer chains.[129] For example, the rates of cyclization were studied for a collection of *m*- and *p*-(ω-bromoalkoxy)phenolate anions, yielding dioxameta- and dioxaparacyclophanes **30** and rates of intramolecular phosphorescence quenching for the *p*-benzoyl benzoate esters of ω-hydroxyalkenes to form oxetane-containing paracyclophanes **31.** In both of these

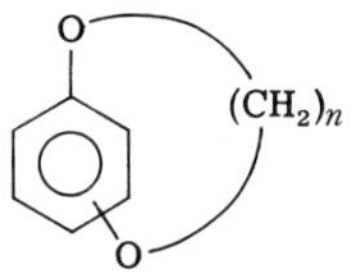

30 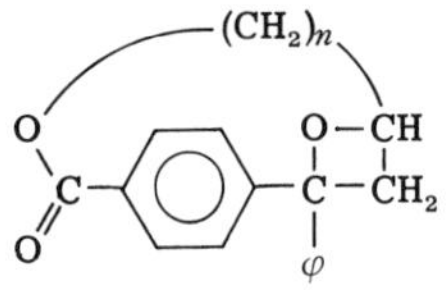 **31**

cases, as well as many others that Winnick cites, the species directly incorporate the essential features of cyclophanes, although the bridging chains are excessively long compared with the the extremely short bridges conventionally desired in these species.

We conclude this section by observing that cross-linked polystyrene, as well as some other polymers, may also be considered to be cyclophane derivatives.

B. Insights about Molecular Rearrangements and Other Thermal Processes

1. Importance and Surprises of Rearrangements

Rearrangements figure prominently in the chemistry of cyclophanes. The reader should recall that this chapter began with a reference to the benzidine rearrangement and the thermochemistry of the rearrangement of [2.2]paracyclophane to [2.2]metaparacyclophane. Furthermore, in the discussion of the systematics of the energetics of neutral cyclophanes, we make conceptual use of the synthetically useful *o*-xylylene ⇌ benzocyclobutene equilibrium, as well as equilibrium between derivatives of benzene and its various valence isomers. The substituent-dependent equilibrium (affected by both steric and electronic factors) of *anti*-[2.2]metacyclophanedienes **32a** and *trans*-15,16-dihydropyrenes **32b** has played an important role in efforts to understand the aromaticity of benzene and bridged [14]annulenes as well as of numerous benzo-annelated derivatives.[130,131]

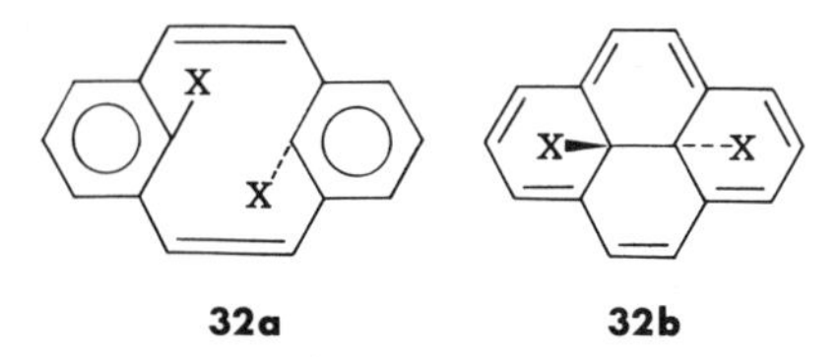

Numerous conventional rearrangements have parallels in cyclophane chemistry but with surprising features. For example, the 2,5,2′,5′-tetracarbinol of [2.2]paracyclophane **(33a)** with four primary alcohol groups quantitatively rearranges to form the 2,5′-dimethyl-5,2′-dicarbaldehyde **33b** via an acid-catalyzed "transannular pinacol rearrangement."[132] Fi-

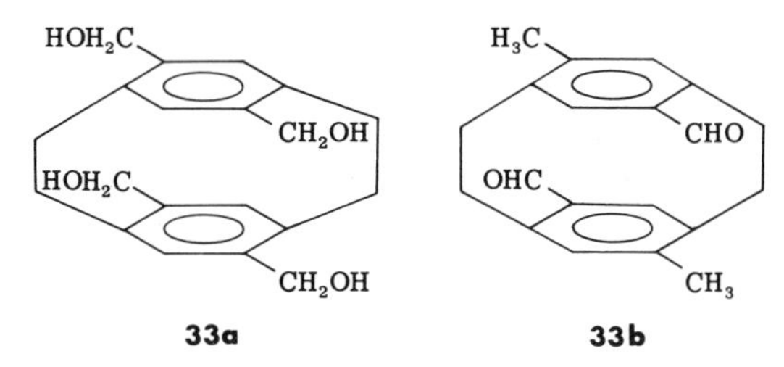

nally, we note the plethora of methods of removing sulfur from dithiacyclophanes, some involving rearrangements (e.g., the Benzyne–Stevens reaction) and others not.[53] (If only sulfur could be removed from coal, and oil, so cleanly, efficiently, and elegantly!) We now turn to some rearrangements that do not happen.

2. *p*-Isotoluene, *p*-Xylylene, and Thermolyses of Cyclophanes

The simple thermal rearrangement of *p*-isotoluene **(34a)** to the aromatic toluene **34b** occurs by way of a bimolecular, cyclic transition state.[133] It appears that the preferred products of a bimolecular encounter are the 3-methylcyclohexadienyl and benzyl radicals **35a** and **35b,** although the isomerization reaction is ~23 (±3) kcal/mol exothermic per mole of toluene.[134] That these findings relate to cyclophane chemistry becomes clear when one recalls that the endocyclic $>C^{H}_{H}$ has "pseudo-π" orbitals analogous to the C=C double bond of a $=CH_2$ group, i.e., hyper-

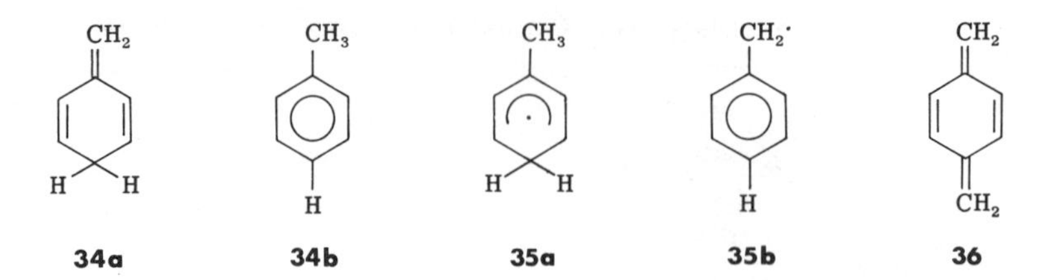

conjugation. As such, the equivalent reaction is for two *p*-xylylene molecules **(36)** to couple initially to form the diradical $\cdot CH_2C_6H_4(CH_2)_2C_6H_4CH_2\cdot$ rather than to form the [2.2]paracyclophane directly. ⟨Although formation of the diradical is obviously entropically favored, it may simply be shown to be enthalpically disfavored. In particular, we can estimate the heat of formation of this diradical by Eq. (8),

$$2[\Delta H_f(C_6H_5CH_2\cdot)] + \Delta H_f(\text{bibenzyl}) - 2[\Delta H_f(C_6H_6)] = 2(48.8) + 34.2 - 2(19.8) = 92.2 \text{ kcal/mol} \quad (8)$$

where the value for benzyl radical was taken from Meot-Ner.[135] By contrast, the heat of formation of *p*-xylylene has been given as 50 kcal/mol,[60] and so the dimerization reaction is ~8 kcal/mol exothermic. For further calibration purposes, using an approach systematically described elsewhere,[26] we can estimate the heat of formation of its diradical by Eq. (9),

$$2[\Delta H_f(C_6H_5CH_2\cdot)] - \Delta H_f(C_6H_6) = 2(48.8) - 19.8 = 77.8 \text{ kcal/mol} \quad (9)$$

whereas the experimental value[60] is a lower bound of 76 kcal/mol.⟩ This corresponds to the experimental observation that, whereas *p*-xylylene and [2.2]paracyclophane (and substituted derivatives) are in thermal equilibrium,[136,137] diradicals are intermediates. The intermediacy of diradicals may be viewed as being a consequence of the dimerization involving $4n$ electrons and so cannot be concerted.[138,139] It is encouraging that the diradical may be trapped in thermolysis reactions of the cyclophane, resulting in *p,p'*-disubstituted derivatives of bibenzyl with hydrogen atom donors[140,141] and of [2.4]paracyclophane with olefins.[140] We remind the reader of analogous thermolysis reactions of [2.2.2](1,2,4)cyclophane[141] and of [2.2.2.2](1,2,3,5)cyclophane.[48] These reactions, in the presence of hydrogen atom donors, result in the formation of cyclophanes with one less bridge.

3. *m*-Xylylenes and Aromatization

We note here a reaction involving *m*-xylylenes and a particularly strong drive to aromatization. In particular, a synthesis of a [2.2]metacyclophane derivative was reported[142] from a compound lacking a preformed six-membered ring, a bismethylene derivative of bicyclo[3.1.0]hexene [**37a**: X = Y = $C(CH_3)_2$]. Although there was spontaneous rearrangement to the *m*-xylylene **38a** in this case, the analogous monoketo compound **37b,** with X = O, Y = $C(CH_3)_2$, seemingly does not rearrange[143] to form the hetero-*m*-xylylene **38b,** even when heated to 150°C. Nor does it form a fulvene plus CO via a cyclopropanone derivative, that is, **37c** with the X and Y transposed.

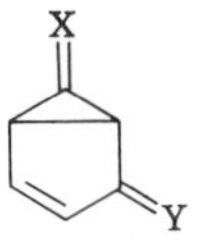

X Y **37a** · $\cdot C(CH_3)_2$ $\cdot C(CH_3)_2$ **38a** · $C(CH_3)_2$ O **37b** · X· Y· **38b** · $C(CH_3)_2$ O **37c**

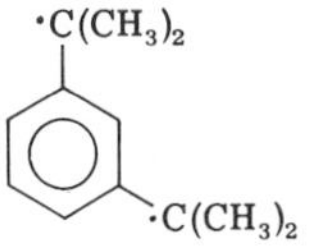

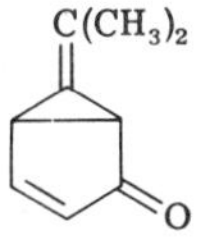

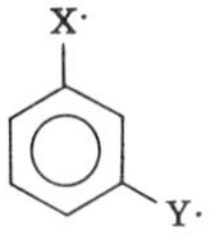

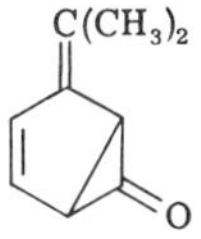

4. Failures in Aromatizing Six-membered Rings

We now consider cases in which six-membered rings fail to aromatize. The reader should recall Sections II,A,1 and II,A,2, in which several cases involving cyclophanes containing benzene valence isomers were shown to be more stable than the corresponding benzenoid case. We now turn to analogous examples in which the two species are related by simple proton transfer or tautomerism. The first case is the [5]metacyclophane precursor **39,** which fails to form the nitrophenol, whereas the higher homologs, that is, [6]- and larger metacyclophane precursors, facilely

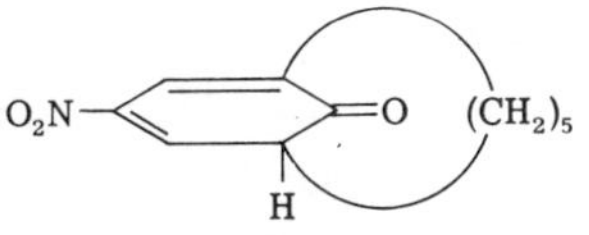

O_2N O $(CH_2)_5$ H **39** · $(CH_2)_8$ H F_2C CF_3 **40** · O OH H **41**

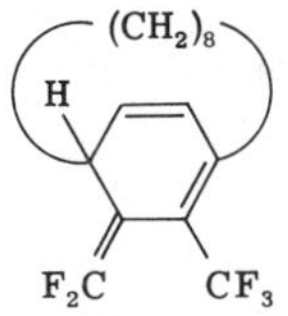

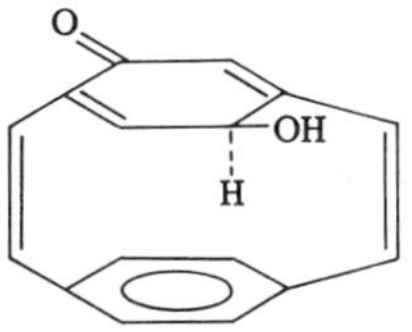

rearrange.[144,145] A more recent example is the failure of an *o*-isotoluene **(40)** to aromatize to an [8]paracyclophane even with strong base.[146] We note the analogous failures of a [2.2]paracyclophanediene precursor **(41)** to form the second benzene ring,[147] a result also shown by a [2.2.2.2](1,2,4,5)cyclophane derivative.[148]

5. [2.2]Paracyclophanediene, "[2.2]Paraxylenophane," and [*m.n.o.p*]Paracyclophanepolyenes

Another example involves the absence of equilibrium, resonance, or rearrangement between the [2.2]paracyclophanediene **42a** and the quite obviously impossible "[2.2]paraxylylenophane" **42b.** This nonprocess

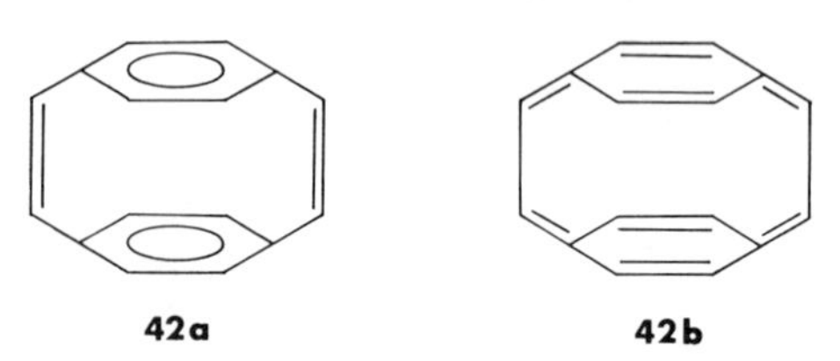

has been referred to as "an extreme example of steric inhibition of resonance in a classically conjugated hydrocarbon."[149] Analogously, the "acyclic" bi-*p*-xylylenyl $[CH_2{=}C_6H_4{=}CH]_2$, formed *in situ* from a preformed [2.2]paracyclophane derivative, fails to recyclize to form the singly unsaturated [2.2]paracyclophane derivative.[150] Making some reasonable assumptions, we deduce that structure **42a** is at least 30 kcal/mol more stable than **42b.** ⟨We consider **42a** to be two styrenes joined at their termini by two new C—C bonds and **42b** to be two *p*-xylylenes likewise attached. We then tacitly assume that all C_{sp^2}—C_{sp^2} bonds are of equal strength, that the interring repulsions should be comparable, and that the twisting energy for a phenyl–vinyl bond in a styrene or stilbene is comparable to that of twisting the interxylylene bond in [2.2]paraxylylenophane. The last assumption is patently untrue[151] and so makes the 30 kcal/mol a significant underestimate because structure **42b** corresponds to a planar species that has been excessively twisted.⟩ It should not be surprising that no resonance occurs and that structure **42b** is a nonentity. However, it is important to note that this resonance could not result in stabilization any more than do the two "Kekulé" structures for the anti-aromatic cyclobutadiene or a planar cyclooctatetraene. Both the hypothetical planar paracyclophanediene and paraxylylenophane are derivatives of [12]annulene. In principle, one could make a cyclophanepolyene large enough that both structures would contribute without twisting of double bonds or spatially overlapping rings. Likewise, in principle, a $4n + 2\pi$ electron annulene analog could also be envisioned. In fact, there are species that fulfill both criteria[152]: structures **43** and **44.** However, except for rather small ring current effects, there is little with which to differentiate these species from the related compounds with only $4n\pi$ electrons (**45a, 45b,** and **45c,** with $m = n = o = p = 2$, $m = n = o = p = 4$, and $m = o = 2$, $n = p = 4$, respectively).

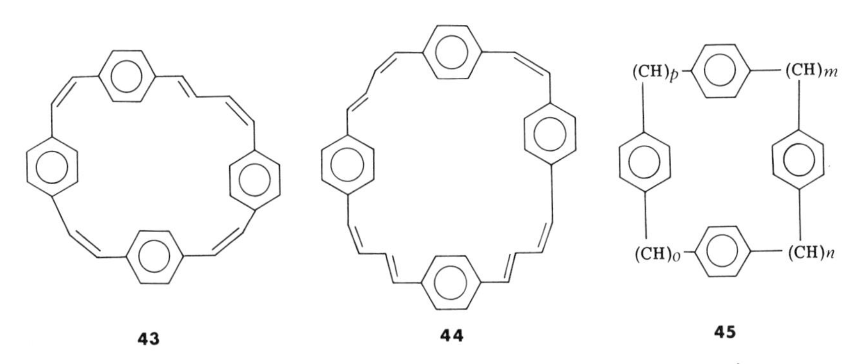

Although perhaps disappointing, this is not surprising because these are derivatives of [26]-, [30]-, [24]-, [28]-, and [32]annulene, respectively, and so one expects little vestige of aromaticity or anti-aromaticity.[153] Furthermore, resonance between structures analogous to that for the [2.2]paracyclophanediene would entail the destruction of all four benzene rings and the formation of four *p*-xylylenes, costing about 60 kcal/mol. As such, why bother? Equivalently, none of these species would be expected to show resonance or other particularly remarkable properties. However, it is well established for annulenes that aromaticity and anti-aromaticity are markedly accentuated in ions. The electrochemical reduction of these five [2.2.2.2]paracyclophane polyenes has been reported.[152,154] The assignments of the aromaticity or anti-aromaticity of the dianions were corroborated by their nmr ring currents and the voltage and reversibility of formation. In particular, the aromatic dianions have diamagnetic ring currents, are formed at lower voltages, are reversibly formed, and even appear to be stable in the presence of a small amount of water. We note the formation of other ions from these species, mono- and dications in the mass spectra of the hydrocarbons,[152] and the tetraanion on further reduction of the simplest species **45a.**[154] ⟨Even though these are alternant hydrocarbons and so have an inherent symmetry relating the chemistry of cations and anions⟩, we are loathe to compare ions with different charges and such different modes of formation. We await more precise data on the cations, for example, electrochemical oxidation potentials, and on the tetraanions, that is, data on more than one species, before drawing any additional conclusions.

Another rearrangement that does not occur is the spontaneous isomerization of *cis*- to *trans*-stilbene when associated with small, simple cyclophanes. It is not surprising that [2.2]paracyclophanediene occurs only as the cis,cis form with no evidence for the cis,trans or trans,trans form (e.g., **46a**). It is interesting that the [2.2.2.2]paracyclophanediene can be obtained[155] in all three forms (e.g., **46b**). Whereas the unconstrained *trans*-stilbene is ~4 kcal/mol more stable than its cis isomer, the

46a **46b**

trans,trans isomer of this [2.2.2.2]paracyclophanediene isomerizes on standing to this less symmetric cis,trans form.

C. Some Relationships with Biochemistry

1. Porphyrins and Natural Cyclophanes

There are numerous compounds of biological origin and biochemical significance that may be considered cyclophane derivatives. For example, Chapter 5 discusses natural [*n*]paracyclophane derivatives, the so-called ansa antibiotics, and Chapters 11 and 12 discuss enzymelike aspects of cyclophane derivatives. A less obvious application, "perhaps of slight consequence,"[156] relates cyclophane chemistry to porphyrins. Although it may be immediately apparent that porphyrins can be described as [1.1.1.1](2,5)pyrrolophanes **(47)** structures **48a–48d** document the alter-

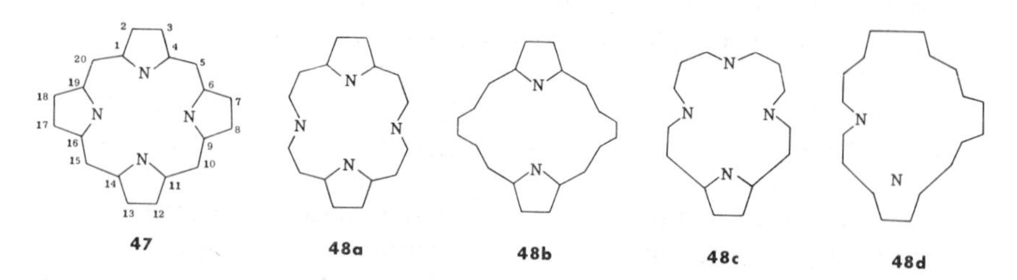

47 **48a** **48b** **48c** **48d**

native descriptions as [5.5]- and [6.6](2,5)pyrrolophanes as well as [11]- and [14](2,5)pyrrolophanes. It is hoped that these alternative descriptions provide some insight into the synthesis of porphyrin analogs as well as a greater sense of the inherent flexibility of this species. Indeed, it is this variability in the number and kind of nitrogens involved in metal binding that accounts in large part for the plethora of structural features that natural and synthetic metalloporphyrins enjoy.[157]

Looking at a porphyrin as a single ring suggests that porphyrins may be components of cyclophanes. We briefly note some of the numerous species that have been made to help us better understand the importance of chlorophyll stacking in photosynthesis and of hydrophobic interactions in the reversibility of binding of oxygen and carbon dioxide by hemoglobin and myoglobin. It is not surprising that there are several different types of porphyrin-containing cyclophanes that may be conceptually distinguished. The first has two porphyrin ring systems in parallel planes, whereas the second has but one porphyrin and one ring of another type. Both are recognizable as [*m.n*]porphyrinophanes. [*n*]Porphyrinophanes define a third class of species of interest, and, noting the importance of metal–metal interactions, we choose a metal-bearing chain [*n*]porphyrinophane as the final class of species to be considered. Bioinorganic chemists have been active in devising species to mimic hemoglobin (and myoglobin), chlorophyll, and the cytochromes. Members of all four classes of compound have been prepared, for example, [4.4](5,15]porphyrinophane,[158] a [4.4](2,12)porphyrino(4,4′)biphenylophane,[159] [13](2,12)-porphyrinophane with a —$(CH_2)_2CONH(CH_2)_4NHCO(CH_2)_2$— bridge,[160] and finally a formal [15]porphyrinophane containing four *o*-(nicotinamidophenyl)phenyl groups (5,10,15,20 substitution) with M = Cu and a bridging Cl.[161]

2. Nucleic Acids

It has long been known that nucleic acid bases and their derivatives have a tendency to "stack," that is, align in parallel planes, whether in the native macromolecule DNA, in synthetic polynucleotides, or even as "monomers." The degree of stacking affects the optical properties of the species, and so spectroscopic measurements are customarily used to determine the degree of disorder or denaturation of nucleic acids. In particular, it is well established that the absorption of a polynucleotide with *n* bases is generally less than *n* times the absorption of the free base. This decrement is referred to as "hypochromism."[162]

Despite the extensive activity of biochemists in investigating this phenomenon, surprisingly little work has been done on synthesizing species that contain suitably joined nucleic acid bases to model this phenomenon. Cyclophane chemistry would appear to provide a suitable framework for studying this problem, and indeed purinophanes have been reported.[163] In particular, compounds **49** (with X = S, $n = 3$; X = NH, $n = 3$; and X = S, $n = 4$) were synthesized and show a hypochromicity of ~30% of that for poly(A), the essentially infinite polymer derived from adenine. Noting that there are four bases of general biochemical interest and that each is multifunctional suggests that numerous analogs of the above compound are desirable synthetic goals. One caveat should be presented: To mimic the naturally occurring environment, the bases should be essentially planar. In particular, the amino group in an adenine analog should lie in the same plane as the purine ring. All hydrocarbon analogs are anticipated to be rather poor predictors of the structural and energy features of the desired species, as illustrated[164] by the extremely different conformational properties of multihetero [3.3]metacyclophane derivatives **50** (X = S or —CH_2—) and the parent all-carbon [3.3]metacyclophane.

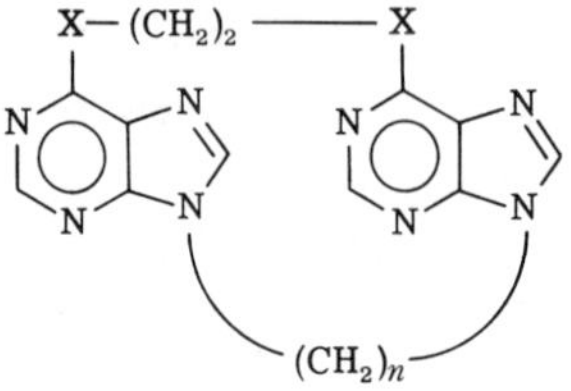

49

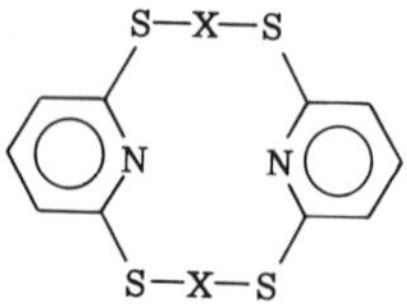

50

Finally, if we consider the Watson–Crick hydrogen-bonded base pairs[18] as but a single ring, then the whole double-stranded helix of DNA can be described as a multilayered [11.11]cyclophane. Whether some significance can be attached to this is left to the reader to decide.

3. Pyridoxal Analogs

The final application to biochemistry that we discuss entails the synthesis of vitamin B_6 or pyridoxal **(51)** analogs containing cyclophanes.[165] In particular, the [8]-, [10]- and [12]pyridinophanes **52a** (n = 4, 6, and 8) and the [3.3]cyclopyridinophane **52b** were synthesized. As determined by racemization rates of monosodium glutamate, the relative activities were [12] > [10] > [8] > natural > [3.3]. Having so demonstrated that hydrophobic interactions are important components for activity, these [n]pyridinophanes being the first synthetic analogs with activity exceeding that

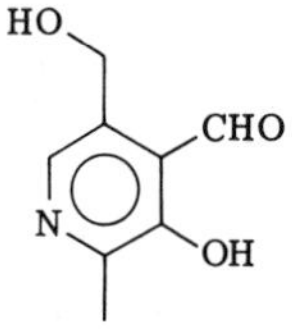

51

$(CH_2)_n$ CHO N OH S S — **52a**; CHO N OH S S — **52b**; C_2H_5S CHO N OH C_2H_5S — **53**

of the natural vitamin, these authors then synthesized the "acyclic" **53** and also found it to have higher activity than pyridoxal.

D. Relationships with Inorganic and Organometallic Chemistry

1. Arene–Metal Complexes, Boranes, and Carboranes

Soon after the initial report of arene–$Cr(CO)_3$ complexes[166] appeared, the first cyclophane complexes were reported.[167] In particular, these new species confirmed that the two rings behaved like a single π system; that is, only monometal complexes were formed for [*m.n*]paracyclophanes with $m, n \leq 4$ (see **54a**). In contrast, [4.5]- and [6.6]paracyclophanes formed both mono- and di-$Cr(CO)_3$ derivatives (see **54b**). In no case was a

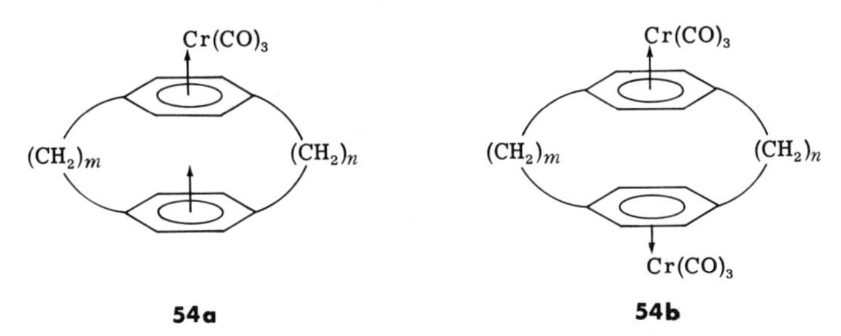

54a **54b**

sandwich compound directly prepared in which the chromium was encapsulated by the two rings, nor could such a species be prepared by thermally decomposing a preexisting chromium complex. ⟨We note that Eq. (10),

$$2\ C_6H_6Cr(CO)_3 \rightarrow Cr(CO)_6 + Cr(C_6H_6)_2 \qquad (10)$$

is nearly thermoneutral.⟩ Only with the development of simple methods of forming metal atoms did it become possible to synthesize such species

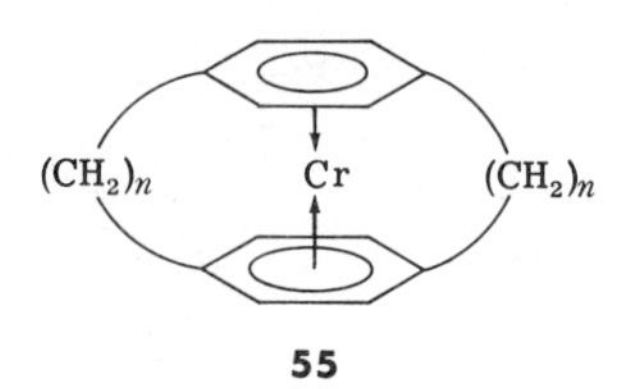

55

(**55**: $n = 2$,[168] $n = 3$[169]) directly. The radical-cation of the $n = 2$ species showed surprising kinetic stability in solution.[168] Furthermore, facile one-electron oxidation of the the $n = 3$ species resulted in stable, crystalline I_3^- and PF_6^- salts in which it was shown[170] that the interring spacing with the metal is smaller than in the parent cyclophane!

Related examples of metal encapsulation are found in multibridged ferrocenophanes (see Chapter 3) and in the Ag^+ complex of [2.2.2]paracyclophane (alternatively called π-prismand).[171] We anticipate analogous encapsulation complexes being formed between the latter cyclophane and some neutral metal with a low ionization potential. The alkali metals come to mind; cf. the cyclophane radical-anion–K^+ or radical-anion–Rb^+ ion pairs discussed in Section II,D,1. Alternatively, the lanthanides and "stable" actinides should be considered.[172] ⟨Numerous orbitals of the three benzene rings, both occupied and unoccupied, have the same symmetry as the partially filled *f* orbitals of the metal. This suggests that these species should enjoy stabilization analogous to more conventional metallocenes containing *d*-orbital transition elements that have but two rings and to the "π-isoelectronic" organoactinides $(C_5H_5)_3MR$.[172a]⟩

Schematically, letting M be some transition metal, Ar some single-ring arene, and Ar{}Ar the two benzene rings of an [*m.n*]paracyclophane, the species described above can be recast as ArM, ArMAr, Ar{}ArM, MAr{}ArM, and {ArMAr}, respectively. Other logical classes of complexes include Ar{}ArMAr, MArM{ArAr}MAr, $[\text{Ar}\{\}\text{ArM}]_x$, and $[\{\text{ArMAr}\}\text{M}]_x$. A surprising "nonexample" of the first class has been reported.[173] In particular, recalling the comparatively simple [2.2]-paracyclophane complexes with $Cr(CO)_3$ and other π-acids (see Chapter 3) suggests a paracyclophane–metal–arene ternary complex. [2.2]-Paracyclophane·$Cr(CO)_3$·1,3,5-$C_6H_3(NO_2)_3$ would appear to be such a compound. However, these authors showed the trinitrobenzene to be associated with the CO groups on the metal. Although this seems to contradict the ArArM geometry given a decade before for the nonparacyclophane analog with anisole,[174] the data were shown to be equally consistent with the same surprising structural feature.

Examples of the second class have also been reported. Such species include **56**[175] and **57**.[176] Synthetic routes to the third class have been suggested.[176] The final class seems more distant; the first MArM species

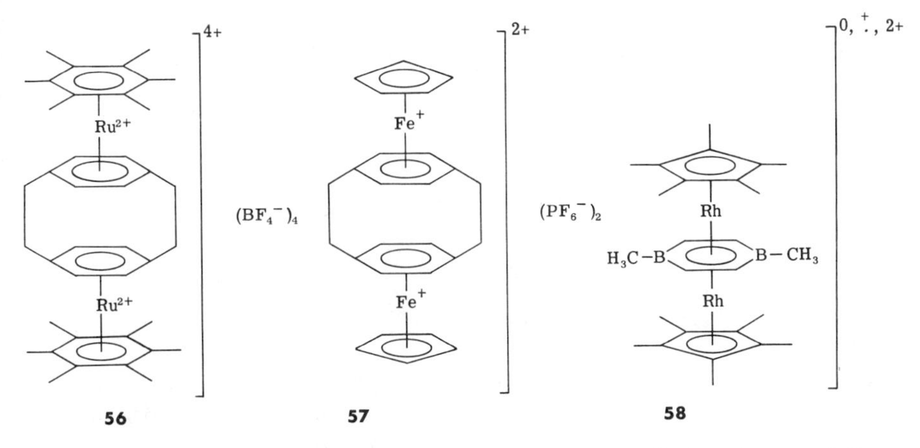

with a six-membered ring, the neutral form, the radical-cation, and the dications of **58,** were just reported.[177]

This series of redox-related complexes recalls the redox chemistry of the central C_4B_2 ring. In particular, whereas the dianion undoubtedly has a benzenoid, 6π, structure, the neutral form need not be a diboracyclohexadiene **(59a)** but rather is probably a pentagonal pyramid **(59b).**[178,179]

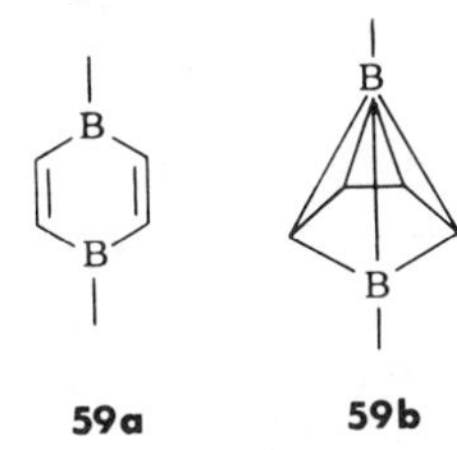

In turn, this suggests that one can consider the chemistry of other boranes and carboranes.[179] For example, one can consider the highly stable $(BH)_{12}^{2-}$ ion,[179] **60a** and the yet unknown "supericosahedral"[180] $(BH)_{14}^{2-}$ **(60b)** as a 6π complex of two :BH groups, two added electrons, and two $(BH)_n$ rings with $n = 5$ and 6, respectively.[181] Derivatives of the latter ion have also been described[182] as related to $(BH)_{12}^{6-}$ **(60c).** Related electron counting and isoelectronic reasoning would describe neutral [*m.n*]cyclophanes as derivatives of $(BH)_{12}^{12-}$ **(60d).**

2. "Cyclomers" and "Dicuprophanes"

So-called cyclomers with stacked rings form remarkably stable complexes with Mg^{2+} and with Li^+. These complexes may be considered derivatives of [3.5](1,4)pyridinophanes **(61a),** much as the parent cyclo-

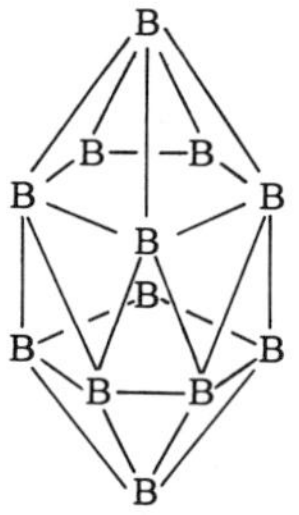

60a

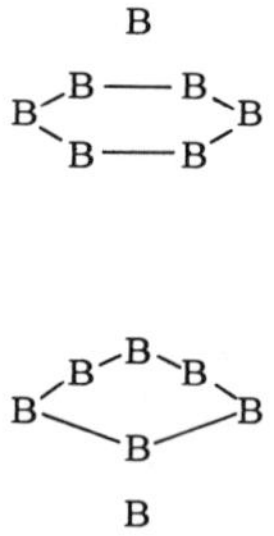

60b

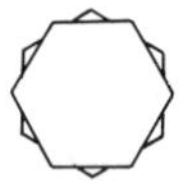

60c

60d

mers may be considered [0.3](1,4)pyridinophanes.[183] Indeed, the conformational properties of these cyclomers and the open diradical **61b** mimic those of cyclophanes and their "acyclic" analogs.

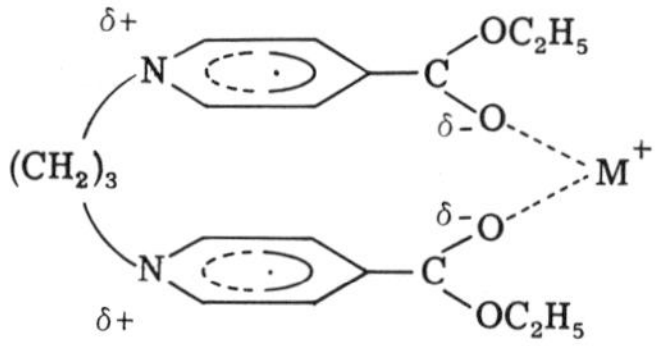

61a

61b

We close this section and this chapter, with an application explicitly noted by the original authors,[184] the synthesis of a "dicuprophane" **(62).**

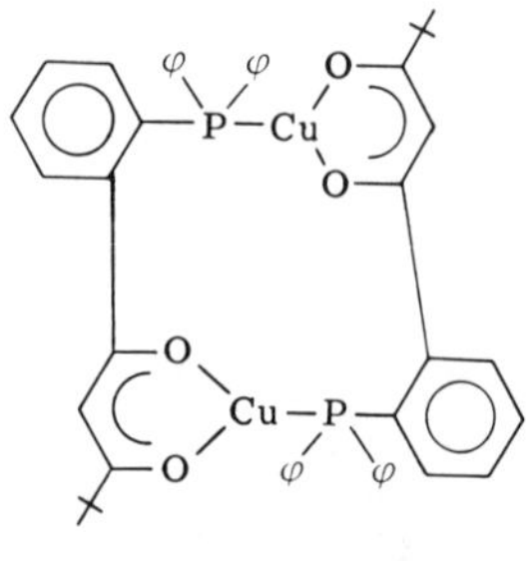

62

It may be tempting to note that numerous other metal complexes qualify as cyclophane analogs and conclude that there is comparatively little novel about this species (except for two spatially proximate trigonally coordinated Cu^+ ions). This conclusion, however, would belie a primary message of this chapter—that interrelationships between cyclophane chemistry and the chemistry of other species are generally instructive.

In summary, it has been seen that the general chemistry of cyclophanes encompasses the structure and energetics of numerous organic, organometallic, and inorganic species that possess a wide variety of molecular features. It has also been shown that the chemistry of cyclophanes and that of other species can be profitably interrelated and so provide a better understanding of general chemical phenomena. However, it has also been shown that there are many gaps in our knowledge, both qualitative and quantitative, that often prevent us from answering even the simplest of questions. More study is clearly indicated, heartily suggested, and warmly welcomed.

Acknowledgments

Thus concludes a cyclophanist named L,
Who thanks those with uncited findings to tell,
But thanks most his spouse, Deborah Van Vechten,
For her love, patience, and style correctin',
So now the chapter reads better, if not well.

REFERENCES

1. P. J. Garratt, "Aromaticity." McGraw-Hill, New York, 1971.
2. A. Greenberg and J. F. Liebman, "Strained Organic Compounds." Academic Press, New York, 1978.
3. D. J. Cram, R. C. Hegelson, D. Lock, and L. A. Singer, *J. Am. Chem. Soc.* **88,** 1324 (1966).
4. S. A. Godleski, P. von R. Schleyer, E. Ōsawa, and W. T. Wipke, *Prog. Phys. Org. Chem.* **13,** 63 (1981).
5. J. S. Chickos, A. S. Hyman, L. H. Ladon, and J. F. Liebman, *J. Org. Chem.* **46,** 4294 (1981).
6. C. G. DeKruif, *J. Chem. Thermo.* **12,** 203 (1980).
7. J. B. Pedley and J. Rylance, "Sussex-N.P.L. Computer Analysed Thermochemical Data: Organic and Organometallic Compounds." University of Sussex, U. K., 1977.
8. J. L. Devlin, III, J. F. Wolf, R. W. Taft, and W. J. Hehre, *J. Am. Chem. Soc.* **98,** 1990 (1976).
9. D. H. Aue and M. T. Bowers, *in* "Gas Phase Ion Chemistry" (M. T. Bowers, ed.), Vol. 2, p. 1. Academic Press, New York, 1979.
10. K. N. Hartman, S. Lias, P. Ausloos, H. M. Rosenstock, S. S. Schroyer, C. Schmidt, D. Martinsen, and G. W. A. Milne, "A Compendium of Gas Phase Basicity and Proton Affinity Measurements," NBSIR 79-1777. U.S. Dept. of Commerce, Washington, D. C., 1979.

11. M. H. Delton, R. E. Gillian, and D. J. Cram, *J. Am. Chem. Soc.* **93,** 2329 (1971).
12. D. J. Cram and H. Steinberg, *J. Am. Chem. Soc.* **73,** 5691 (1951), footnote 5.
13. Prof. Donald J. Cram, personal communication to the author.
14. A. Greenberg and J. F. Liebman, "Strained Organic Compounds," p. 174. Academic Press, New York, 1978.
15. M. J. S. Dewar, *J. Chem. Soc.* p. 777 (1966).
16. M. Meot-Ner (Mautner), P. Hamlet, E. P. Hunter, and F. H. Field, *J. Am. Chem. Soc.* **100,** 5466 (1978).
17. F. A. L. Anet and J. M. Muchowski, *Proc. Chem. Soc., London* p. 219 (1961).
18. J. D. Watson and F. H. C. Crick, *Nature* (*London*) **171,** 737 (1953).
19. Prof. Frank A. L. Anet, personal communication to the author.
20. F. Fehér, K. Glinka, and F. Malcharek, *Angew. Chem., Int. Ed. Engl.* **10,** 413 (1971).
21. N. L. Allinger, J. T. Sprague, and T. Liljefors, *J. Am. Chem. Soc.* **96,** 5100 (1974).
22. H. Schmidt, A. Schweig, G. Thiel, and M. Jones, Jr., *Chem. Ber.* **111,** 1958 (1978).
23. S. L. Kammula, L. D. Iroff, M. Jones, Jr., J. W. van Stratten, W. H. deWolf, and F. Bickelhaupt, *J. Am. Chem. Soc.* **99,** 5815 (1977).
24. K. Weinges, P. Günther, W. Kasei, C. Hubertal, and P. Günther, *Angew. Chem., Int. Ed. Engl.* **20,** 981 (1981).
25. A. Greenberg and J. F. Liebman, "Strained Organic Compounds," pp. 232–245. Academic Press, New York, 1978.
26. J. Dannacher, H. M. Rosenstock and J. F. Liebman, *Radiat. Phys. Chem.,* **20,** 7 (1982).
27. A. Greenberg and J. F. Liebman, "Strained Organic Compounds," pp. 117–133. Academic Press, New York, 1978.
28. W. F. Maier and P. von R. Schleyer, *J. Am. Chem. Soc.* **103,** 1891 (1981).
29. H. Wynberg, W. L. Nieuwpoort, and H. T. Jonkman, *Tetrahedron Lett.* p. 4623 (1973).
30. J. A. Pople, *J. Am. Chem. Soc.* **97,** 5306 (1975).
31. W. J. Hehre, *J. Am. Chem. Soc.* **97,** 5308 (1975).
32. M. J. S. Dewar, *J. Am. Chem. Soc.* **97,** 6591 (1975).
33. W. J. Hehre and J. A. Pople, *J. Am. Chem. Soc.* **97,** 6941 (1975).
34. I. Shavitt, *Mod. Theor. Chem.,* **3,** 189 (1977).
35. E. Wasserman, R. S. Hutton, and F. B. Bramwell, *J. Am. Chem. Soc.* **98,** 7429 (1976).
36. A. Greenberg and J. F. Liebman, "Strained Organic Compounds," pp. 343–369. Academic Press, New York, 1978.
37. J. W. Van Stratten, W. H. deWolf, and F. Bickelhaupt, *Tetrahedron Lett.* p. 4667 (1977).
38. L. A. M. Turkenburg, P. M. L. Blok, W. H. deWolf, and F. Bickelhaupt, *Tetrahedron Lett.* **22,** 3317 (1981).
39. L. A. M. Turkenburg, P. M. L. Blok, W. H. deWolf, and F. Bickelhaupt, *J. Am. Chem. Soc.* **102,** 3256 (1980).
40. W. E. Parnham, D. R. Johnson, C. T. Hughes, M. K. Meilahn, and J. K. Rinehart, *J. Org. Chem.* **35,** 1048 (1971).
41. E. Vogel, W. Wiedermann, H. D. Roth, J. Eimer, and H. Guenther, *Justig Liebigs Ann. Chem.* **759,** 1 (1972).
42. S. Hirano, T. Hiyama, and H. Nozaki, *Tetrahedron* **32,** 2381 (1970).
43. C.-F. Shieh, D. McNally, and R. H. Boyd, *Tetrahedron* **25,** 3653 (1969).
44. H. J. Lindner, *Tetrahedron* **32,** 753 (1976).
45. R. H. Boyd, *Tetrahedron* **22,** 119 (1966).
45a. K. Nishiyama, N. Sakiyama, S. Seki, H. Horita, T. Otsubo, and S. Misumi, *Bull. Chem. Soc. Jpn.,* **53,** 869 (1980).

46. W. R. Roth, M. Biermann, H. Dekker, R. Jochems, C. Mosselman, and H. Hermann, *Chem. Ber.* **111,** 3892 (1978).
47. S. W. Benson and H. E. O'Neal, *Natl. Stand. Ref. Data Ser.* (*U.S., Natl. Bur. Stand.*) **NSRDS-NBS 21** (1970).
48. H. Hopf, J. Kleinschroth, and A. E. Murad, *Isr. J. Chem.* **20,** 291 (1980).
49. B. Neuschwander and V. Boekelheide, *Isr. J. Chem.* **20,** 288 (1980).
50. L. H. Ladon and J. F. Liebman, unpublished results.
51. B. E. Eaton, E. D. Laganis, and V. Boekelheide, *Proc. Natl. Acad. Sci. U.S.A.* **78,** 6564 (1981).
52. Y. Sekine, M. Brown, and V. Boekelheide, *J. Am. Chem. Soc.* **101,** 3126 (1979).
53. V. Boekelheide, *Acc. Chem. Res.* **13,** 65 (1980).
54. Y. Sekine and V. Boekelheide, *J. Am. Chem. Soc.* **103,** 1777 (1981).
55. V. V. Kane, A. D. Wolf, and M. Jones, Jr., *J. Am. Chem. Soc.* **96,** 2643 (1974).
56. H. H. Wasserman and P. M. Keehn, *J. Am. Chem. Soc.* **91,** 2374 (1969).
57. G. Kaupp and I. Zimmerman, *Angew. Chem., Int. Ed. Engl.* **15,** 441 (1976).
58. J. Juriew, T. Skorochodowa, W. Merauschew, W. Winter, and H. Meier, *Angew. Chem., Int. Ed. Engl.* **21,** 269 (1981).
59. C. I. Lin, P. Singh, M. Maddox, and E. F. Ullman, *J. Am. Chem. Soc.* **102,** 3261 (1980).
60. S. K. Pollack, B. C. Raine, and W. J. Hehre, *J. Am. Chem. Soc.* **103,** 6308 (1981).
61. D. J. Cram and H. Steinberg, *J. Am. Chem. Soc.* **73,** 5691 (1951).
62. D. J. Cram, N. L. Allinger, and H. Steinberg, *J. Am. Chem. Soc.* **76,** 6132 (1954).
63. D. J. Cram and H. Steinberg, *J. Am. Chem. Soc.* **73,** 5691 (1951).
64. K. C. Janda, J. C. Hemminger, J. S. Winn, S. E. Novick, S. J. Harris, and W. Klemperer, *J. Chem. Phys.* **63,** 1419 (1975).
65. J. M. Steed, T. A. Dixon, and W. Klemperer, *J. Chem. Phys.* **70,** 4940 (1979).
66. J. B. Hopkins, D. E. Powers, and R. E. Smalley, *J. Phys. Chem.* **85,** 3739 (1981).
67. P. R. R. Langridge-Smith, D. V. Brumbaugh, C. A. Hayman, and D. H. Levy, *J. Phys. Chem.* **85,** 3742 (1981).
68. J. Ferguson, *Chem. Phys. Lett.* **79,** 198 (1981).
69. D. J. Cram and M. F. Antar, *J. Am. Chem. Soc.* **80,** 3103 (1958).
70. D. J. Cram and L. A. Singer, *J. Am. Chem. Soc.* **85,** 1084 (1963).
71. N. E. Blank and M. W. Haenel, *Chem. Ber.* **114,** 1520 (1981).
72. N. E. Blank and M. W. Haenel, *Chem. Ber.* **114,** 1531 (1981).
73. H. H. Wasserman and P. M. Keehn, *J. Am. Chem. Soc.* **94,** 298 (1972).
74. D. J. Cram and R. H. Bauer, *J. Am. Chem. Soc.* **81,** 5971 (1959).
75. T. L. Cairns, R. A. Carboni, D. D. Coffman, M. Engelhardt, R. E. Heckert, W. J. Middleton, R. M. Scribner, G. W. Theobald, and H. E. Winberg, *J. Am. Chem. Soc.* **80,** 2775 (1958).
76. R. E. Merrifield and W. D. Phillips, *J. Am. Chem. Soc.* **80,** 2778 (1958).
77. J. Bruhin, F. Gerson, and H. Ohya-Nishiguchi, *J. Chem. Soc., Perkin Trans. 2* p. 1045 (1980).
78. J. Bruhin, F. Gerson, and H. Ohya-Nishiguchi, *Helv. Chim. Acta* **60,** 2471 (1977).
79. F. Gerson, G. Kaupp, and H. Ohya-Nishiguchi, *Angew. Chem., Int. Ed. Engl.* **16,** 657 (1977).
80. T. Sato, K. Torizuka, R. Komaki, and H. Atobe, *J. Chem. Soc., Perkin Trans. 2* p. 561 (1980).
81. C. Kröhnke, V. Enkelmann, and G. Wegner, *Angew. Chem., Int. Ed. Engl.* **20,** 981 (1981).

82. A. F. Murad, J. Kleinschroth, and H. Hopf, *Angew. Chem., Int. Ed. Engl.* **19,** 389 (1980).
83. B. Kovač, M. Mohraz, E. Heilbronner, V. Boekelheide, and H. Hopf, *J. Am. Chem. Soc.* **102,** 4314 (1980).
84. B. Kovač, M. Allan, E. Heilbronner, J. P. Maier, R. Gleiter, M. W. Haenel, P. M. Keehn, and J. A. Reiss, *J. Electron Spectrosc. Relat. Phenom.* **19,** 167 (1980).
85. B. Kovač, M. Mohraz, E. Heilbronner, S. Itô, Y. Fukazawa, and P. M. Keehn, *J. Electron. Spectrosc. Relat. Phenom.* **22,** 327 (1981).
86. B. Kovač, M. Allan, and E. Heilbronner, *Helv. Chim. Acta* **64,** 430 (1981).
87. Y. Zhong-zhi, B. Kovač, E. Heilbronner, S. Eltamany, and H. Hopf, *Helv. Chim. Acta* **64,** 1991 (1981).
88. H. M. Rosenstock, K. Draxl, B. W. Steiner, and J. T. Herron, *J. Phys. Chem. Ref. Data* **6,** Suppl. 1 (1977).
89. R. Levin and S. G. Lias, *Natl. Stand. Ref. Data Ser.* (*U.S. Natl. Bur. Stand.*) **NSRDS-NBS 71** (1982).
90. Y. Zhong-zhi, B. Kovač, E. Heilbronner, S. Eltamany, and H. Hopf, *Helv. Chim. Acta* **64,** 1991 (1981).
91. R. Gleiter, H. Hopf, M. Eckert-Macksić, and K.-L. Noble, *Chem. Ber.* **113,** 3401 (1980).
92. S. I. Weissman, *J. Am. Chem. Soc.* **80,** 6462 (1958).
93. F. Gerson and W. B. Martin, Jr., *J. Am. Chem. Soc.* **91,** 1883 (1969).
94. F. Gerson, W. B. Martin, Jr., and C. Wydler, *Helv. Chim. Acta* **59,** 1565 (1976).
95. J. C. Moore, C. Thornton, W. B. Collier, and J. P. Devlin, *J. Phys. Chem.* **85,** 350 (1981).
96. K. D. Jordan, J. A. Michejda, and P. D. Burrow, *J. Am. Chem. Soc.* **98,** 1295 (1976).
97. J. McHale and J. Simons, *J. Chem. Phys.* **72,** 425 (1981).
98. C. Eischenbroisch, F. Gerson, and J. A. Reiss, *J. Am. Chem. Soc.* **99,** 60 (1977).
99. T. Sato, K. R. Torizuka, R. Kumaki, and H. Otabe, *J. Chem. Soc., Perkin Trans. 2* p. 56 (1980).
100. B. Natsume, N. Nishikawa, T. Kaneda, Y. Sakata, and S. Misumi, *Chem. Lett.* p. 600 (1981).
101. F. Gerson, J. Lopez, and V. Boekelheide, *J. Chem. Soc., Perkin Trans. 2* p. 1298 (1981).
102. Prof. Virgil Boekelheide, personal communication to the author.
103. J. Dale, *Top. Sterochem.* **9,** 199 (1976).
104. Cycloalkanes and Bicycloalkanes: A. Greenberg and J. F. Liebman, "Strained Organic Compounds," pp. 65–70 and 70–76 resp. Academic Press, New York, 1978.
105. T. Olsson, D. Tanner, O. Wennerström, and T. Liljefors, *Tetrahedron* **37,** 3743 (1981).
106. T. Olsson, D. Tanner, B. Thulin, and O. Wennerström, *Tetrahedron* **37,** 3485 (1981).
107. M. Nakazaki, K. Yamamoto, and T. Toya, *J. Org. Chem.* **46,** 1611 (1981).
108. A. Greenberg and J. F. Liebman, "Strained Organic Compounds," p. 368. Academic Press, New York, 1978.
109. N. L. Allinger and A. Y. Myers, *Tetrahedron* **31,** 1807 (1975).
110. E. M. Engler, J. D. Andose, and P. von R. Schleyer, *J. Am. Chem. Soc.* **95,** 8005 (1973).
111. A. Greenberg and J. F. Liebman, unpublished results.
112. M. Saunders, *Tetrahedron* **23,** 2105 (1967).
113. W. Leupin and J. Wirz, *Helv. Chim. Acta* **61,** 1663 (1978).
114. N. L. Allinger and D. J. Cram, *J. Am. Chem. Soc.* **76,** 2362 (1956).
115. F. A. L. Anet and A. K. Cheng, *J. Am. Chem. Soc.* **97,** 2420 (1975).

116. R. B. Woodward, T. Fukunaga, and R. C. Kelly, *J. Am. Chem. Soc.* **86,** 3162 (1964).
117. L. A. Paquette, D. W. Balogh, R. Usha, D. Kountz, and G. G. Christoph, *Science* **211,** 575 (1981).
118. L. A. Paquette and D. W. Balogh, *J. Am. Chem. Soc.* **104,** 774 (1982).
119. P. E. Eaton, *Tetrahedron* **39,** 2189 (1979).
120. L. A. Paquette, I. Itoh, and W. B. Farnham, *J. Am. Chem. Soc.* **97,** 7280 (1975).
121. S. P. Roberts and G. Shoham, *Tetrahedron Lett.* **22,** 4895 (1981).
122. P. E. Eaton, Y. S. Or, and S. J. Branca, *J. Am. Chem. Soc.* **103,** 2134 (1981).
123. T. J. Katz and N. Acton, *J. Am. Chem. Soc.* **95,** 2738 (1973).
124. P. E. Eaton and T. W. Cole, Jr., *J. Am. Chem. Soc.* **86,** 3157 (1964).
125. Prof. Philip E. Eaton, personal communication to the author.
126. N. P. Adams, R. D. Perkins, F. P. Wilgis, and J. F. Liebman, unpublished results.
127. M. J. S. Dewar, *Bull. Chem. Soc. Belg.* **88,** 957 (1979).
128. Prof. Arthur Greenberg, personal communication to the author.
128a. V. I. Minkin and R. M. Minyaeev, *Russ. J. Org. Chem.* (*Engl. Transl.*) **17,** 175 (1982).
129. M. A. Winnick, *Chem. Rev.* **81,** 491 (1981).
130. V. Boekelheide and J. B. Phillips, *J. Am. Chem. Soc.* **89,** 1695 (1967).
131. R. H. Mitchell and R. Mahadevan, *Tetrahedron Lett.* **22,** 5131 (1981).
132. R. Bray and V. Boekelheide, *J. Am. Chem. Soc.* **101,** 2129 (1979).
133. J. J. Gajewski and A. M. Gortva, *J. Am. Chem. Soc.* **104,** 334 (1982).
134. J. E. Bartmess, *J. Am. Chem. Soc.* **104,** 335 (1982).
135. M. Meot-Ner (Mautner), *J. Am. Chem. Soc.* **104,** 5 (1982).
136. J. R. Schaefgen, *J. Polym. Sci.* **15,** 203 (1955).
137. W. F. Gorham, *J. Polym. Sci., Part A-1* **4,** 3027 (1966).
138. R. B. Woodward and R. Hoffmann, *Angew. Chem., Int. Ed. Engl.* **8,** 789 (1969).
139. R. Luhowy and P. M. Keehn, *J. Am. Chem. Soc.* **99,** 3794 (1977).
140. H. J. Reich and D. J. Cram, *J. Am. Chem. Soc.* **91,** 351 (1969).
141. D. J. Cram, R. B. Hornsby, E. A. Truesdale, H. J. Reich, M. H. Delton, and J. M. Cram, *Tetrahedron* **30,** 1757 (1974).
142. J. J. Gajewski, M. J. Chang, P. J. Stang, and T. E. Fish, *J. Am. Chem. Soc.* **102,** 2096 (1980).
143. M. Rule, A. R. Martin, E. F. Hilinski, A. D. Dougherty, and J. A. Berson, *J. Am. Chem. Soc.* **101,** 5098 (1979).
144. V. Prelog and K. Wiesner, *Helv. Chim. Acta* **30,** 1465 (1947).
145. V. Prelog, K. Wiesner, W. Ingold, and O. Hafliger, *Helv. Chim. Acta* **31,** 1325 (1948).
146. P. G. Gassman, S. R. Korn, and R. P. Thummel, *J. Am. Chem. Soc.* **96,** 6948 (1974).
147. I. Erden, P. Gölitz, R. Nüder, and A. de Meijere, *Angew. Chem., Int. Ed. Engl.* **20,** 583 (1981).
148. R. Gray and V. Boekelheide, *J. Am. Chem. Soc.* **101,** 2128 (1979).
149. K. C. Dewhirst and D. J. Cram, *J. Am. Chem. Soc.* **80,** 3115 (1958).
150. D. J. Cram and R. H. Bauer, *J. Am. Chem. Soc.* **81,** 5983 (1959).
151. A. Greenberg and J. F. Liebman, "Strained Organic Compounds," pp. 117–133. Academic Press, New York, 1978.
152. B. Thulin, *J. Chem. Soc., Perkin Trans. 1* p. 664 (1981).
153. F. Sondheimer, *Acc. Chem. Res.* **5,** 81 (1972).
154. W. Huber, K. Müllen, and O. Wennerström, *Angew. Chem., Int. Ed. Engl.* **19,** 624 (1980).
155. D. Tanner and O. Wennerström, *Tetrahedron Lett.* **22,** 2313 (1981).
156. B. H. Smith, "Bridged Aromatic Compounds," p. 177. Academic Press, New York, 1964.

157. K. Tatsumi and R. Hoffmann, *J. Am. Chem. Soc.* **103,** 3328 (1981).
158. J. P. Collman, A. O. Chong, G. B. Jameson, R. T. Oakley, E. Rose, E. R. Schmittou, and J. A. Ibers, *J. Am. Chem. Soc.* **103,** 517 (1981).
159. H. Diekmann, C. K. Chang, and T. G. Traylor, *J. Am. Chem. Soc.* **93,** 4068 (1971).
160. B. Ward, C.-B. Wang, and C. K. Chang, *J. Am. Chem. Soc.* **103,** 5236 (1981).
161. M. J. Gunter, L. N. Mander, G. M. McLaughlin, K. S. Murray, K. J. Berry, P. E. Clark, and D. A. Buckingham, *J. Am. Chem. Soc.* **102,** 1470 (1980).
162. A. M. Michelson, *in* "Molecular Associations in Biology" (B. Pullman, ed.), p. 83. Academic Press, New York, 1968.
163. F. Hama, Y. Sakata, and S. Misumi, *Tetrahedron Lett.* **22,** 1123 (1981).
164. F. Bottino and S. Pappalardo, *Chem. Lett.* p. 1781 (1981).
165. M. Iwata and H. Kuzubara, *Chem. Lett.* p. 5 (1981).
166. E. Fischer and K. Ofele, *Chem. Ber.* **90,** 2532 (1957).
167. D. J. Cram and D. I. Wilkinson, *J. Am. Chem. Soc.* **82,** 5721 (1960).
168. C. Eischenbroich, R. Möckel, and U. Zennech, *Angew. Chem., Int. Ed. Engl.* **17,** 531 (1978).
169. A. R. Koray, M. L. Ziegler, N. E. Blank, and M. W. Haenel, *Tetrahedron Lett.* p. 2465 (1979).
170. R. Benn, N. E. Blank, M. W. Haenel, J. Klein, A. R. Koray, K. Weidenhammer, and M. L. Ziegler, *Angew. Chem., Int. Ed. Engl.* **19,** 44 (1980).
171. J.-L. Pierre, P. Baret, P. Chautemps, and M. Armand, *J. Am. Chem. Soc.* **103,** 2986 (1981).
172. J. S. Vincent and J. F. Liebman, unpublished results.
172a. T. J. Marks, *Science* **217,** 989 (1982).
173. H. Kobayashi, K. Kobayashi, and Y. Kaizu, *Angew. Chem., Int. Ed. Engl.* **20,** 4135 (1981).
174. O. L. Carter, A. T. McPhail, and G. A. Sim, *J. Chem. Soc. A* p. 822 (1966).
175. E. D. Laganis, R. G. Finke, and V. Boekelheide, *Tetrahedron Lett.* p. 4405 (1980).
176. E. D. Laganis, R. G. Finke, and V. Boekelheide, *Proc. Natl. Acad. Sci. U.S.A.,* **78,** 2657 (1981).
177. G. E. Herbereich, B. Hessier, G. Huttner, and L. Zsolnai, *Angew. Chem., Int. Ed. Engl.* **20,** 472 (1981).
178. A. Greenberg and J. F. Liebman, "Strained Organic Compounds," pp. 375–385. Academic Press, New York, 1978.
179. See, for example, E. L. Muetterties, ed., "Boron Hydride Chemistry." Academic Press, New York, 1975.
180. M. F. Hawthorne, *J. Organomet. Chem.* **100,** 97 (1975).
181. R. L. Kellert, S. D. Frans, and J. F. Liebman, unpublished results.
182. R. N. Grimes, J. R. Pipal, and E. Sinn, *J. Am. Chem. Soc.* **101,** 4172 (1979).
183. J. Hermolin and E. M. Kosower, *J. Am. Chem. Soc.* **103,** 4813 (1981).
184. B. Rauchfuss, S. R. Wilson, and D. A. Wrobleski, *J. Am. Chem. Soc.* **103,** 6769 (1981).

CHAPTER 3

Crystal Structure of Cyclophanes

PHILIP M. KEEHN

Department of Chemistry
Brandeis University
Waltham, Massachusetts

I. INTRODUCTION

Since the discovery of cyclophanes, numerous analytical methods have been used to probe a variety of physical and chemical properties of these

* See this section (pp. 78–79) for a listing of the contents of the tabulation.

ISBN 0-12-403001-7

molecules. Although uv and nmr spectroscopic methods have been used to define and understand some of the effects of molecular distortion on the properties of these unusual systems, X-ray crystallographic analysis has afforded the most direct description of the structural features that are responsible for the unique phenomena associated with these compounds.

From the outset, when the structure of [2.2]paracyclophane was determined,[1] the distortion of the aromatic ring atoms from planarity and the further displacement of the bridging benzylic carbon atoms from the aromatic plane (as compared with *p*-xylene) became the hallmarks by which these molecules were identified. Additional structural characteristics that are typical of these strained and contorted systems are angle strain in the bridge and at the aromatic ring atom bound to the bridge as well as bond stretching in the bridge and around the ring atom bound to the bridge. The overall geometry, which is quite distorted in the lower [*m*.*n*]meta- and [*m*.*n*]paracyclophanes, is the result of a compromise between the maintenance of ideal hybridization geometry in the individual moieties of the cyclophane and the attempt to reduce the transannular π–π and other nonbonded interactions within the molecule.

In this chapter an attempt is made to collate the known X-ray crystal structure determinations of cyclophanes (through the early part of 1982) in order to place the crystallographic and structural data of these systems in a form that is useful for making comparisons, for acquiring a greater understanding of the overall geometries and deviations therefrom, and for applying the data to other areas of cyclophane chemistry. The approach taken was to present the salient features of these structures in noncrystallographic terms so the information could be of general use. The next section provides a very qualitative overview of the trends observed in the structural chemistry of cyclophanes. Packing analyses and other details have not been considered in a critical or definitive way.

II. GENERAL STRUCTURAL FEATURES AND GEOMETRIC RELATIONSHIPS IN CYCLOPHANES

A. Bond Lengths

Although, in general, bond lengths do not deviate dramatically from normal values, changes are observed in the bond lengths of cyclophane macrocycles. Specifically, the bond lengths in the aromatic ring around the atom bound to the bridge (*e* in the table at the end of the chapter) and the bond lengths of the atoms in the bridge (*a* and *b* in the table) deviate from their normal values. The bridging bond lengths (*b*) are generally the longest and in some instances have been found to be greater than 1.6 Å.

B. Bond Angles

The molecular distortion in cyclophanes is reflected largely in the deviation of the bond angles from the normal values of 109.5, 120, and 180° for sp^3-, sp^2-, and *sp*-hybridized atoms, respectively. These deviations are observed primarily in the internal angle of the bridging atom in the aromatic ring (λ in the table) and in the angle of the bridge at the atom once removed from the aromatic ring (γ in the table). The former values (sp^2) have been observed to be as low as 115°, whereas the latter (sp^3) have been found to be as high as 118°. Angle γ also deviates from 120° when the bridging atom is sp^2-hybridized.

C. Nonplanarity of Aromatic Rings and Displacement of Appended Atoms from the Ring Planes

In the meta- and paracyclophanes, in which fewer than four atoms are present in each bridge, the aromatic ring atoms are displaced from planarity and the rings are distorted into boat shapes. In the paracyclophanes the 1 and 4 atoms of the aromatic rings are displaced out of the plane of the other four atoms *toward* the cyclophane cavity (α in the table). In the metacyclophanes the 2 and 5 atoms of the aromatic ring are displaced from the plane of the other four atoms *away* from the cyclophane cavity (β and δ, respectively, in the table). The 2 atom of the metacyclophane aromatic ring is generally more affected by the transannular π–π interaction, and thus angle β is usually larger than δ. In some instances, as in the multibridged and multilayered cyclophanes, chair geometries of the aromatic rings are also observed (see Fig. 1).

In addition to the distortion from planarity of the aromatic rings the atom in the bridge once removed from the aromatic ring (benzylic) is also displaced from the aromatic plane. In the lower [*m.n*]- and [*n*]cyclophanes these atoms are displaced from the aromatic plane toward the cyclophane cavity (β in the paracyclophane case and α in the metacyclophane case in the table).

D. Proximity of Nonbonded Atoms and Relationships of Interplanar Angles

The ideal nonbonded contact distance between aromatic rings in crystals containing these groups is 3.4 Å, as observed in graphite. In the lower homologs of [*m.n*]para- and [*m.n*]metacyclophanes, however, mean intramolecular aromatic ring separations are near 3 Å, and specific C · · · C nonbonded distances of the corresponding 1,4 and 1,3 atoms, which eclipse one another transannularly, are frequently less than 2.8 Å (p and q in the table). These close proximities cause distortions in the geometry of the ring sp^2 atoms such that the π density grows on the outer face of the aromatic ring. This rehybridization characteristic is observed crystallographically as the "turning-in" of the H atoms bound to the aromatic ring and has been observed in most structures in which the H atoms were found during the structure refinement.

Another aspect of intraannular π–π interactions sometimes found in cyclophane structures is that the mean aromatic planes are not parallel but are inclined with respect to one another (θ in the table). In addition, in many cases the aromatic ring atoms do not perfectly eclipse the corresponding atoms in the other aromatic ring but are staggered somewhat by a rotation about an axis normal to the aromatic planes. A lateral or slipping motion of the aromatic rings relative to one another (ω in the table) has also been observed, although mainly in systems in which there is greater capacity to do so because the bridging aliphatic chains are longer. All of these nonbonded interactions and interplanar relationships have been observed in the cyclophane structures. Whereas the former features are the result of the inability of the aromatic rings to move any farther away from one another, the latter features all help to reduce the transannular nonbonded interactions.

The interaction of forces in the four categories just delineated gives rise to the unique deviations from the norm that are observed in cyclophane structures. As the aliphatic bridges are lengthened the conditions requiring nonideal atomic geometries are removed, and the distortions in the categories are no longer observed.

III. TABLE OF CYCLOPHANE CRYSTAL STRUCTURE DETERMINATIONS WITH IMPORTANT CRYSTALLOGRAPHIC DATA AND STRUCTURAL FEATURES

A. Description of the Table

1. General Description

The table presented in this chapter is divided into eight categories of cyclophanes: (1) para ring phanes; (2) meta ring phanes; (3) condensed aromatic and nonbenzenoid ring phanes; (4) heterophanes; (5) ortho ring phanes including *o*-thymotides, cycloveratrils, and related structures; (6) multilayered phanes; (7) ferrocenophanes; and (8) miscellaneous phanes. Entries in the first two categories are divided into subsections according to structural similarities and include (a) general structures containing at least one para or meta ring; (b) cyclophanes with three or more atoms in the aliphatic bridge; (c) cyclophanes with heteroatoms in the bridge; (d) cyclophanes with unsaturation in the bridge; (e) multibridged phanes of the $[m_y]$ type; and (f) multibridged phanes of the $[m^y]$ type.

In some cases category lines are crossed because it seemed important to list certain structures together. Thus, all cyclophanes with sulfur atoms in the bridge are listed together despite the fact that some have three or more atoms in the bridge and could be placed in subsection (b). As another example, multilayered phanes that also contain a heteroaromatic ring are listed with the layered phanes rather than with the heterophanes because comparison with layered phanes is of greater interest.

An attempt was made to include all published X-ray structure determinations of cyclophane molecules through early 1982. Structural information that was referred to in published work was also included even though a complete crystallographic publication has not yet appeared. Whatever information was available from footnotes and discussions in these articles was incorporated into the table even for a single datum. (The absence of an entry implies that information was not available.) In addition, preliminary reports, if they were available (crystallographic meetings), and some as yet unpublished results were also incorporated if the information was known to the author, if it was complete, or if the available data were significant. Although orthocyclophanes and some large macro-, bicyclic, and crownlike structures are generally not distorted in the classic manner of cyclophanes, and not directly related to the para- and metacyclophane systems, some orthocyclophane, tri-*o*-thymotide, cycloveratril, and bicyclophane structures were included for completeness and because of the manner in which their cavities (formed by the aromatic rings and bridges) resemble the other cyclophanes.

2. Explanation of Table Format and Symbols

A table of cyclophane X-ray determinations is presented beginning on p. 80. It contains the following entry categories:

1. Structure: A two-dimensional representation of the compound is given. Attempts were made to give syn and anti as well as other geometric relationships concerning the three-dimensional structure, but not in every case. When necessary, a numbering system is given to the atoms in the structure so that descriptive ambiguity does not arise. The compound number is given in boldface type below each structure.
2. Name: Standard cyclophane nomenclature as described by Smith* and Vögtle† is used. The superscripted number following the structure name and boldface structure number refers to the reference at the end of the chapter, and boldface numbers in parentheses indicate the compound number. In multibridged cyclophanes the $[m_y]$ notation refers to y,m-atom bridges bridging two aromatic rings that are layered, whereas the $[m^y]$ notation refers to y,m-atom bridges bridging y aromatic rings in a cycle. Nine columns appear below the name with the following information:
3. Crystal system: The crystal system is given, and if known the solvent or method of crystallization is given in parentheses.
4. Space group: The space group determined crystallographically is given. Space groups in parentheses indicate that a unique space group was not determined.
5. Radiation: The radiation used for the collection of intensity data is given.
6. Cell dimensions: The unit cell dimensions are given in angstroms for a, b, and c and in degrees for α, β, and γ. Values in parentheses are standard deviations in the least significant digit. In those cases in which data are given at more than one temperature, the first entry is at the temperature at which the intensity data were collected.
7. Cell volume: The unit cell volume is given in cubic angstroms.
8. Density: The calculated density is given first. The measured density is given in parentheses. Both values are in grams per cubic centimeter.

* "Bridged Aromatic Compounds," B. H. Smith, Academic Press, New York 1964.
† Vögtle, F., Neumann, P., *Tetrahedron,* **26,** 5847 (1970).

9. Z: The number of molecules per unit cell is given. For entries that are clathrates or inclusion compounds this value is for the host only. The host/guest ratio (given with the name) gives the number of guest molecules per unit cell.

10. R: The final refinement index is given.*

11. Significant bond lengths, bond and interplanar angles, and nonbonded interactions: In this column all the relevant information about bond lengths (given in angstroms), bond angles (given in degrees), interplanar angles (given in degrees), and nonbonded interactions (given in angstroms) is given. These are generally mean values averaged over all equivalent lengths and angles with standard deviations given in parentheses.

In this column letters *a*, *b*, *c*, *d*, *e*, *f*, *g*, *i*, and *j* denote bond lengths within the cyclophane structure, as described below. Letters *u*, *v*, *w*, *x*, *y*, and *z* denote a functional group bond length and represent the bond length of the link between the atom in the cyclophane and the atom in the functional group that is bound to the cyclophane framework. These letters specifically describe the following types of functional group bonds: *u*, C—C; *v*, C—O (or C=O); *w*, C—N; *x*, C—F; *y*, C—Br; *z*, C—Cl.

Letters *p*, *q*, *r*, *s*, and *t* denote transannular nonbonded distances, and letters *m*, *n*, and *o* denote distances in ferrocenophane structures, as described below.

Bond angles and interplanar angles are denoted by the Greek letters α, β, γ, δ, ε, θ, λ, μ, φ, and ω and are specifically used as described below. Additional angles associated with ferrocenophanes are described using the Greek letters π, ρ and σ, and their use is described below.

12. Comments: Below the nine columns of crystallographic and geometric data, the important and relevant descriptions, as discussed in the literature, are summarized. In certain instances comparisons with other entries of the table are made.

3. Expanded Explanation of Symbols Used for Data Entries in Last Column of Table

The bond length and bond angle information in the last column of the table is given relative to three general types of cyclophane structures by which each of the cyclophanes can be described. These are given in Fig. 1 for a para- (a), meta- (b), and ortho-type (c) cyclophane structure.

In all cyclophane types, *a*, *b*, *c*, and *d* denote the bond distances in the

* In most cases this value was for nonzero reflections.

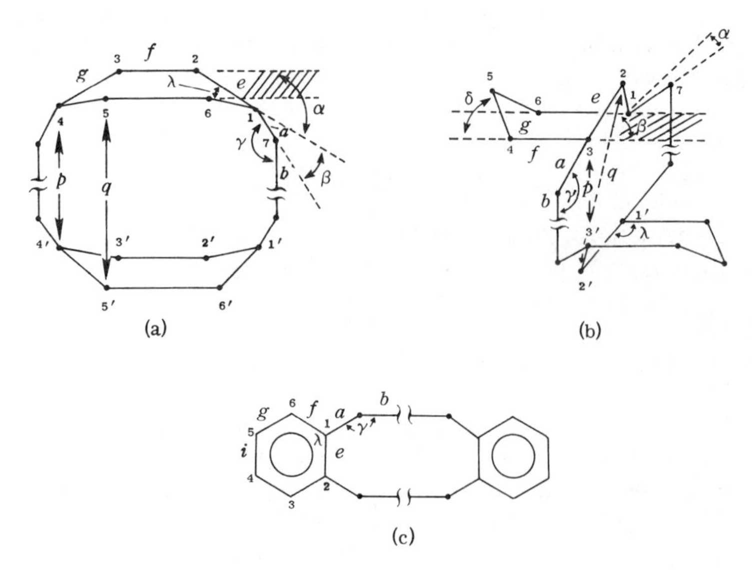

Fig. 1. The three general types of cyclophane structures used to describe the crystallographic data: (a) para, (b) meta, and (c) ortho.

bridge (Fig. 1). The first bond removed from the aromatic ring is denoted by *a*, the second by *b*, the third by *c*, and the fourth by *d*. These values specify any atom–atom bond length, whether C—C or otherwise, which is found in these specified positions. In general, these represent C—C bond lengths. However, sulfur, nitrogen, and oxygen atoms are frequently found in cyclophane bridges, and if the bridge contains three atoms in the following sequence, $\mathrm{C}\underset{\text{aromatic}}{\text{——}}\mathrm{C}$—S—C, *a* denotes a C—C bond and *b* denotes a C—S bond. (If ambiguities arise specific atoms with their associated numbers are given, e.g., C-1—C-2 = 1.54 Å).

Letters *e*, *f*, *g*, *i*, and *j* specifically denote the bond lengths in the aromatic rings. For the para-type ring (Fig. 1a) *e* describes the 1,2(1,6) bonds (1 being the atom bound to the bridge) and, when related by symmetry, the 3,4(4,5) bonds. If the latter bonds are unrelated to the former because of dissymmetry the value is denoted by the letter *g*. The 2,3(5,6) bonds are denoted by *f*.

In the meta-type ring (Fig. 1b) *e* denotes the 1,2(2,3) bonds, *f* denotes the 3,4(6,1) bonds, and *g* denotes the 4,5(5,6) bonds. In the ortho-type ring (Fig. 1c) *e* denotes the 1,2 bond, *f* denotes the 2,3(6,1) bonds, *g* denotes the 3,4(5,6) bonds, and *i* denotes the 4,5 bond. The letter *j* is reserved for larger rings and is specified when used. (As in the case of the bridging bond lengths, ambiguities are dealt with by specifying the numbered atoms.)

Letters p and q denote the nonbonded distances as described in Fig. 1a,b. For the para-type structure p represents the distance between the 1 and 1′ (4 and 4′) atoms, and q the distance between the 2 and 2′ or 6 and 6′ (3 and 3′ or 5 and 5′) atoms. For the meta-type structure p represents the distance between the 1 and 1′ (3 and 3′) atoms, and q denotes the distance between 2 and 2′ atoms in either a syn or anti structure. Letters r, s, and t also denote nonbonded distances and are defined wherever needed.

The important bond angles given in the table are denoted by γ and λ and are pictured in Fig. 1. In all three ring types the angle in the bridge once removed from the aromatic ring ($\text{atom}\overline{_{\text{aromatic ring}}}\text{atom}\overline{_{\alpha \text{ to ring}}}\text{atom}_{\beta \text{ to ring}}$) is described by the γ, and the internal aromatic ring angle at the atom attached to the bridge (the 6–1–2 angle) is denoted by λ.

Distortions associated with the aromatic rings are described by angles α, β, and δ. Thus, in the para-type structure, α denotes the interplanar angle between the 2,3,5,6 plane and the 2,1,6 plane, and β denotes the angle between the 2,1,6 plane and the 1–7 bond vector. In the meta-type structure, α denotes the angle between the 1,3,4,6 plane and the 1–7 bond vector, β denotes the interplanar angle between the 1,2,3 and the 1,3,4,6 planes, and δ denotes the interplanar angle between the 4,5,6 plane and the 1,3,4,6 plane. (These angles are absent in the ortho-type phanes because the aromatic ring atoms and the atom once removed from the ring are all coplanar.) An asterisk appended to these letters (e.g., α^*) indicates that the values given are the perpendicular distances (angstroms) that the appropriate atoms are displaced from the adjacent plane. Thus, for the para-type structure, α^* indicates the distance that atom 1 is displaced from the 2,3,5,6 plane, and β^* indicates the distance that atom 7 is displaced from the 1,2,6 plane. For the meta-type structure, α^*, β^*, and δ^*, respectively, indicate the distance that atom 7, atom 2, and atom 5 are displaced from the 1,3,4,6 plane.

Greek letters ε and μ are used when necessary and are defined when needed. The letter θ describes the angle of inclination of two aromatic ring planes (least-squares planes) (Fig. 2a). Letter φ describes the angle that the 1,2,3,4 plane and the 4,5,6,1 plane make with one another in a para-type structure, and ω defines the angle found between the least-squares aromatic plane and the line drawn between the aromatic ring atoms of each ring bound to the same bridge (Fig. 2b).

For ferrocenophanes letters a, b, c, and d denote bridge bond lengths (as described above), and letters e, f, and g denote ring bond lengths (cyclopentadienide), as pictured in Fig. 3. Depending on the number of bridges, these phanes are compared with general ortho-, meta-, or para-type cyclophane structures, and then the α, β, γ, δ, and λ values correspond to those described above. Thus, a monobridged ferrocenophane is

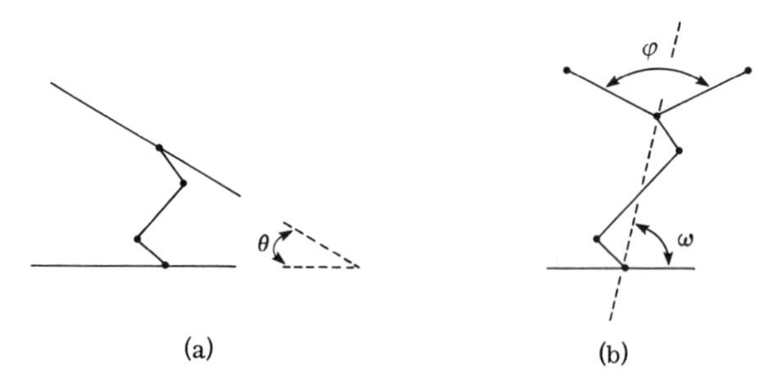

Fig. 2. Definition of angles θ (a) and φ and ω (b).

compared with the para-type structure, a (1,1′)(2,2′)-dibridged ferrocenophane is compared with the ortho-type structure, and a (1,1′)(3,3′)-dibridged ferrocenophane is compared with the meta-type structure described above.

In addition to the letters describing bond lengths and bond and interplanar angles, the letter n is used in ferrocenophanes to describe the perpendicular distance of the iron atom to the cyclopentadienyl ring plane, the letter o (with subscripts to describe particular atoms) is used to describe C—Fe bond lengths in the ferrocenoid moiety, and the letter m is used to denote Fe · · · Fe nonbonded distances in those structures containing more than one ferrocenoid unit. The Greek letter π denotes the angle of inclination between the two cyclopentadienyl rings (related to θ in Fig. 2a), and ρ denotes the angle of rotation that the two rings make with one another about a normal to both rings (staggered structure, $\rho = 36°$; eclipsed structure, $\rho = 0°$; see Fig. 3b). Finally, σ denotes the twisting of two ferrocenoid units with respect to one another (twisting of the two normals through each ferrocenoid cyclopentadienyl unit and iron atom) in a ferrocenophane containing more than one ferrocenoid moiety.

In this last column of the table describing lengths and angles a double dagger (‡) indicates that the value is described more fully in the last section, entitled "Comments."

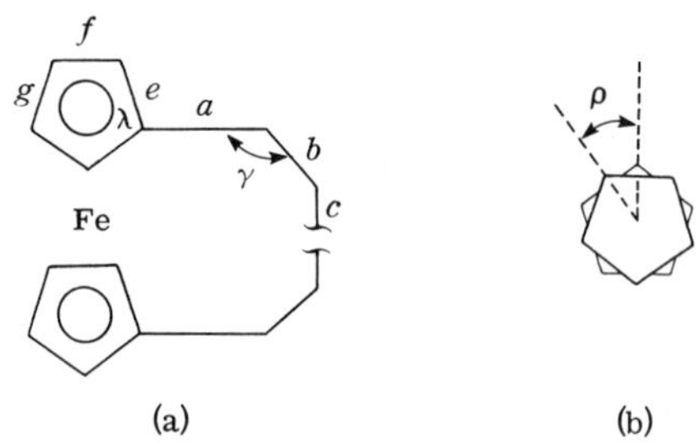

Fig. 3. (a) Bond lengths and angles in a monobridged ferrocenophane; (b) definition of rotational angle ρ.

B. Tabulation of Crystallographic Determinations

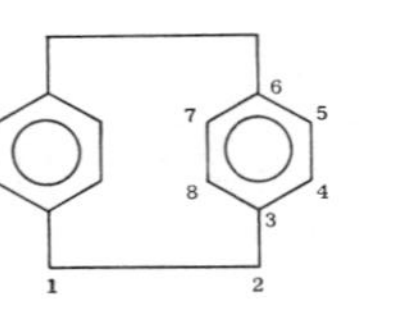

1

Crystal system	Space group	Radiation	Cell dimensions (Å)	Cell volume (Å³)	Density (g/cm³), calculated (measured)	Z	R	Significant bond lengths and nonbonded distances (Å), and bond and interplanar angles (deg)

I. PARA RING PHANES

A. STRUCTURES CONTAINING AT LEAST ONE PARA RING

[2.2]Paracyclophane (1)[1,2]

Crystal system	Space group	Radiation	Cell dimensions (Å)	Cell volume (Å³)	Density (g/cm³), calculated (measured)	Z	R	Significant bond lengths and nonbonded distances (Å), and bond and interplanar angles (deg)
Tetragonal (pyridine)	$P4_2/mnm$	CuKα	$a = 7.82$ $c = 9.33$	—	(1.23)	2	0.14	$a = 1.54$, $\alpha = 14$ $b = 1.55$, $\gamma = 114.37$ $e = 1.39$, $\lambda = 118.6$ $f = 1.40$ $p = 3.09$ $q = 2.83$

Comments: The aromatic rings are not flat. The C-3 and C-6 ring atoms are 0.13 Å out of the C-4, C-5, C-7, C-8 plane.

[2.2]Paracyclophane (1)[3,4]

Crystal system	Space group	Radiation	Cell dimensions (Å)	Cell volume (Å³)	Density (g/cm³), calculated (measured)	Z	R	Significant bond lengths and nonbonded distances (Å), and bond and interplanar angles (deg)
Tetragonal	$P4_2/mnm$	CuKα	291K $a = 7.79(1)$ $c = 9.30(1)$	—	1.229	—	—	291K $a = 1.547$, $\alpha = 14.0$ $b = 1.630$, $\gamma = 111.2$ $e = 1.380$, $\lambda = 119.7$ $f = 1.421$ $p = 2.751$ $q = 3.087$

			93K	—	—	—	—	93K
			$a = 7.69(1)$ $c = 9.23(1)$					$a = 1.534$, $b = 1.558$, $e = 1.388$, $f = 1.415$ $\alpha = 14.0$, $\gamma = 112.9$, $\lambda = 118.4$ $p = 2.749$, $q = 3.093$

Comments: The molecule is 0.03 Å shorter and the intermolecular distances are 0.05 Å shorter at 93K. The values of α, p, and q are not temperature dependent, but b decreases by 0.07 Å at 93K. The data indicate a concertina vibration of the benzene rings toward and away from one another accompanied by a stretching of the bridge C—C bonds (b) and twisting of the benzenoid rings with respect to one another about an axis perpendicular to the rings and passing through their centers.

[2.2]Paracyclophane (1)[5]

Tetragonal	$P4_2/mnm$	CuKα	$a = 7.781(1)$ $c = 9.290(2)$	—	—	2	0.029	$a = 1.511$, $b = 1.593$, $e = 1.386$, $f = 1.387$ $\alpha = 12.6$, $\beta = 11.2$, $\gamma = 113.7$ $p = 2.78$, $q = 3.09$

Comments: In comparison with the above two earlier determinations this one is the most precise. The substituent atoms on the bridge (H) show a large degree of motional anisotropy normal to the bridge–substituent bond, which is explained in terms of a disorder of these atoms. A dynamic disorder exists in which a twist (3°) of the aromatic rings occurs in opposite directions about a normal common to both rings.

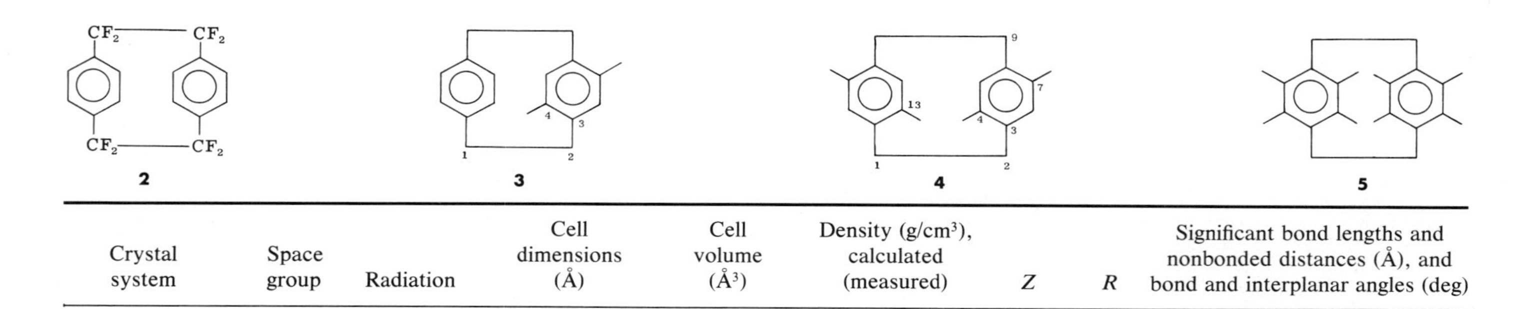

Crystal system	Space group	Radiation	Cell dimensions (Å)	Cell volume (Å^3)	Density (g/cm^3), calculated (measured)	Z	R	Significant bond lengths and nonbonded distances (Å), and bond and interplanar angles (deg)
1,1,2,2,9,9,10,10-Octafluoro[2.2]paracyclophane (2)[5]								
Monoclinic (CH_2Cl_2/ether, 1 : 1)	$P2_1/n$	CuKα	24°C a = 7.994(2) b = 7.986(2) c = 10.855(3) β = 97.84(1)	—	(1.704)	2	0.037	a = 1.496, b = 1.597, e = 1.382, f = 1.373, x = 1.35 p = 2.80, q = 3.09 α = 11.8, β = 12.6, γ = 114.4

Comments: The substituent atoms on the bridge (F) show a large degree of motional anisotropy normal to the bridge–substituent bond, which is explained in terms of a disorder of these atoms. A dynamic disorder exists in which a twist (3°) of the aromatic rings occurs in opposite directions about a normal common to both rings.

Crystal system	Space group	Radiation	Cell dimensions (Å)	Cell volume (Å^3)	Density (g/cm^3), calculated (measured)	Z	R	Significant bond lengths and nonbonded distances (Å), and bond and interplanar angles (deg)
4,7-Dimethyl[2.2]paracyclophane (3)[6]								
Monoclinic	$P2/c$	—	a = 14.974(6) b = 7.739(1) c = 11.518(2) β = 94.83(3)	1,330.0(6)	1.180 (1.172)	4	—	—

Comments: A combination of molecular conformational analysis and molecular packing analysis was applied to the crystal structure determination. The rings are twisted 9.9(10)° with respect to one another about an axis perpendicular to the planes of the rings.

4,7,13,16-Tetramethyl[2.2]paracyclophane (4)[6]

Monoclinic	*C*2/*c*	—	a = 8.157(2)	1,547(5)	1.137	4	—		—
			b = 15.846(3)		(1.133)				
			c = 12.026(2)						
			β = 95.58(2)						

Comments: A combination use of molecular conformation analysis and molecular packing analysis was applied to the crystal structure determination. The rings are twisted 19.3(1)° with respect to one another about an axis perpendicular to the planes of the rings.

4,5,7,8,12,13,15,16-Octamethyl[2.2]paracyclophane (5)[7]

Monoclinic	*I*2/*c*	MoKα	a = 24.539(12)	3,645.35	1.168	8	0.085	a = 1.525	α = 15.8
			b = 8.750(4)		[1.13(2)]			b = 1.561	β = 8.3
			c = 16.981(8)					e = 1.390	γ = 114.5
			β = 91.16(4)					f = 1.397	λ = 118.5
								u = 1.551	
								p = 2.74	
								q = 3.16	
								$H_3C \cdots CH_3$ = 3.52	

Comments: The rings are displaced 0.62 Å from being directly above one another, which causes a torsional angle (defined by the two bridging atoms and the two atoms to which they are bound in the two rings) of 15.3° in the bridge. This twisting and lateral displacement decreases the interring π–π repulsion and nonbonded $H_3C \cdots CH_3$ interaction.

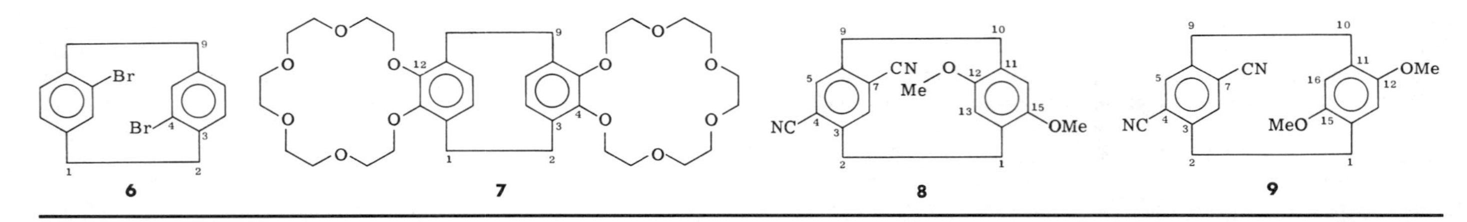

Crystal system	Space group	Radiation	Cell dimensions (Å)	Cell volume (Å³)	Density (g/cm³), calculated (measured)	Z	R	Significant bond lengths and nonbonded distances (Å), and bond and interplanar angles (deg)
			4,12-Dimbromo[2.2]paracyclophane (6)[8]					
Monoclinic	$P2_1/c$	—	$a = 7.797(6)$ $b = 7.737(5)$ $c = 11.165(7)$ $\beta = 100.26(4)$	—	—	2	0.054	Average distance after correction for thermal motion $a = 1.523$ $b = 1.577$ $e,f = 1.392$

Comments: Bromine substitution enlarges the adjacent angle (β) but does not substantially change bond distances a,b and e,f from those found in [2.2]paracyclophane or 1,1,2,2,9,9,10,10-octafluoro[2.2]paracyclophane. There is no evidence for molecular disorder.

Crystal system	Space group	Radiation	Cell dimensions (Å)	Cell volume (Å³)	Density (g/cm³), calculated (measured)	Z	R	Significant bond lengths and nonbonded distances (Å), and bond and interplanar angles (deg)
			Bis(4,5;12,13)-18-crown-6[2.2]paracyclophane (7)[9]					
Orthorhombic	*Pbca*	—	$a = 8.347$ $b = 16.939$ $c = 24.330$	—	—	4	—	—

Comments: The molecular geometry of the [2.2]paracyclophano portion of the molecule is similar to [2.2]paracyclophane, but the twisting of the aromatic rings relative to an axis normal to the rings, which is pronounced in the latter compound, is absent in the ether and is ascribed to the bulk of the polyether rings attached to the cyclophane. The crown ether portion of the molecule departs significantly from sixfold symmetry.

4,7-Dicyano-12,15-dimethoxy[2.2]paracyclophane (pseudo-ortho isomer) (8) [10,11]

								Di-CN ring	Di-OMe ring
Monoclinic	*C*2/*c*	—	$a = 9.334(1)$	—	—	—	—		
			$b = 11.444(1)$						
			$c = 14.915(4)$					$\alpha = 12.3$	$\alpha = 11.1$
			$\beta = 99.05$					$\beta = 10.8$	$\beta = 12.0$
								$p = 2.755$	
								$q_{5,12} = 3.054$	
								$q_{4,13} = 3.049$	

Comments: The intermolecular distance between donor and acceptor portions of neighboring phane molecules is 8.394 Å.

4,7-Dicyano-12,15-dimethoxy[2.2]paracyclophane (pseudo-gem isomer) (9) [10,11]

								Di-CN ring	Di-OMe ring
Monoclinic	*C*2/*c*	—	$a = 9.248(1)$	—	—	—	—		
			$b = 11.348(2)$						
			$c = 15.505(3)$					$\alpha = 13.4$	$\alpha = 12.0$
			$\beta = 97.94$					$\beta = 9.7$	$\beta = 11.2$
								$p = 2.747$	
								$q_{5,16} = 3.046$	
								$q_{4,15} = 3.089$	

Comments: The intermolecular distance between donor and acceptor portions of neighboring phane molecules is 8.282 Å. The methoxy-O · · · C≡N nonbonded distance is 3.197 Å.

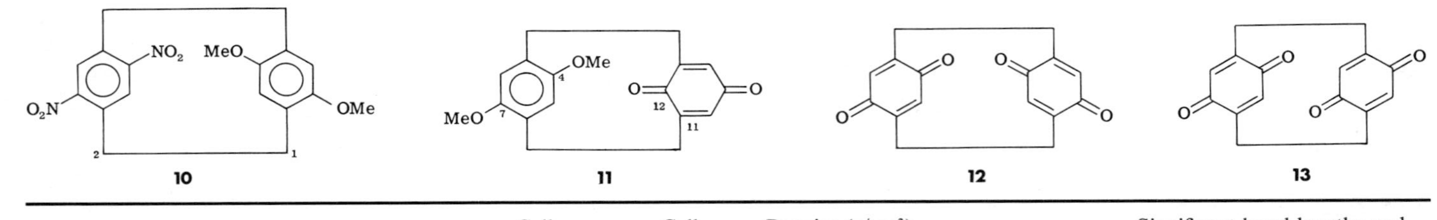

Crystal system	Space group	Radiation	Cell dimensions (Å)	Cell volume (Å³)	Density (g/cm³), calculated (measured)	Z	R	Significant bond lengths and nonbonded distances (Å), and bond and interplanar angles (deg)
4,7-Dimethoxy-12,15-dinitro[2.2]paracyclophane (pseudo-ortho isomer) (10)[11]								
Monoclinic	$C2/c$	—	$a = 9.172$ $b = 11.522$ $c = 15.530$ $\beta = 93.81$	—	—	4	0.06	$b = 1.575$ $p = 2.746$ $q_{4,3} = q_{5,12} = 3.054$

Comments: The aromatic rings are deformed into boat shapes. The dihedral angle in the bridge (around the C-1—C-2 bond) is 22°. The intermolecular distance between donor and acceptor parts of neighboring molecules is ~8.4 Å, but the molecules are not layered in the crystal.

Crystal system	Space group	Radiation	Cell dimensions (Å)	Cell volume (Å³)	Density (g/cm³), calculated (measured)	Z	R	Para ring	Meta ring
4,7-Dimethoxy[2.2](2,6)*p*-benzoquinonoparacyclophane (11)[12]									
Orthorhombic	$P2_12_12_1$	—	—	—	—	—		$\alpha = 11$	$\beta = 21$ $\delta = 7$ $\theta = 19$

Comments: Both rings are distorted into boat conformations. The inner carbonyl C atom (C-12) is 2.87 Å from the benzenoid plane.

[2.2](2,5)*p*-Benzoquinonophane (pseudo-ortho isomer) (12)[11]

Monoclinic	$P2_1/c$	—	$a = 9.115$ $b = 23.632$ $c = 11.457$ $\beta = 90.55$	—	—	8	0.06		$\alpha = 16$ $\beta = 6$ $p = 2.65$ $q = 3.04$

Comments: The quinone rings are deformed into boat shapes. The C=O bonds make an angle of 9° with the plane of the four nonbridged atoms.

[2.2](2,5)Benzoqunionophane (pseudo-gem isomer) (13)[11,13]

Monoclinic (dioxane)	$P2_1/n$	MoKα	$a = 7.930(2)$ $b = 9.280(2)$ $c = 8.024(2)$ $\beta = 93.30(2)$	—	1.51	2	0.051	$a = 1.494$ $b = 1.581$ $v = 1.213$	$\alpha = 16$ $\beta = 65$ $\gamma = 112.5$ $\lambda = 117.2$ $p = 2.73$ $q = 3.11$

Comments: The molecule deviates from 2/*m* symmetry because of a parallel shifting of the two rings with respect to one another (0.24 Å). The rings are boat-shaped due to transannular repulsion. The C=O bonds make an angle of 5.5° with the plane of the four nonbridged C atoms of the rings in a direction opposite to the proximate aromatic ring. The nonbonded distance between the pseudogeminal carbonyl C atoms and carbonyl O atoms are 3.17 and 3.37 Å, respectively.

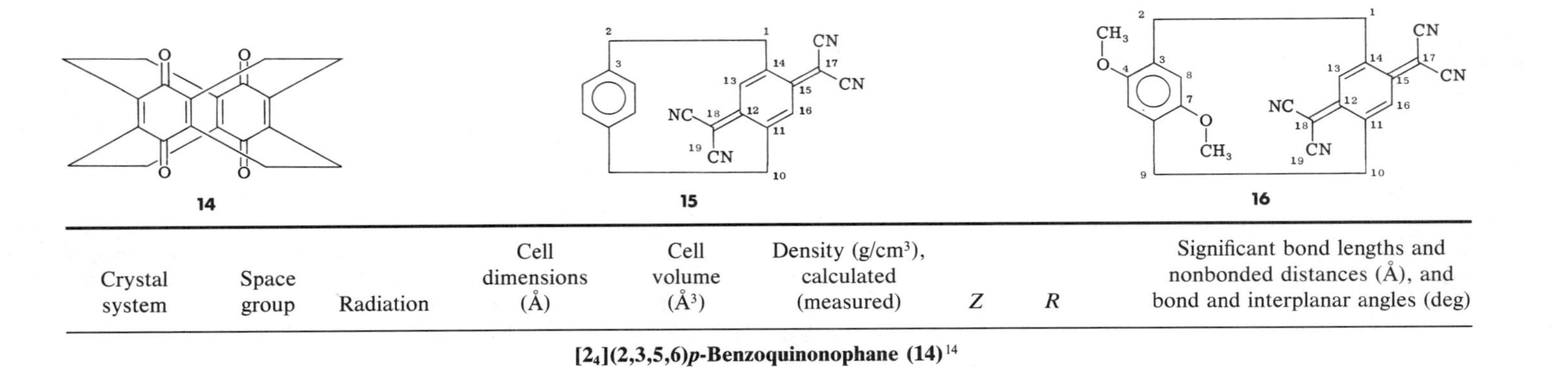

Crystal system	Space group	Radiation	Cell dimensions (Å)	Cell volume (Å³)	Density (g/cm³), calculated (measured)	Z	R	Significant bond lengths and nonbonded distances (Å), and bond and interplanar angles (deg)
[2₄](2,3,5,6)*p*-Benzoquinonophane (14) [14]								
—	—	—	—	—	—	—	—	Meta type: $b = 1.556$; $\beta = \delta = 23$; $p = 2.67$

Comments: The quinono rings are distorted into boat shapes with eclipsing of the corresponding atoms in the two rings. The C=O vectors make an angle of 16° with the plane of the four bridging C atoms of its ring toward the opposite ring. The nonbonded distances of the transannular carbonyl groups are 3.95 Å for the C=O · · · O=C and 3.26 Å for the O=C · · · C=O interactions.

[2](2,5)(17,17,18,18-Tetracyanoquinodimethano)[2]paracyclophane · benzene (15) [15]

Crystal system	Space group	Radiation	Cell dimensions (Å)	Cell volume (Å³)	Density (g/cm³), calculated (measured)	Z	R	Para ring	Quninoid ring
Triclinic	$P\bar{1}$	MoKα	$a = 9.772(2)$	1,107.8(4)	1.236	2	0.088	$a = 1.503(7)$	$a = 1.514(6)$
			$b = 18.477(4)$		(1.240)			$b = 1.570(7)$	
			$c = 6.569(1)$					$e = 1.394(6)$	$e_{11,12} = 1.464(5)$
			$\alpha = 90.38(2)$						$e_{11,16} = 1.349(6)$
			$\beta = 107.92(2)$						
			$\gamma = 100.32(2)$						

Para ring	Quinoid ring
$f = 1.385(7)$	$f = 1.445(5)$
	$u_{12,18} = 1.378(5)$
	$u_{18,19} = 1.429(6)$
$\alpha = 12.1; 9.7$	$\alpha = 15.2; 15.7$
$\beta = 12.8; 11.9$	$\beta = 7.7; 11.3$
$\gamma = 111.8(4)$	$\gamma = 111.5(4)$
$\lambda = 116.8(4)$	$\lambda = 115.8(4)$

Comments: The paracyclophano moiety has the usual boat structure, and the quinodimethano moiety has a twisted-boat form. Whereas the C-11 and C-14 atoms are bent toward the benzenoid ring ($\alpha = 15.5°$), the two $C(CN)_2$ groups are bent away from that ring. The C-12—C-18 and C-15—C-17 vectors make angles of about 12.5° with the C-12, C-13, C-15, C-16 plane. The two rings are twisted with respect to one another about an axis that is normal to both rings and passing through their centers. This causes torsional angles of from 5 to 21° around the C-1—C-2 and C-9—C-10 bonds.

4,7-Dimethoxy[2](2,5)(17,17,18,18-tetracyanoquinodimethano)[2]paracyclophane (16)[15]

								Para ring	Quinoid ring
Orthorhombic	*Fdd*2	MoKα	$a = 25.618(3)$ $b = 31.895(5)$ $c = 9.808(4)$	8,014(3)	1.307 (1.30)	16	0.109	$a = 1.496(16)$	
								$b = 1.544(17)$	
								$e_{3,4} = 1.401(17)$	$e_{11,12} = 1.461(14)$
								$e_{3,8} = 1.400(16)$	$e_{11,16} = 1.354(14)$
								$f = 1.406(17)$	$f = 1.431(14)$
								$\alpha = 8.2; 12.4$	$\alpha = 16.9; 16.4$
								$\beta = 15.3; 13.0$	
								$\gamma = 113.4(10)$	$\gamma = 110.6(10)$
								$v = 1.374(16)$	$u_{12,18} = 1.389(15)$
									$u_{18,19} = 1.438(17)$
								$\lambda = 117.9(11)$	$\lambda = 115.3(9)$

Comments: The structure is similar to that of [2](2,5)(17,17,18,18-tetracyanoquinodimethano)[2]paracyclophane, and both rings are boat-shaped. The torsional angles around the C-1—C-2 and C-9—C-10 bonds range from 4.0 to 23.8°.

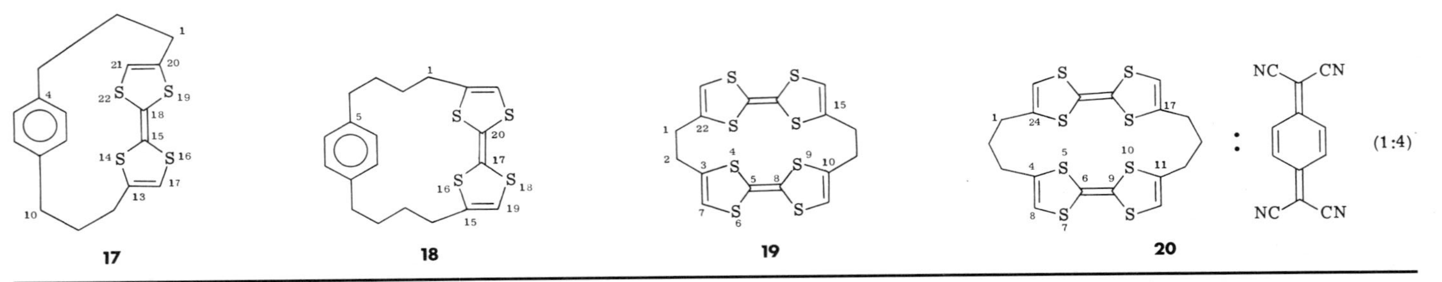

Crystal system	Space group	Radiation	Cell dimensions (Å)	Cell volume (Å^3)	Density (g/cm^3), calculated (measured)	Z	R	Significant bond lengths and nonbonded distances (Å), and bond and interplanar angles (deg)	
[3.3](13,20)Tetrathiafulvalenoparacyclophane (17)[16]									
Monoclinic	$P2_1/n$	—	$a = 14.193(4)$	—	—	4	0.034	Para ring	Tetrathiafulvaleno ring
			$b = 12.804(3)$					$a = 1.515(4)$	$a = 1.493(4)$
			$c = 9.832(2)$					$b = 1.526(4)$	$b = 1.531(5)$
			$\beta = 105.5(1)$					$e = 1.385(4)$	C-15—C-18 = 1.327(3)
								$f = 1.385(4)$	S-14—C-15 = 1.757(3)
								$\alpha_{\text{inverted}} = 2^{\ddagger}$	S-16—C-15 = 1.761(2)
								$\beta_{\text{inverted}} = 4^{\ddagger}$	S-14—C-13 = 1.769(3)
									S-16—C-17 = 1.749(3)
									C-13—C-17 = 1.317(4)

Comments: The central double bond of the tetrathiafulvalenoid (TTF) moiety is almost exactly centered above, and parallel to, the benzenoid ring, being 3.73 Å from the least-squares benzenoid plane. Although the benzenoid ring is only slightly boat-shaped, the para C atoms are bent *away* rather than toward the TTF system. The α and β angles corresponding to this inverted benzenoid ring are 2 and 4°, respectively. The long axis of the TTF moiety and the para axis of the benzenoid ring are twisted 15° with respect to one another. The TTF group deviates dramatically from planarity. The two S atoms of each of the five-membered rings of the TTF group are approximately 4° displaced from the C-15═C-18 double-bond plane, and there exists a torsional twist around this double bond of 7°. The S—C-20—C-21—S plane is inclined ~30° to the C-15═C-18 double-bond plane.

[4.4](15,22)Tetrathiafulvalenoparacyclophane (18)[16]

Monoclinic	$C2/c$	—	$a = 18.12(1)$ $b = 9.137(5)$ $c = 23.58(2)$ $\beta = 100.2(1)$	—	—	8	0.060	—

Comments: The benzenoid ring is displaced sideways to the TTF moiety. The benzenoid plane is inclined 63.5° to the C-17═C-20 double-bond plane. The tetrathiaalkene unit (C-17, C-20, and the four S atoms) of the TTF group is planar, but the S-16, C-15, C-19, S-18 plane is displaced 22° from the alkene plane toward the benzenoid moiety.

[2](3,10)[2](15,22)Tetrathiafulvalenophane (19)[17]

Monoclinic (CS_2)	$P2_1/c$	—	$a = 10.013(2)$ $b = 9.731(2)$ $c = 11.185(2)$ $\beta = 121.00(3)$	—	—	2	0.043	—

Comments: The two TTF units are in a stepped-anti conformation. The TTF units deviate from planarity. The S-4, C-3, C-7, S-6 plane (like all similar planes) is displaced 20° from the alkene plane defined by C-5═C-8 and the four attached S atoms.

[3](4,11)[3](17,24)Tetrathiafulvalenophane : Tetracyanoquinodimethane (1 : 4) (20)[17]

Triclinic (CS_2/CH_3CN)	$P\bar{1}$	—	$a = 7.80$ $b = 13.93$ $c = 28.73$ $\alpha = 92.9$ $\beta = 88.9$ $\gamma = 91.9$	—	1.39	—	—	—

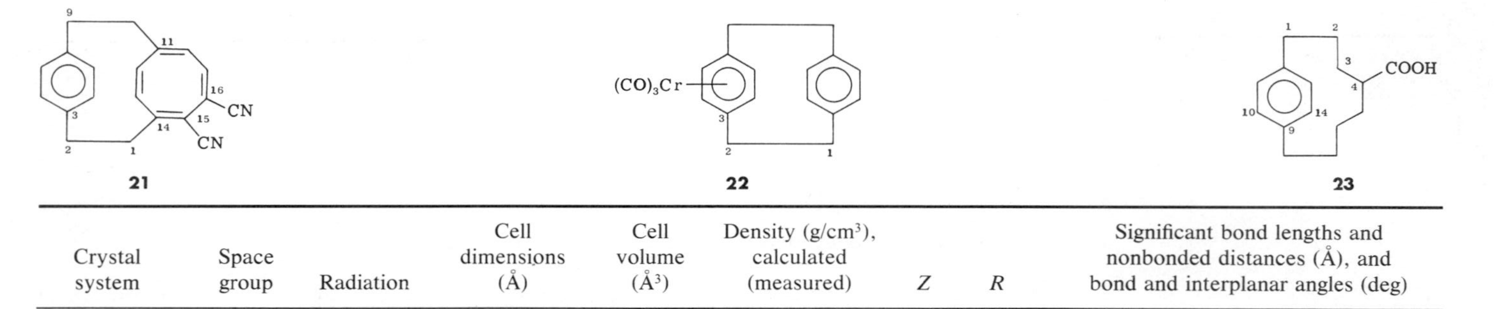

Crystal system	Space group	Radiation	Cell dimensions (Å)	Cell volume (Å^3)	Density (g/cm^3), calculated (measured)	Z	R	Significant bond lengths and nonbonded distances (Å), and bond and interplanar angles (deg)
			15,16-Dicyano[2.2](11,14)cyclooctatetraenoparacyclophane (21)[18]					
Orthorhombic	$P2_12_12_1$	CuKα	a = 11.897(1) b = 15.490(1) c = 8.0605(6)	1,485.4	1.271	4	0.036	Para ring: a = 1.501(4); e = 1.385(4); f = 1.382(5); α^* = 0.11 Cyclooctatetraeno ring (para type): a = 1.554(4); b = 1.541(4); e = 1.473(4); f = 1.331(4); C-14—C-15 = 1.343(3); C-18—C-11 = 1.331(4); C-15—C-16 = 1.493(3); C-17—C-18 = 1.464(4); C-16—C-17 = 1.328(4); u = 1.440(4)

Comments: The benzenoid ring is boat-shaped, whereas the cyclooctatetraenoid ring is slightly distorted from a tub-shaped conformation. The torsional angles around the C—C single bonds and the C═C double bonds in the cyclooctatetraene ring are from 0.4 to 5.4° and 53.0 to 56.9°, respectively.

Tricarbonyl(3,4,5,6,7,8-η^6-[2.2]paracyclophane)chromium (22)[19]

								Complexed ring	Uncomplexed ring
Orthorhombic	$P2_12_12_1$	MoKα	−150°C	1,536.5	—	—	0.067		
			a = 10.651(3)					a = 1.510(2)	a = 1.519(2)
			b = 13.031(3)					b = 1.595(4)	
			c = 11.071(4)					e = 1.401(6)	e = 1.397(4)
								f = 1.402(6)	f = 1.392(1)
								α = 12.2(1)	α = 11.4(1)
								β = 8.9(1)	β = 11.2(8)
								γ = 110.9(1)	γ = 112.8(3)
			20°C	1,573.3	1.453 (1.445)	4	—	λ = 118.3	λ = 116.9
			a = 10.694(2)					p = 2.723(5)	
			b = 13.115(2)					q = 3.017(2)	
			c = 11.217(2)						

Comments: The rings are essentially parallel. The Cr atom is situated centrally above one ring with three types of Cr—C bonds to the ring: short, 1.19 Å; medium, 2.23 Å; long, 2.34 Å. A twist of 3.8(7)° exists between the two rings around an axis perpendicular to the planes of the two rings. The interannular distance q is reduced in comparison with [2.2]paracyclophane due to the coordination of the $Cr(CO)_3$ group to one of the rings.

B. STRUCTURES CONTAINING THREE OR MORE ATOMS IN THE BRIDGE

4-Carboxy[8]paracyclophane (23)[20]

Monoclinic (hexane)	$P2_1/c$	CuKα	a = 12.807(3)	—	1.21 (1.21)	4	0.05	a = 1.503	α = 9.1
			b = 6.144(1)					b = 1.550	β = 7.8
			c = 16.354(4)					c = 1.541	γ = 110.7
			β = 98.07(4)					d = 1.558	λ = 117.5
								u = 1.511	

Comments: The benzene ring is not planar but has a boat-shaped form. The aromatic ring H atoms are bent toward the aliphatic chain by an average of 0.09 Å. Ignoring the —COOH group the molecule has C_2 symmetry along an axis perpendicular to the ring and bisecting the central C—C bond of the aliphatic chain. The nonbonded distance between C-14 of the aromatic ring and C-5 of the aliphatic chain is 3.362(3) Å. The average C—C—C angle in the bridge (excluding the benzylic C atoms) is 114.5°, about 4° larger than γ.

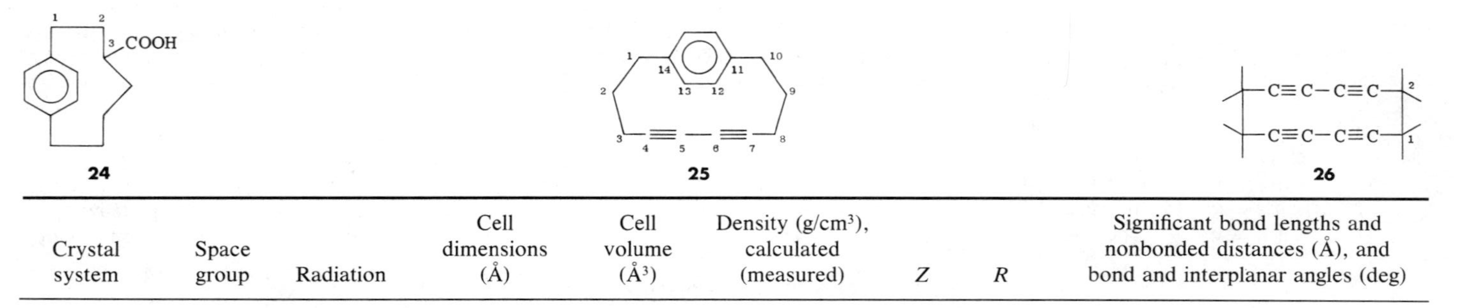

Crystal system	Space group	Radiation	Cell dimensions (Å)	Cell volume (Å³)	Density (g/cm³), calculated (measured)	Z	R	Significant bond lengths and nonbonded distances (Å), and bond and interplanar angles (deg)	
				3-Carboxy[7]paracyclophane (24)[21]					
Monoclinic (hexane)	$P2_1/c$	CuKα	a = 12.36 b = 7.81 c = 12.74 β = 99.19	1,215	1.20 (1.19)	4	0.163	a = 1.48 b = 1.56 c = 1.56 d = 1.57 u = 1.52	α = 16.5 β = 6.7 γ = 108 λ = 118

Comments: The benzenoid ring is more distorted than it is in 4-carboxy[8]paracyclophane. The average nonbonded distance between the C-3, C-4, and C-5 bridge C atoms and the aromatic ring C atoms is 3.60, 3.28, and 3.55 Å, respectively. The C—C—C angles in the aliphatic chain range from 111° (C-2—C-3—C-4) and 115° (C-1—C-2—C-3) to 117° (C-3—C-4—C-5). The widening of these angles (compared with normal values of 109°) results from a compromise to separate the aliphatic chain from the aromatic ring and lengthen the side chain to facilitate bridging. The molecules form racemic dimers in the crystal.

Crystal system	Space group	Radiation	Cell dimensions (Å)	Cell volume (Å³)	Density (g/cm³), calculated (measured)	Z	R	Significant bond lengths and nonbonded distances (Å), and bond and interplanar angles (deg)	
				[10]Paracyclophane-4,6-diyne (25)[22,23]					
Monoclinic (petroleum spirit)	$P2_1/a$	CuKα	10°C a = 14.509(2) b = 8.539(1) c = 10.623(1) β = 112.37(5)	—	1.04 (1.05)	4	0.087	a = 1.518(5) b = 1.501(7) c = 1.545(6) d = 1.476(6) C-4—C-5 = 1.183(5) C-5—C-6 = 1.380(5) $p_{4,14}$ = 3.09 [illegible] = 3.47	e = 1.382(5) f = 1,379(5) α = 1.8 γ = 116 λ = 117.5

Comments: The benzenoid ring is boat-shaped but less so than [3.3]- and [2.2]paracyclophanes. The diacetylene unit is bent into a bow shape (see below). The projection of the C-4, C-5, C-6, and C-7 atoms onto the aromatic plane is linear, but the axis of the diacetylene unit is rotated 2.5° from the diagonal (C-14,C-11) of the benzenoid ring. Whereas the C-1,C-2,C-3 trimethylene bridge is ordered (and has a bridge geometry similar to that of [3,3]paracyclophane) the C-8,C-9,C-10 trimethylene bridge is disordered and is explained in terms of hyperconjugative contributions of the bridge and acetylene units.

1,1,2,2,7,7,8,8-Octamethyl-3,5,9,11-cyclododecatetrayne (26)[24]

Monoclinic	$P2_1/n$	MoKα	−160°C a = 7.053 b = 10.436 c = 10.821 β = 92.91	—	—	—	0.060	—
			Ambient temperature a = 6.974 b = 10.725 c = 11.279 β = 90.62	—	—	—	—	—

Comments: At ambient temperature a disorder arises from changes in the ring dihedral angles and large shifts in the methyl group positions. Disorder was no longer apparent at −160°C. The diacetylene units are nearly planar (linear), and the saturated bridges are about 20° from a gauche conformation. Transannular interaction between the diacetylene units and methyl steric repulsions lead to angle and bond distortions in the saturated bridge. The molecule is quite similar to [2.2]paracyclophanes.

Crystal system	Space group	Radiation	Cell dimensions (Å)	Cell volume (Å³)	Density (g/cm³), calculated (measured)	Z	R	Significant bond lengths and nonbonded distances (Å), and bond and interplanar angles (deg)	
12,13,15,16-Tetramethyl[10]paracyclophane-4,6-diyne (27)[25]									
Monoclinic	$P2_1/c$	Zr-Filtered MoKα	$a = 9.170(4)$ $b = 9.208(3)$ $c = 18.232(7)$ $\beta = 94.0(1)$	1,535.7	1.143	4	0.058	$a = 1.530(6)$ $b = 1.535(6)$ $c = 1.516(6)$ $d = 1.465(6)$ C-4—C-5 = 1.184(5) C-5—C-6 = 1.388(5) $p_{4,14} = 3.203$ $q_{5,13} = 3.283$	$u_{12,17} = 1.519(6)$ $\alpha = 4.2$ $\gamma = 114.7(4)$ $\lambda = 119.4(3)$

Comments: The benzenoid ring is boat-shaped, and the diacetylene unit is bow-shaped. The molecule has C_s symmetry, and the nonbonded distances between the aromatic ring and the diacetylene unit seem to be smaller than those in [10]paracyclophane-4,6-diyne. The C-5 and C-6 atoms of the diacetylene unit nearly perfectly eclipse the C-13 and C-12 atoms of the aromatic ring when viewed down the normal to the ring. Unlike most cyclophanes, which exhibit the H atoms of the aromatic bent toward the inner part of the cyclophane macrocyle, in this molecule the methyl groups bound to the ring are bent away from the cyclophane cavity. A similar displacement is observed in tetramethyl quadruple-layered cyclophane. (See Chapter 10, Fig. 16, for triple-layered paracyclophanediyne.)

Crystal system	Space group	Radiation	Cell dimensions (Å)	Cell volume (Å³)	Density (g/cm³), calculated (measured)	Z	R	Tetramethylene side	Trimethylene side
[11]Paracyclophane-5,7-diyne (28)[25]									
Monoclinic	$P2_1/c$	Ni-Filtered CuKα	$a = 10.938(3)$ $b = 9.173(2)$ $c = 14.988(4)$	1,350.5	1.094	4	0.085	$a = 1.513(8)$	[illegible]

$c = 1.481(9)$ $d = 1.467(10)$ C-4—C-5 = 1.454(9) C-5—C-6 = 1.163(7) C-6—C-7 = 1.367(7) $\alpha = 0.6$ $\beta^* = 0.004$ $\gamma = 112.8(5)$ $\lambda = 117.4$ $p_{5,15} = 3.729$ $q_{6,14} = 3.769$	$c = 1.498$‡ $d = 1.497(9)$ C-8—C-7 = 1.178(7) $\alpha = 0.2$ $\beta^* = 0.055$ $\gamma = 114.0(6)$ $\lambda = 116.7$ $p_{8,12} = 3.250$

Comments: The aromatic ring is almost perfectly planar, and no deviation of the aromatic H atoms toward the center of the macrocyclic cavity is observed. Atoms C-1 and C-11 are displaced 0.004 and 0.055 Å, respectively, from the C-13,C-14,C-16,C-17 plane. The bow shape deformation is smaller than that found in 12,13,15,16-tetramethyl[10]paracyclophane-4,6-diyne and [10]paracyclophane-4,6-diyne. The projection of the axis of the diacetylene unit onto the aromatic plane shows that the axis makes an angle of 4° with the C-12,C-15 axis of the aromatic ring. Whereas the tetramethylene bridge is ordered, the trimethylene bridge is disordered, like the trimethylene unit in [10]paracyclophane-4,6-diyne. This disorder gives values of 1.4 to 1.7 Å for the b and c bonds, which are less reliable.

[12]Paracyclophane-5,7-diyne : TCNE (2 : 1) (29)[22,26]

Monoclinic	$C2/c$	—	$a = 18.29$ $b = 10.80$ $c = 20.03$ $\beta = 115.3$	—	—	4	—	—

Comments: The projection of the diacetylene unit onto the benzenoid plane is linear, but the axis of the diacetylene group is found to be rotated 10° from the diagonal (C-13,C-16) of the aromatic ring. The TCNE unit is sandwiched between the aromatic rings of two [12]paracyclophanediyne units, with an interplanar distance of 3.27 Å.

1,3;10,12-Biscyclopropano[2.2]paracyclophane (30a)[26a,26b]

Monoclinic	$P2_1/n$	MoKα	$a = 9.327(2)$ $b = 8.032(2)$ $c = 8.089(2)$ $\beta = 92.28(2)$	609.81	1.268 (1.250)	2	0.038	$a = 1.495$ $b = 1.558$ $c_{2,3} = 1.499$ $e = 1.393$ $f = 1.385$ $p = 2.81$ $q = 3.14$	$\alpha = 12.7$ $\beta = 12.3$ $\gamma = 115.0$ $\lambda = 116.6$

Comments: The molecule has Ci symmetry.

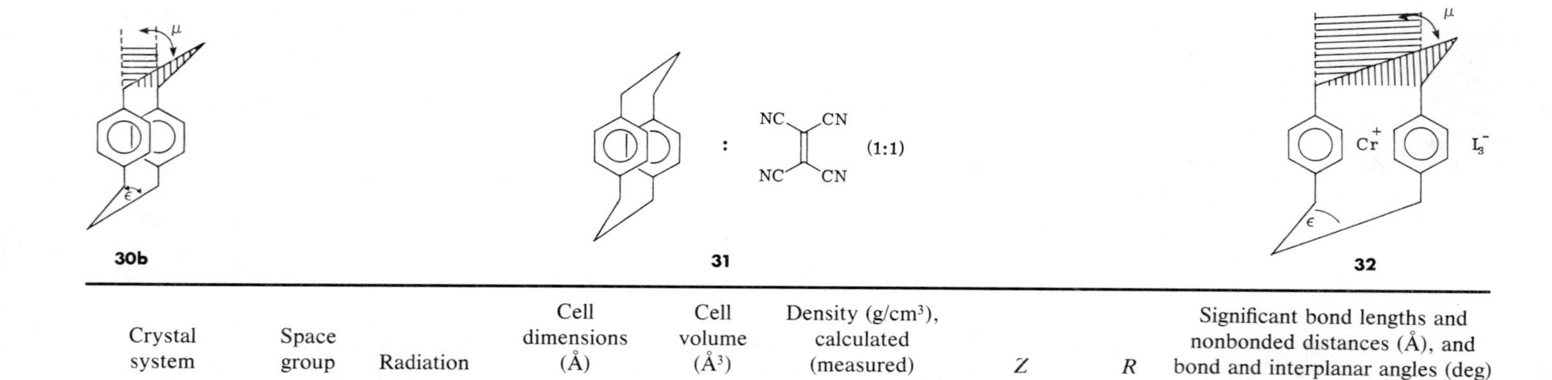

Crystal system	Space group	Radiation	Cell dimensions (Å)	Cell volume (Å³)	Density (g/cm³), calculated (measured)	Z	R	Significant bond lengths and nonbonded distances (Å), and bond and interplanar angles (deg)
				[3.3]Paracyclophane (30b)[27]				
Monoclinic (benzene)	$P2_1/n$	CuKα	$a = 9.175$ $b = 8.138$ $c = 8.524$ $\beta = 90.69(3)$	—	1.165 (1.156)	2	0.123	$a = 1.512$ $\alpha = 6.4$ $b = 1.525$ $\beta = 2.5; 4.6$ $e = 1.388$ $\gamma = 113.6; 115.9$ $f = 1.383$ $\lambda = 116.9$ $\mu = 65; 70$ $\varepsilon = 116.8$ $p = 3.14$ $q = 3.30$

Comments: The molecule is less distorted than [2.2]paracyclophane. The strain energy (the major contribution being from out-of-plane distortion of aromatic rings and p substituents) is estimated to be about 20–25% of [2.2]paracyclophane. The two 3-carbon atom bridges are oriented anti to each other, and the two aromatic rings are displaced about 0.5 Å from being directly above one another. The difference in β and γ angles at the two ends of the molecule is related to this parallel displacement of the rings and the effects of nonbonded C · · · C repulsions. The aromatic H atoms are displaced toward the inside of the molecule by about 0.08 Å, an effect (observed in [2.2]paracyclophane and more pronounced in [2.2]paracyclophane-1,9-diene) that seems to be due to the increased π density on the outer side of the aromatic rings because of transannular π–π interaction.

[3.3]Paracyclophane : TCNE (1 : 1) (31)[28,29]

Triclinic	$P\bar{1}$	Ni-Filtered CuKα	$a = 8.533$ $b = 8.538$ $c = 7.705$ $\alpha = 103.35$ $\beta = 110.78$ $\gamma = 104.01$	447.23	1.268 (1.256)	1 (complex unit)	0.055	$a = 1.513$ $b = 1.539$ $e = 1.397$ $f = 1.387$	$\alpha = 6.3$ $\beta = 2.3; 5.0$ $\gamma = 113.4; 115.8$ $\lambda = 117.2$ $\mu = 63; 70$ $\varepsilon = 117.1$ $p = 3.13$ $q = 3.20$

Comments: The two molecules of the complex are alternately stacked in columns above one another along the [111] direction. The TCNE and aromatic planes are essentially parallel (2°) to one another, with the TCNE molecule disordered by rotation about an axis through its center and normal to its plane. The cyclophane portion is not significantly different from the uncomplexed parent [3.3]paracyclophane, although the two aromatic rings are slightly more displaced with respect to one another, causing a decrease in the perpendicular distance between the aromatic planes. Aromatic H atoms are displaced slightly toward the macrocyclic cavity (by about 0.09 Å) from the plane formed by the C atoms to which they are bound.

(η^{12}-[3.3]Paracyclophane)chromium(I) triiodide (32)[30,31]

Monoclinic (toluene/H_2O)	$P2_1/c$	MoKα	$a = 7.482(2)$ $b = 7.642(2)$ $c = 16.701(3)$ $\beta = 102.90(2)$	930.82	2.493 (2.54)	2	0.057	$a = 1.506$ $b = 1.541$ $e = 1.409$ $f = 1.400$	$\alpha = 6.51; 4.69$ $\beta = 2.21; 4.36$ $\gamma = 115.7; 114.1$ $\lambda = 116.4$ $\mu = 57.0; 66.6$ $\varepsilon = 115.8$ $p = 3.071$ $q = 3.222$

Comments: The two 3-carbon bridges are oriented anti to one another. Unlike the case of [3.3]paracyclophane, the C atoms of the aromatic rings are almost ideally eclipsed. The Cr atom is centrally located between the aromatic rings (perpendicular distance 1.58 and 1.61 Å) with an average C—Cr bond distance of 2.12 Å. The H atoms on the aromatic ring are inclined toward the Cr atom by an average of 0.10 Å from the plane defined by the C atoms to which they are bound. The linear I_3^- lies above the aromatic ring and is almost parallel (1.6°) to the plane of nonbridging atoms of the aromatic ring, the perpendicular distance to that plane being 3.73 Å. The line between the two central atoms of the I_3^- units and the Cr atom is almost perpendicular (87.0°) to the aromatic planes. An angle of 22.52° exists between the I_3^- axis and the *p* axis of the aromatic ring.

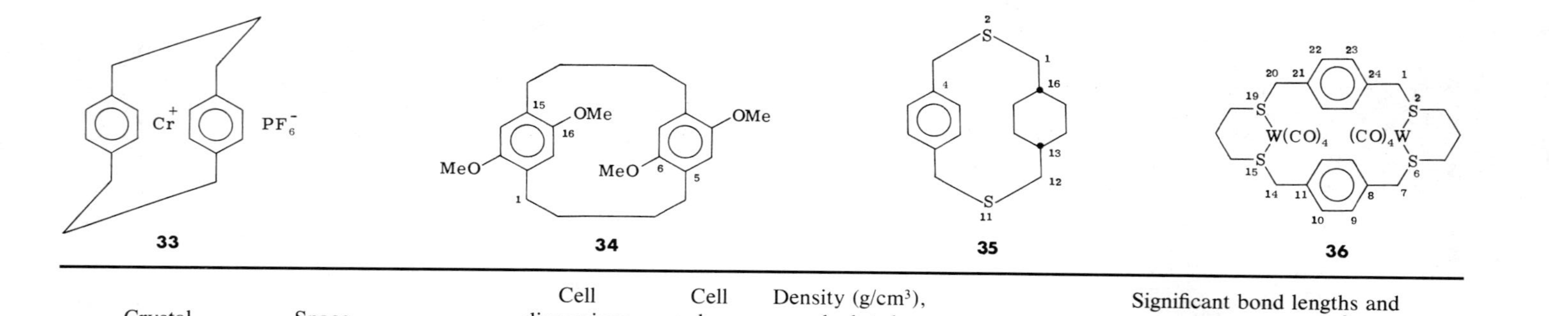

Crystal system	Space group	Radiation	Cell dimensions (Å)	Cell volume (Å³)	Density (g/cm³), calculated (measured)	Z	R	Significant bond lengths and nonbonded distances (Å), and bond and interplanar angles (deg)	
(η^{12}-[3.3]Paracyclophane)chromium(I) hexafluorophosphate (33)[31,32]									
Tetragonal	*P4/mbm*	MoKα	$a = 11.385(4)$ $c = 6.580(3)$	852.89	1.687 (1.74)	2	0.083	$a = 1.61$ $b = 1.56$ $e = 1.37$ $f = 1.43$	$\alpha = 4.5$ $\beta = 4.6$ $\gamma = 114.2$ $\mu = 58.6$ $\varepsilon = 113.0$

Comments: The structure is similar to the **32** above except that due to the statistical distribution of the central bridge C atom, unequivocal assignment of the anti orientation of the bridges was not made. The average C—Cr distance is 2.07 Å.

Crystal system	Space group	Radiation	Cell dimensions (Å)	Cell volume (Å³)	Density (g/cm³), calculated (measured)	Z	R	Significant bond lengths and nonbonded distances (Å), and bond and interplanar angles (deg)	
6,9,16,19-Tetramethoxy[4.4]paracyclophane (34)[33]									
Triclinic (acetone)	$P\bar{1}$	MoKα	$a = 7.889(2)$ $b = 7.924(2)$ $c = 18.431(4)$ $\alpha = 83.79(1)$ $\beta = 76.24(1)$ $\gamma = 71.07(1)$	—	1.21	2	0.048	$a = 1.51$ $b = 1.52$ $c = 1.52$ $e = 1.38$ $f = 1.38$ $v = 1.38$ $p = 3.98$ $q = 4.01$	$\alpha = 0$ $\beta = 0$ $\gamma = 111.5$ $\lambda = 117.8$

Comments: The aromatic rings are planar and parallel to one another but twisted 9.5° with respect to one another about an axis perpendicular and common to both rings. Torsion angles of 76 and 143° arise between the first and second and the second and third bridging C atoms, respectively, due to the twisting of the rings. The methoxy groups of each ring are twisted about 10° in opposite directions out of the aromatic plane.

C. STRUCTURES CONTAINING HETEROATOMS IN THE BRIDGE

2,11-Dithia[3.3](13,16)-*cis*-1,4-cyclohexanoparacyclophane (35)[34]

Monoclinic (benzene/hexanes)	$P2_1/c$	CuKα	a = 11.749(3) b = 7.959(2) c = 16.097(4) β = 100.72(1)	1,479	1.25	4	0.055	Para benzenoid ring α = 6.0 β = 4.4

Comments: The benzenoid ring is slightly distorted into a boat shape, and although the aromatic ring is directly above the cyclohexanoid ring it is tipped 25° with respect to a plane passing through the C-13 and C-16 atoms of, and bisecting, the cyclohexanoid ring. The cyclohexanoid ring is somewhat flattened from an idealized chair conformation. The S—C—C bond angles at C-1 and C-12 are 114.0(3) and 117.6(3)°, respectively, and are slightly larger than normal. The C—S—C bond angles at the S atoms are 102.7(2)°, about 3° larger than normal.

2,6,15,19-Tetrathia[7.7]paracyclophane ditungsten octacarbonyl (36)[35]

Monoclinic	$P2_1/n$	MoKα	a = 8.273(9) b = 11.569(8) c = 16.815(13) β = 92.85(10)	1,607.38	—	2	0.093	a = 1.55(5) b = 1.80(4) c = 1.80(3) d = 1.55(5) e = 1.38(6) f = 1.41(5)	S—W = 2.527(10) α^* = 0.015; 0.018 γ = 117(2) λ = 122(3)

Comments: The molecule has a center of symmetry. The two S atoms of each bridge take up two coordination sites on the two W atoms, which are oriented anti to one another with respect to the four-S-atom plane. The three C atoms linking the two S atoms in each bridge are also anti to one another. The aromatic rings are absolutely parallel to one another and lie essentially above one another. The benzylic atoms C-1, C-7, C-14, and C-20 are displaced slightly from their aromatic planes toward the opposite aromatic ring (α^*).

37 **38** **39**

Crystal system	Space group	Radiation	Cell dimensions (Å)	Cell volume (Å^3)	Density (g/cm^3), calculated (measured)	Z	R	Significant bond lengths and nonbonded distances (Å), and bond and interplanar angles (deg)	
2,6,15,19-Tetrathia[7.7]paracyclophane dimolybdenum octacarbonyl · 2 benzene (37)[35]									
Monoclinic	$P2_1/a$	MoKα	a = 17.866(6) b = 13.059(6) c = 10.251(4) β = 112.47(3)	2,210.01	—	2	0.078	a = 1.516(16) b = 1.832(14) c = 1.811(11) d = 1.505(20) e = 1.375(17) f = 1.384(17)	S—Mo = 2.570(3) α^* = 0.046; 0.088 γ = 117.7(11) λ = 117.5(11)

Comments: The molecular geometry is similar to that of the ditungsten derivative (previous entry), with two S atoms occupying two coordination sites on each Mo atom. The Mo atoms, and the three-C-atom chain linking the two S atoms in each bridge are oriented anti to one another. The aromatic rings are parallel to one another and lie essentially above one another. The benzylic atoms C-1, C-7, C-14, and C-20 are displaced slightly from their aromatic planes toward the opposite ring (α^*). There are two benzene solvate molecules, which lie on the periphery of the cyclophane macrocycle between the bridges. Their planes are perpendicular to the benzenoid planes of the cyclophane.

2,6,15,19-Tetrathia[7.7]paracyclophane ditungsten decacarbonyl (38)[35]

Triclinic	$P\bar{1}$	MoKα	$a = 6.659(3)$ $b = 9.915(4)$ $c = 14.703(8)$ $\alpha = 76.28(4)$ $\beta = 81.94(4)$ $\gamma = 86.56(4)$	933.38	—	1	0.087	$a = 1.524(18)$ $b = 1.814(12)$ $c = 1.825(14)$ $d = 1.508(17)$ $e = 1.406(20)$ $f = 1.395(17)$	S—W = 2.560(3) $\alpha^* = 0.019; 0.041$ $\gamma = 111.6(7)$ $\lambda = 120.0(11)$

Comments: The molecule has a center of symmetry. The W atoms are linked to a single S atom in each bridge and directed toward the outside of the cavity of the cyclophane macrocycle. The aromatic rings are essentially planar and lie above each other and are parallel to one another. They are, however, laterally displaced by 3.13 Å with respect to one another along their 1,4 axes. The benzylic atoms at C-1, C-7, C-14, and C-20 are displaced slightly from their aromatic planes away from the opposite aromatic ring (α^*).

6,9,17,20-Tetraethoxy-1,2,3,4,11,12,13,14-octathia[4.4]paracyclophane (pseudo-ortho isomer) (39)[36]

Monoclinic	$P2_1/c$	MoKα	$a = 10.198(5)$ $b = 9.454(5)$ $c = 29.845(16)$ $\beta = 115.17(8)$	2,604.3	1.49 (1.48)	4	0.063	$a = 1.765(8)$ $b = 2.028(5)$ $c = 2.068(5)$ $e,f = 1.390(10)$ $v = 1.365(10)$	$\alpha = 0$ $\beta = 0$ $\gamma = 105.6$ $p = 3.599$ $q = 3.657$

Comments: The benzenoid rings are planar and have normal bond lengths and bond angles. They are also very nearly parallel and twisted with respect to the normal to both rings so that there is no eclipsing of the C atoms of the rings. The closest ring contact is 3.41 Å (C-5 · · · C-19). The C—S bonds and the S—S bonds show a slight degree of π-bond character.

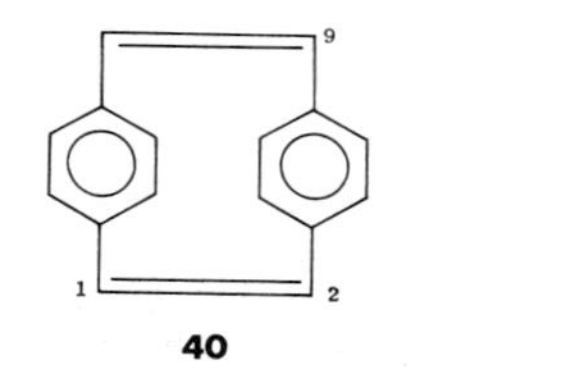

40

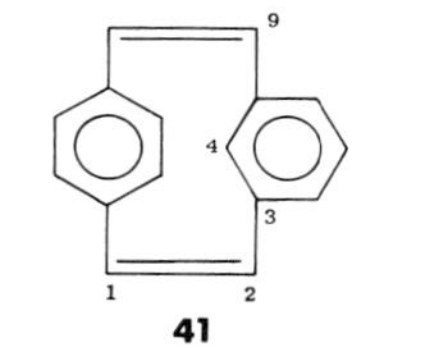
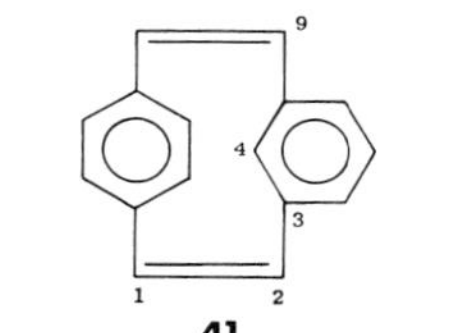

41

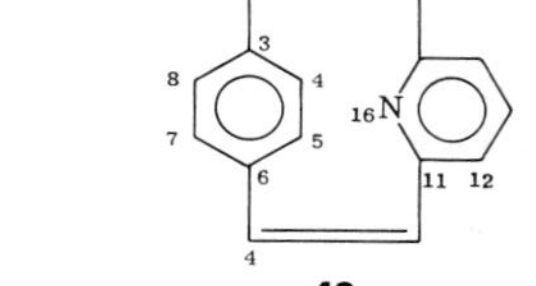

42

Crystal system	Space group	Radiation	Cell dimensions (Å)	Cell volume (Å³)	Density (g/cm³), calculated (measured)	Z	R	Significant bond lengths and nonbonded distances (Å), and bond and interplanar angles (deg)
D. STRUCTURES WITH UNSATURATION IN THE BRIDGE								
[2.2]Paracyclophane-1,9-diene (40)[37]								
Monoclinic	$P2_1/c$	CuKα	a = 6.866 b = 11.636 c = 7.640 β = 116.9	—	1.246 (1.25)	2	0.100	a = 1.511 α = 13.7 b = 1.336 β = 15.2 e = 1.384 γ = 118.9 f = 1.396 λ = 117.4 φ = 163.6 p = 3.14 q = 2.80

Comments: The angle distortions in the aromatic rings and aliphatic bridges are greater than in [2.2]paracyclophane. The H atoms of the aromatic ring are all significantly bent toward the inner part of the macrocycle.

Crystal system	Space group	Radiation	Cell dimensions (Å)	Cell volume (Å³)	Density (g/cm³), calculated (measured)	Z	R	Significant bond lengths and nonbonded distances (Å), and bond and interplanar angles (deg)
[2.2]Metaparacyclophane-1,9-diene (41)[38]								
Orthorhombic	$Pbca$	CuKα	a = 14.91(1) b = 18.51(1) c = 8.133(5)	2,245	1.21 (1.20)	8	0.046	Para ring / Meta ring a = 1.486 / a = 1.484 b = 1.346

$e = 1.405$	$e = 1.400$
$f = 1.381$	$f = 1.413$
	$g = 1.376$
$\alpha = 18.4$	$\alpha^* = 0.226$
$\beta = 18.0$	$\beta^* = 0.108$
$\gamma = 120.8$	$\gamma = 125.7$
	$\delta^* = 0.026$
$\lambda = 116.0$	$\lambda = 116.9$

$\theta = 41$

Comments: The molecule does not have *mm*2 symmetry. Both rings are distorted into boat shapes, although the distortion is more severe for the para-bridged ring. Angles α and β are more than 3° greater than in [2.2]paracyclophane-1,9-diene, and the aromatic rings are inclined 41° (θ) with respect to one another. The C-4 atom and its appended H atom lie 2.71 and 2.16 Å, respectively, from the mean plane of the para-bridged ring.

[2.2](2,6)Pyridinoparacyclophane-1,9-diene (42)[39]

								Para ring	Meta ring
Orthorhombic	*Fdd*2	MoKα	$a = 13.85(4)$	2,130	1.28	8	0.056		
			$b = 17.38(6)$		(1.27)				
			$c = 8.84(2)$					$a = 1.446(8)$	$a = 1.490(8)$
								$b = 1.324(12)$	
								$e = 1.392(6)$	$e = 1.340(6)$
								$f = 1.375(6)$	$f = 1.405(8)$
									$g = 1.361(9)$
								$\alpha^* = 0.222(18.2°)$	
								$\beta = 17.1$	
								$\gamma = 119.5$	$\gamma = 128.2$
								$\lambda = 117.5$	$\lambda = 120.2$

$\theta = 90$

Comments: Whereas the para ring is distorted severely into a boat shape, the meta ring is essentially planar and perpendicular to the para ring. The N atom lies 2.51 Å from the C-4,C-5,C-7,C-8 plane of the para ring.

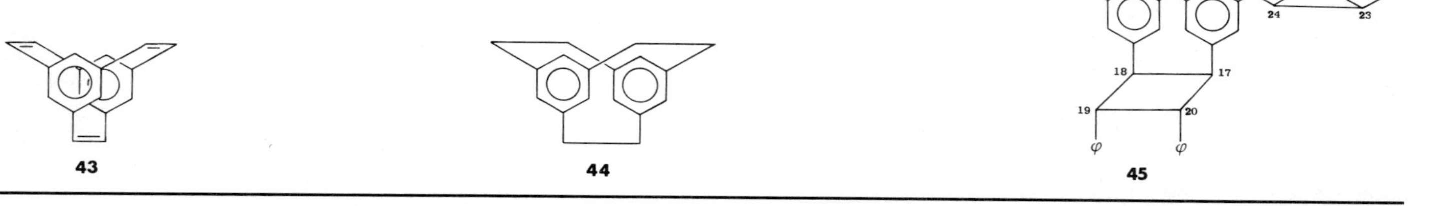

Crystal system	Space group	Radiation	Cell dimensions (Å)	Cell volume (Å^3)	Density (g/cm^3), calculated (measured)	Z	R	Significant bond lengths and nonbonded distances (Å), and bond and interplanar angles (deg)
			[2.2.2](1,3,5)Cyclophane-1,9,17-triene (43)[40,40a]					
Triclinic	$P\bar{1}$	CuKα	a = 7.332(5) b = 11.663(10) c = 7.224(5) α = 87.73(5) β = 100.24(5) γ = 105.52(5)	586	1.29 (1.29)	2	0.043	a = 1.502(1) α = 24.2 b = 1.340(4) β = 6.1(*0.024) e,f,g = 1.397(1) γ = 118.1 λ = 118.4 p = 2.759 q = 2.859

Comments: The molecule has almost exact $\bar{6}m2$ symmetry. The two sets of aromatic ring C atoms eclipse one another, with a mean interplanar spacing of 2.809 Å. Each aromatic ring is chair-shaped. The three nonbridged atoms of each ring lie 0.024 Å above the mean aromatic plane and directed away from the second aromatic ring. The C—H bonds of aromatic rings are displaced from coplanarity (with the nearest three C atoms of the aromatic ring) an average of 0.23 Å toward the other aromatic ring. This corresponds to a 13° angle between the plane and the C—H bond and a transannular nonbonded H · · · H distance of 2.52 Å.

$[2_3](1,3,5)$Cyclophane (44)[41]

Trigonal	$P3_12$ $(P3_221)$	CuKα	a = 8.005(1) c = 16.973(2)	941.9	1.239 (1.24)	3	0.31	a = 1.507 b = 1.600 e,f,g = 1.382 p = 2.75 q = 2.83

Comments: The benzenoid rings are distorted into boat shapes (adjacent atoms lie alternately 0.020 Å above and below the mean aromatic plane), and the atoms of each benzenoid ring eclipse the corresponding atoms in the proximate ring.

19,20,21,22,23,24-Hexaphenyl-$\Delta^{1,2}$, $\Delta^{9,10}$, $\Delta^{17,18}$-tricyclobuta[2.2.2](1,3,5)cyclophane (45)[42]

Orthorhombic	$Pccn$	—	a = 9.633(1) b = 18.320(2) c = 24.170(3)	—	—	4	0.137	b = 1.64(2)

Comments: A symmetry distortion in the crystal causes a complicated disorder within the unit cell. The molecule adopts a "paddlewheel" geometry, with the metacyclo rings forming the wheels and the bridging C atoms with their appended cyclobuta rings providing the paddle. Two of the three cyclobutanoid rings are planar, and the third is slightly folded. They are all directed in the same sense (clockwise) around the periphery of the cyclophane and are perpendicular to the aromatic planes. The six phenyl groups on the cyclobutanoid rings are also directed in the same sense but counterclockwise around the periphery. The average distance between the cyclophane aromatic planes is 2.85(6) Å, which is similar to that found in $[2_3]$(1,3,5)cyclophane.

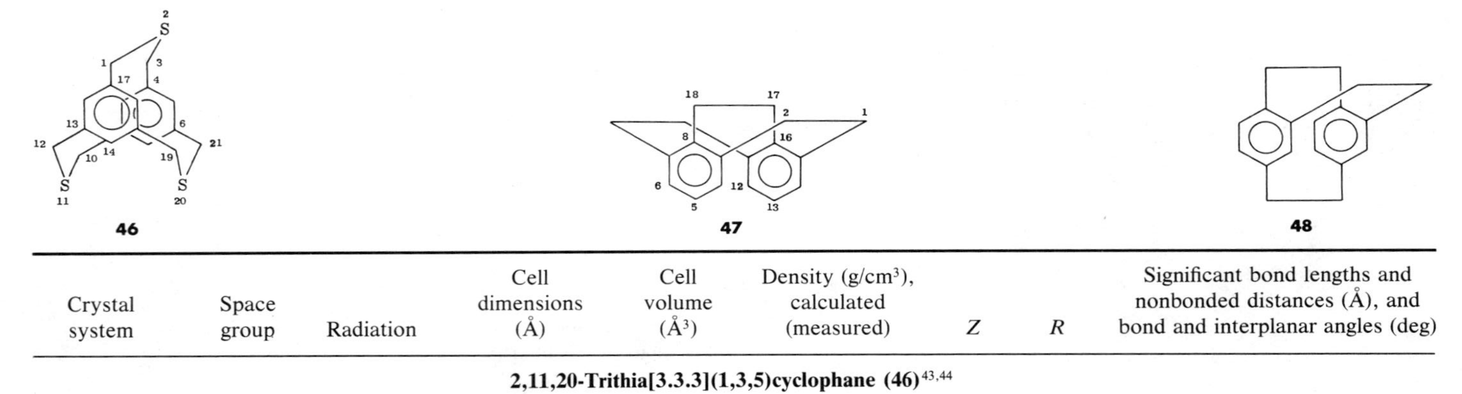

Crystal system	Space group	Radiation	Cell dimensions (Å)	Cell volume (Å³)	Density (g/cm³), calculated (measured)	Z	R	Significant bond lengths and nonbonded distances (Å), and bond and interplanar angles (deg)
			2,11,20-Trithia[3.3.3](1,3,5)cyclophane (46)[43,44]					
Monoclinic	$P2_1/c$ (C_2/c)	CuKα	$a = 13.19(2)$ $b = 33.35(5)$ $c = 7.06(1)$ $\beta = 93.18(5)$	3,102	1.41 [1.39(2)]	8	0.064	$a = 1.50$ $\alpha^* = 0.163$ $b = 1.82$ $\gamma = 115.4$ $e = 1.38$

Comments: Although the structure is disordered, the essential features of the molecule were obtained. The benzenoid rings are parallel, and the atoms of each aromatic ring eclipse the corresponding atoms of the proximate ring with an interplanar separation of 3.19 Å. There is evidence that one of the C—S—C bridges is rapidly flipping.

[2_3](1,2,3)Cyclophane (47)[45]

								1 and 3 bridges	2 bridge
Orthorhombic	*Pbca*	CuKα	$a = 14.499(1)$	2,562.0	1.214	8	0.041		
			$b = 12.282(1)$		(1.22)				
			$c = 14.387(1)$						
								$a = 1.510$	$a = 1.502$
								$b = 1.564$	$b = 1.541$
								$e = 1.401$	
								$f = 1.387$	
								$g = 1.374$	
								$\beta^* = 0.14$	
								$\gamma = 114.2$	$\gamma = 109.9$
								$\delta^* = 0.05$	
								$p = 2.564$	
								$q = 2.806$	
								$\theta = 42$	

Comments: The C-17—C-18 bond (b) is normal because the angle of inclination between the rings allows for minimal stretching of this bond. The aromatic rings are distorted into boat shapes, the bow and stern of each facing away from each other. The nonbonded distances C-6 · · · C-12 and C-5 · · · C-13 are 3.808 and 4.405 Å, respectively.

[2.2.2](1,2,4)Cyclophane (48)[24,26]

Triclinic	$P\bar{1}$	CuKα	$a = 12.709$	—	1.25	4	0.052	$\theta = 13.3$
			$b = 12.710$					
			$c = 7.784$					
			$\alpha = 94.95$					
			$\beta = 94.95$					
			$\gamma = 89.96$					

Comments: No evidence for disorder was observed. The aromatic rings are twisted 3° about the normal to both rings and are inclined 13.3° with respect to one another. The rings are distorted into ''twist-boat'' shapes.

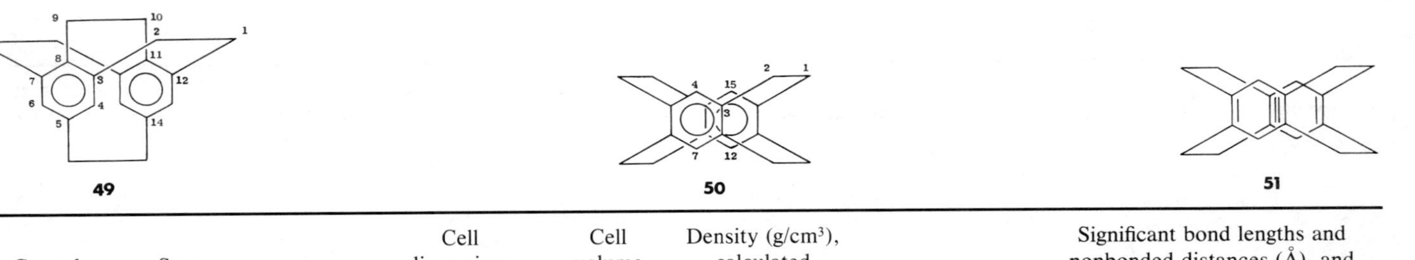

Crystal system	Space group	Radiation	Cell dimensions (Å)	Cell volume (Å³)	Density (g/cm³), calculated (measured)	Z	R	Significant bond lengths and nonbonded distances (Å), and bond and interplanar angles (deg)	
				$[2_4](1,2,3,5)$Cyclophane (49)[47]					
Orthorhombic	*Pbca*	MoKα	$a = 15.557(3)$ $b = 22.516(3)$ $c = 7.585(1)$	—	1.3	8	0.057	1,3 bridge (meta type) $a = 1.504(5)$ $b = 1.592(6)$ $e = 1.405(5)$ $f = 1.393(5)$ $g = 1.381(5)$ $\alpha = 19.6$ $\beta = 0$ $\gamma = 111.2(3)$ $\delta = 9.1$ $\lambda = 119.1$ $p = 2.677(5)$ $\theta = 5$	2 bridge (para type) $a = 1.506(6)$ $b = 1.561(6)$ $\alpha = 0$ $\beta = 15.5$ $\gamma = 109.7(3)$ $\lambda = 119.7$ $p = 2.574(4)$ 5 bridge (para type) $a = 1.505(6)$ $b = 1.594(6)$ $\alpha = 9.1$ $\beta = 17.5$ $\gamma = 112.5(4)$ $\lambda = 117.2$ $p = 2.745(5)$ $q = 2.871(5)$

Comments: The aromatic rings are distorted from planarity into half-boat shapes. Whereas the mean planes of the two benzenoid rings are inclined 5° with respect to one another, the planes excluding the C-5 and C-14 atoms are inclined 8.1° with respect to one another (C-3, C-4, C-6, C-7, and C-8 are coplanar). The H atoms at C-4 and C-6 are bent out of the plane of the nearest three C atoms toward the other aromatic ring. Whereas the angle that the C-4—H vector makes with its plane is 20.7°, the angle that the C-6—H vector makes with its plane is only 8.4°.

$[2_4](1,2,4,5)$Cyclophane (50)[48]

Monoclinic	$P2_1/c$	Ni-Filtered CuKα	20°C	—	1.312 (1.310)	2	0.033	$a = 1.514$	$\alpha^* = 0.547$
			$a = 8.787(9)$					$b = 1.591$	$\beta^*,\delta^* = 0.13$
			$b = 10.91(1)$					$e,g = 1.395$	$\gamma = 111.3$
			$c = 7,619(8)$					$f = 1.411$	$\lambda = 118.3$
			$\beta = 115.6(1)$						$\varphi = 12.25$
								$p = 2.688$	
								$q = 2.950$	

Comments: The molecule has almost exact *mmm* symmetry, and the aromatic rings are distorted into a boat shape. The H atoms on the C-4 and C-15 (C-7 and C-12) are bent toward the opposite aromatic ring and are 2.86 Å from one another.

4,7,12,15-Tetrahydro$[2_4](1,2,4,5)$cyclophane (51)[48]

Triclinic	$P\bar{1}$	—	20°C	—	1.229 (1.227)	2	0.068	$a = 1.514$	$\alpha^* = 0.641$
			$a = 7.464(7)$					$b = 1.519$	$\beta^*,\delta^* = 0.454$
			$b = 16.147(16)$					$e,g = 1.516$	$\gamma = 115.2$
			$c = 7.292(7)$					$f = 1.343$	$\lambda = 118.2$
			$\alpha = 109.1(1)$						$\varphi = 40.0$
			$\beta = 69.1(1)$					$p = 2.809$	
			$\gamma = 117.4(1)$					$q = 3.724$	

Comments: There are two independent molecules (A and B) that are slightly different. The reduced aromatic rings are boat-shaped but are much less strained than the boat-shaped aromatic rings in $[2_4](1,2,4,5)$cyclophane.

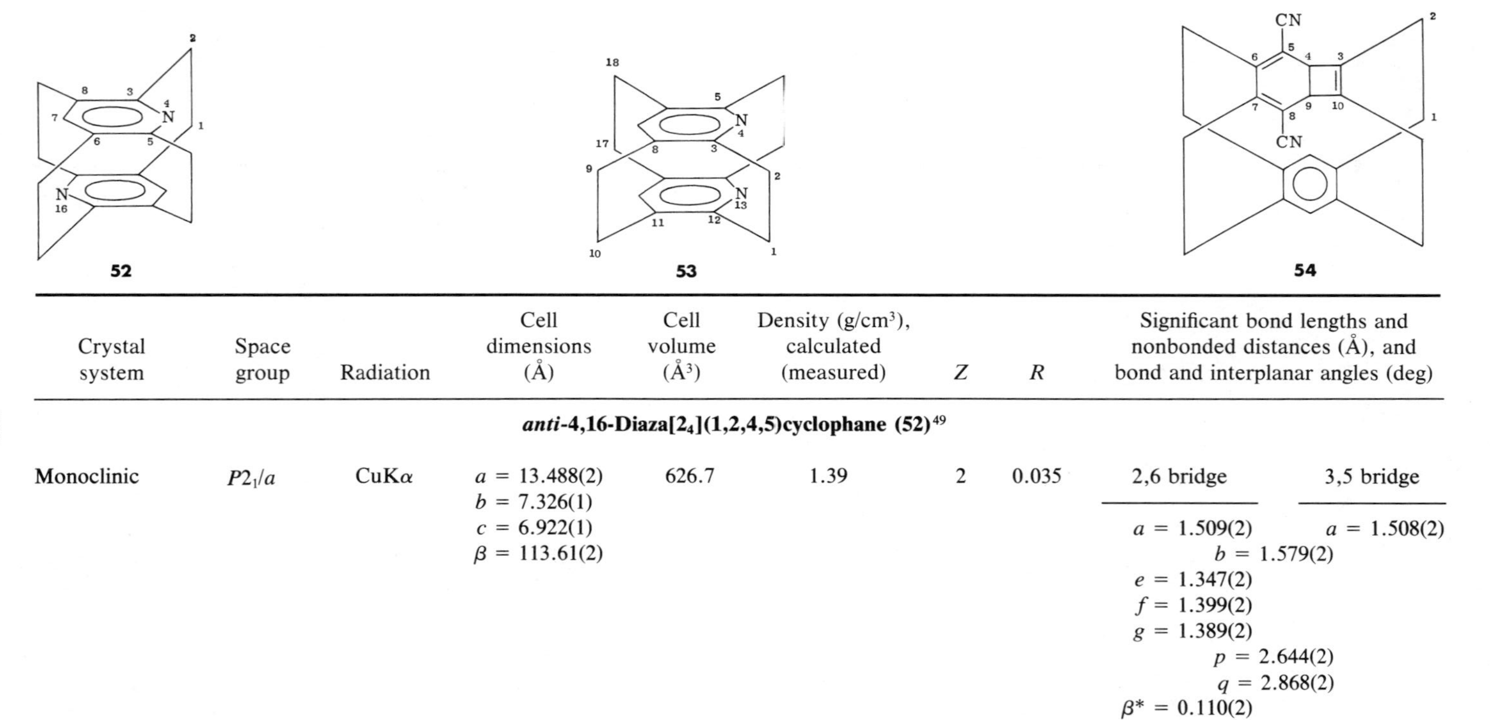

52 **53** **54**

Crystal system	Space group	Radiation	Cell dimensions (Å)	Cell volume (Å³)	Density (g/cm³), calculated (measured)	Z	R	Significant bond lengths and nonbonded distances (Å), and bond and interplanar angles (deg)
			***anti*-4,16-Diaza[2₄](1,2,4,5)cyclophane (52)**[49]					
Monoclinic	$P2_1/a$	CuKα	a = 13.488(2) b = 7.326(1) c = 6.922(1) β = 113.61(2)	626.7	1.39	2	0.035	2,6 bridge: a = 1.509(2); 3,5 bridge: a = 1.508(2) b = 1.579(2) e = 1.347(2) f = 1.399(2) g = 1.389(2) p = 2.644(2) q = 2.868(2) β^* = 0.110(2) δ^* = 0.114(2)

Comments: The molecule is similar in structure to $[2_4]$(1,2,4,5)cyclophane. The aromatic rings are boat-shaped, with the N-4 and C-7 atoms lying 0.112(2) and 0.114(2) Å from the C-3,C-5,C-6,C-8 plane. The H atom at C-7 lies closer to this plane than the C-7 atom alone and points toward the opposite ring.

syn-4,13-Diaza[2$_4$](1,2,4,5)cyclophane (53)[50]

B form, orthorhombic	*Pbca*	CuKα	a = 13.361(1) b = 14.891(1) c = 12.839(1)	2,554.4	1.364	8	0.036	B form, pyridino ring (meta type)	
								$a_{2,3}$ = 1.507 $b_{1,2}$ = 1.581 $e_{C,N}$ = 1.346 f = 1.395 g = 1.389	$a_{8,9}$ = 1.512 $b_{9,10}$ = 1.587
A form, monoclinic	$P2_1/a$	CuKα	a = 13.240(5) b = 7.489(2) c = 7.010(3) β = 114.73(3)	631.3	1.380	2	—	β^* = 0.12 γ = 110.2 δ^* = 0.12 θ = 2.0	$p_{3,12}$ = 2.622 $p_{8,11}$ = 2.670 $q_{N,N}$ = 2.835 $q_{C,C}$ = 2.938

Comments: The A form is disordered, and no complete determination was carried out. The B form is strained, with the bridging C—C bonds longer than usual and the pyridino ring boat-shaped.

5,8-Dicyano[2.2.2.2](3,6,7,10)bicyclo[4.2.0]octa-3^{10},5,7-trienylo(1,2,4,5)cyclophane (54)[51]

Orthorhombic	$P2_12_12_1$	CuKα	a = 13.642(1) b = 14.175(2) c = 8.914(1)	1,723.7	1.296 (1.298)	4	0.046	Benzenoid ring (meta type)	
								Cyclohexadiene side	Cyclobutene side
								a = 1.528(5) b = 1.514(6) e,g = 1.371(6) f = 1.382(6)	a = 1.540(7) b = 1.508(7) e,g = 1.386(6) f = 1.417(6)

Comments: The molecule has an approximate (noncrystallographic) plane of symmetry normal to the N · · · N vector.

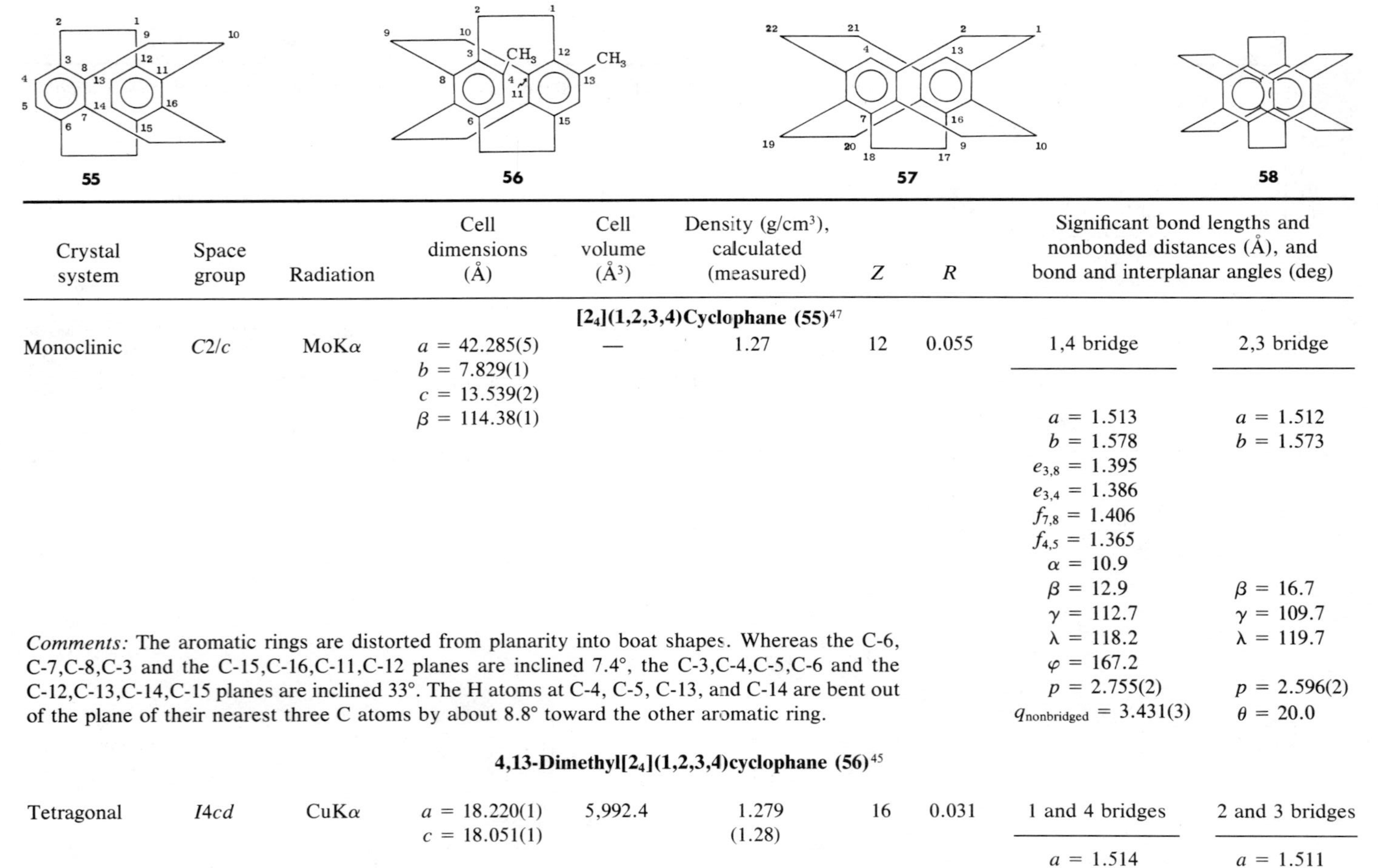

Crystal system	Space group	Radiation	Cell dimensions (Å)	Cell volume (Å³)	Density (g/cm³), calculated (measured)	Z	R	Significant bond lengths and nonbonded distances (Å), and bond and interplanar angles (deg)	
[2₄](1,2,3,4)Cyclophane (55)[47]									
Monoclinic	$C2/c$	MoKα	a = 42.285(5)	—	1.27	12	0.055	1,4 bridge	2,3 bridge
			b = 7.829(1)						
			c = 13.539(2)						
			β = 114.38(1)					a = 1.513	a = 1.512
								b = 1.578	b = 1.573
								$e_{3,8}$ = 1.395	
								$e_{3,4}$ = 1.386	
								$f_{7,8}$ = 1.406	
								$f_{4,5}$ = 1.365	
								α = 10.9	
								β = 12.9	β = 16.7
								γ = 112.7	γ = 109.7
								λ = 118.2	λ = 119.7
								φ = 167.2	
								p = 2.755(2)	p = 2.596(2)
								$q_{nonbridged}$ = 3.431(3)	θ = 20.0

Comments: The aromatic rings are distorted from planarity into boat shapes. Whereas the C-6, C-7,C-8,C-3 and the C-15,C-16,C-11,C-12 planes are inclined 7.4°, the C-3,C-4,C-5,C-6 and the C-12,C-13,C-14,C-15 planes are inclined 33°. The H atoms at C-4, C-5, C-13, and C-14 are bent out of the plane of their nearest three C atoms by about 8.8° toward the other aromatic ring.

Crystal system	Space group	Radiation	Cell dimensions (Å)	Cell volume (Å³)	Density (g/cm³), calculated (measured)	Z	R	Significant bond lengths and nonbonded distances (Å), and bond and interplanar angles (deg)	
4,13-Dimethyl[2₄](1,2,3,4)cyclophane (56)[45]									
Tetragonal	$I4cd$	CuKα	a = 18.220(1)	5,992.4	1.279	16	0.031	1 and 4 bridges	2 and 3 bridges
			c = 18.051(1)		(1.28)				
								a = 1.514	a = 1.511
								b = 1.576	b = 1.576

								$e = 1.397; 1.400$ $f = 1.392; 1.406$ $\alpha^* = 0.14$ $\gamma = 112.8$ $p = 2.754$ $q_{8,11} = 2.594$ $q_{4,13} = 3.460$	$\gamma = 109.6$

Comments: The rings are distorted into boat shapes, with the nonbridged atoms of the ring constituting one side of the boat and the C-3,C-12 and C-6,C-15 being the bow and stern. The C atoms of the appended methyl group lie 0.16 Å from the plane of the nearest three C atoms, although the methyl groups point toward rather than away from one another.

[2₅](1,2,3,4,5)Cyclophane (57)[45,45a,45b]

Orthorhombic	*Pbca*	CuKα	$a = 17.243(2)$ $b = 14.189(2)$ $c = 11.409(1)$	2,791.3	1.363 (1.35)	8	0.043	1 and 5 bridges $a = 1.514$ $b = 1.601$ $e = 1.388$ $f = 1.403$ $g = 1.405$ $\beta^* = 0.10$ $\gamma = 111.5$ $p = 2.712$ $q = 2.973$	2 and 4 bridges $a = 1.515$ $b = 1.575$ $\beta^* = 0.05$ $\gamma = 109.6$ $p = 2.592$ $q = 2.627$ 3 bridge $a = 1.516$ $b = 1.592$ $\gamma = 109.9$

Comments: The aromatic rings are boat-shaped. One end is attached to the middle bridge (3 bridge; C-17—C-18), and the other end is free. Both ends are bent away from the opposite ring. The free end (C-4 and C-13) is displaced more from the plane ($\beta^* = 0.10$) than the bridged end (C-7 and C-16; $\beta^* = 0.05$).

[2₆](1,2,3,4,5,6)Cyclophane ("superphane") (58)[45,45c]

Monoclinic	$P2_1/n$	CuKα	$a = 8.807(1)$ $b = 11.529(1)$ $c = 7.694(1)$ $\beta = 106.60(2)$	748.7	1.386 (1.39)	2	0.034	$a = 1.518$ $b = 1.580$ $e,f,g = 1.406$ $p,q = 2.624$	$\alpha = 20.3$ $\gamma = 110.1$

Comments: The aromatic rings are planar and not distorted. The molecule has D_{6h} symmetry, with all the atoms of one ring eclipsing the corresponding atoms of the other ring. This causes the π orbitals to be deflected from perpendicularity by about 10°, giving the π cloud a bowl shape. The molecule is strained mainly because of the σ framework.

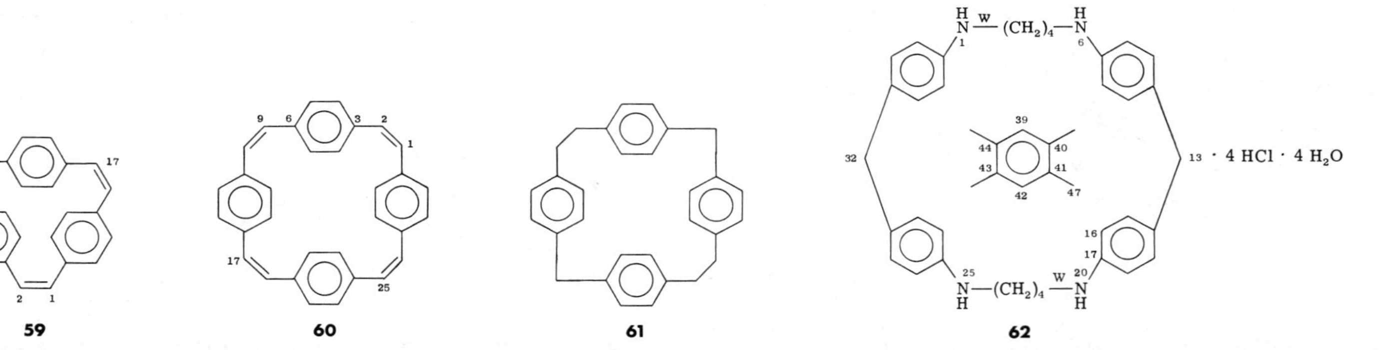

Crystal system	Space group	Radiation	Cell dimensions (Å)	Cell volume ($Å^3$)	Density (g/cm^3), calculated (measured)	Z	R	Significant bond lengths and nonbonded distances (Å), and bond and interplanar angles (deg)
F. MULTIBRIDGED STRUCTURES OF THE $[m^y]$TYPE								
[2[3]]Paracyclophane-1,9,17-triene (59)[52]								
Monoclinic	$P2_1/n$	MoKα	113K a = 16.942 b = 10.057 c = 10.289 β = 105.87	—	—	4	0.076	—
Monoclinic	$C2/c$	MoKα	298K a = 17.111 b = 10.132 c = 10.581 β = 106.95	—	—	4	0.087	—

[2⁴]Paracyclophane-1,9,17,25-tetraene (60)[53]

Orthorhombic (HOAc)	$Pcab$	CuKα	$a = 8.251(1)$ $b = 22.370(4)$ $c = 25.131(5)$	4,640	1.17	8	0.054	$a = 1.470$ $b = 1.312$ $e_{outer} = 1.393$ $e_{inner} = 1.407$	$f_{outer} = 1.358$ $f_{inner} = 1.368$ $\gamma = 131.9$ $\lambda = 116.3$

Comments: The olefinic units and the 1,4 atoms of each ring are planar, with the largest deviation of these atoms from this plane being no more than 0.2 Å. The benzenoid rings are relatively planar and are inclined an average of 33.4° to the above plane such that opposite rings are tilted in a disrotatory manner with respect to one another so as to cap the upper and lower surface of the macroring cavity. Although the planar array of the macroring causes angle distortion of the inner bond angles ($\gamma = 131.9°$), a tub shape would generate more severe steric interactions between the internal C and H atoms. (This is probably the cause for the rather planar macroring structure.)

[2.1.2.1]Paracyclophane (61)[54]

Monoclinic	$P2_1/b$	CuKα	$a = 24.44(4)$ $b = 17.22(6)$ $c = 9.70(4)$ $\beta = 141.3(3)$	2,242	0.976 (0.98)	4	—	—

1,6,20,25-Tetraaza[4.1.4.1]paracyclophane · 4 HCl · durene · 4 H_2O (62)[55]

Monoclinic (H_2O)	$P2_1/n$	CuKα	$a = 14.552(7)$ $b = 22.582(12)$ $c = 7.238(4)$ $\beta = 97.23(4)$	2,359.6	—	2	0.065	—

Comments: The cyclophane forms a macrocyclic cavity with rectangularly shaped open ends approximately 3.5 × 7.9 Å and a depth of 6.5 Å. The four *p*-benzenoid rings are perpendicular to the mean plane of the entire macrocycle (i.e., π clouds facing the center of the macroring), and the bridging chain atoms adopt a trans-planar conformation except for the w bonds (see Structure **62**), which adopt a gauche conformation. The durene molecule (guest) fits into the cavity such that the durene aromatic plane is nearly parallel to the aliphatic "walls," with the methyl groups partly protruding from the cavity. The closest contacts between host and guest are as follows: C-42 · · · C-17 = 3.799 Å; C-41 · · · C-21 = 3.765 Å; C-16 · · · C-47 = 3.592 Å (see Chapter 11, Fig. 6).

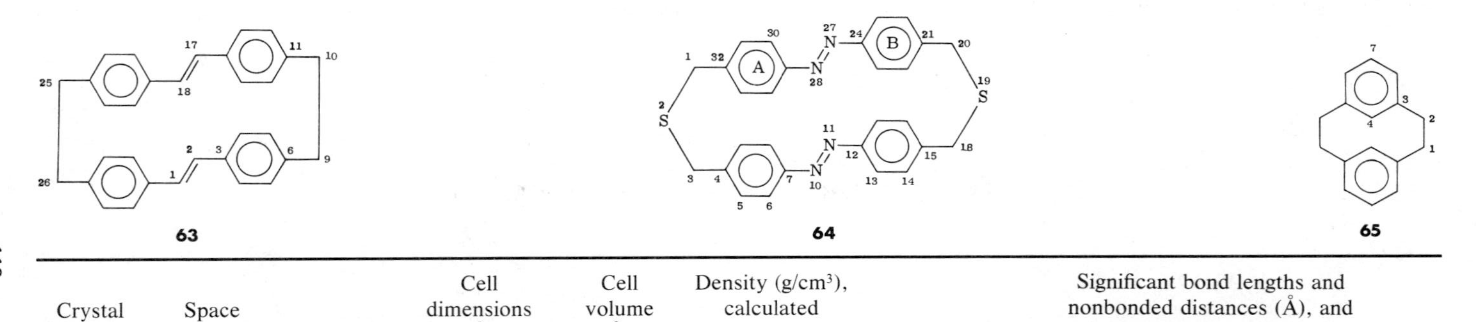

Crystal system	Space group	Radiation	Cell dimensions (Å)	Cell volume (Å^3)	Density (g/cm^3), calculated (measured)	Z	R	Significant bond lengths and nonbonded distances (Å), and bond and interplanar angles (deg)
			(E,E)[2^4]Paracyclophane-1,17-diene (63)[56]					
Monoclinic ($CHCl_3$)	$P2_1/c$	MoKα	a = 11.721(5) b = 8.298(4) c = 12.483(4) β = 102.36(3)	—	1.16	2	0.13	—

Comments: The large thermal vibration of the molecule did not allow for better refinement. The molecule is centrosymmetric (center of the molecule coincides with crystallographic inversion center). The macrocyclic ring is ellipsoidal, with the alkanic bridges at the extremes of the major axis (12.1 Å apart) and the alkenic bridges at the extremes of the minor axis (4.4 Å apart). The two (E)-styrenoid units face one another and are thus perpendicular to the macrocyclic plane (87–90°). The alkanic bridges (C-9—C-10, C-25—C-26) have their respective H atoms eclipsed with a torsion angle of only 3° (C-6—C-9—C-10—C-11).

2,19-Dithia[3.3](4,4′)azobenzenophane (64)[57]

Triclinic	$P\bar{1}$	MoKα	a = 6.301(3) b = 7.555(4) c = 12.324(6) α = 91.46(4) β = 93.50(4) γ = 101.71(4)	—	1.393	1	—	a = 1.519 b = 1.832 e = 1.395 f = 1.397 g = 1.395	C—N = 1.438 N═N = 1.272 γ = 115.1 λ = 119.9

Comments: Both azobenzenoid units are slightly deformed. One of the phenyl units (A) of the azobenzenoid moiety is flat (within error limits), the other (B) is slightly but distinctly boat-shaped, with C-24 and C-21 an average of 0.014 Å out of its mean aromatic plane. The azobenzenoid units are essentially parallel to one another, with a slight parallel displacement (slide) with respect to one another. The azo units are transoid, and the thia bridges are anti with respect to one another. The C_{ring}—N═N—C_{ring} group is not planar but twisted 3.6° from planarity. Rings A and B are also slightly twisted with respect to one another, a 10° angle being formed between the normals to both ring planes. The shortest distances between planes are C-30 · · · C-7, 3.63 Å, and C-5 · · · C-32, 3.34 Å.

II. META RING PHANES

A. STRUCTURES CONTAINING AT LEAST ONE META RING

[2.2]Metacyclophane (65)[58]

Monoclinic (EtOH)	$P2_1/a$	—	a = 12.22 b = 8.28 c = 5.82 β = 93.18	—	1.176 (1.178)	2	0.163	a = 1.535 b = 1.559 e = 1.389 f = 1.383 g = 1.387	α^* ≅ 0.4(15.4°) β^* = 0.143(11.8°) γ = 110.18 δ^* = 0.042(3.4°) λ = 117.3 p = 2.295 q = 2.689

Comments: The molecule has an anti-step structure, with the aromatic rings adopting a boat conformation. The internal ring C atom (C-4) is raised above the mean plane of the two ortho and meta C atoms by 0.143 Å, whereas C-7 is raised from the same plane by 0.042 Å. The C-2 atom is lowered from that plane by 0.368 Å (see next entry).

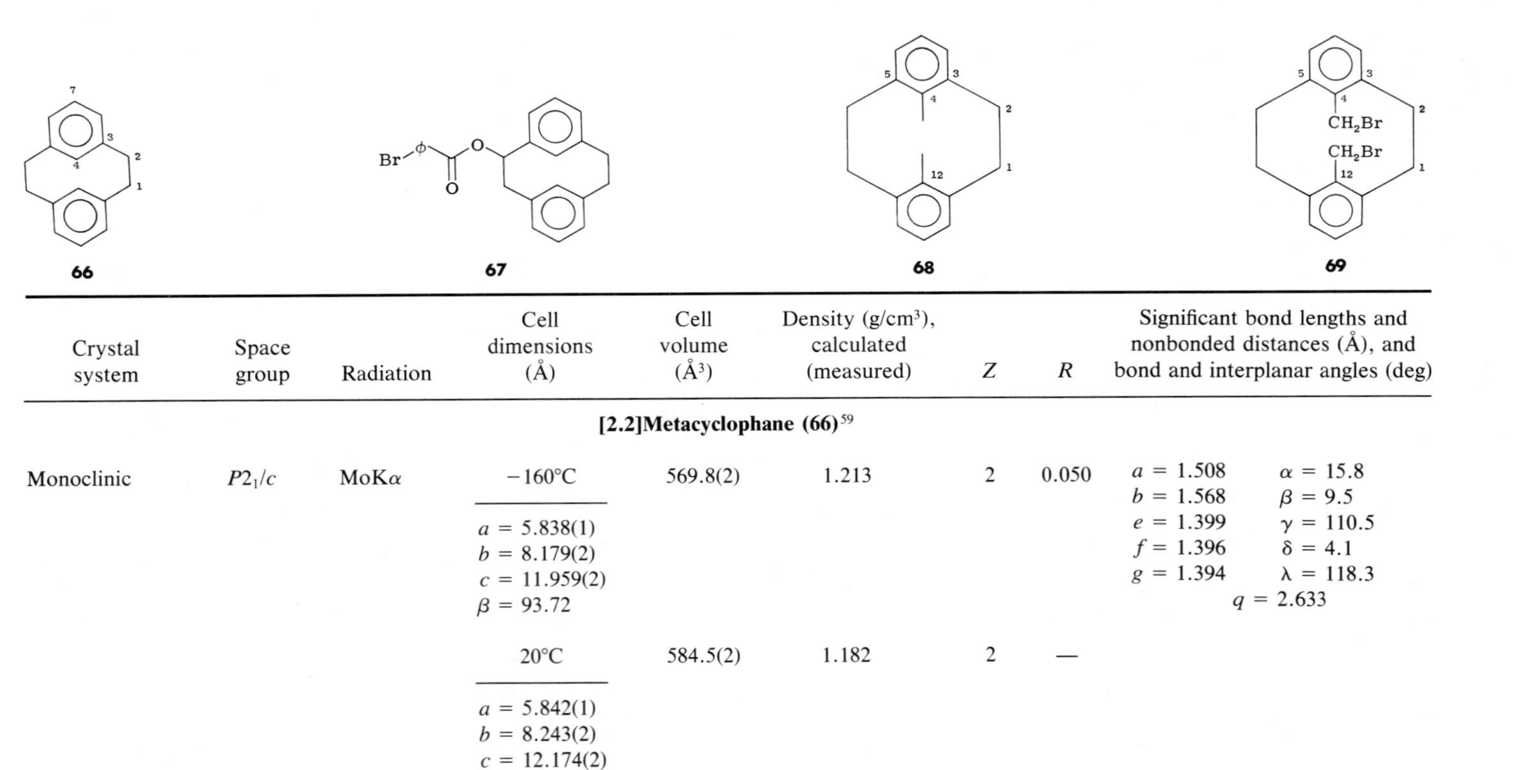

Crystal system	Space group	Radiation	Cell dimensions (Å)	Cell volume (Å^3)	Density (g/cm^3), calculated (measured)	Z	R	Significant bond lengths and nonbonded distances (Å), and bond and interplanar angles (deg)
[2.2]Metacyclophane (66)[59]								
Monoclinic	$P2_1/c$	MoKα	−160°C: $a = 5.838(1)$, $b = 8.179(2)$, $c = 11.959(2)$, $\beta = 93.72$	569.8(2)	1.213	2	0.050	$a = 1.508$, $b = 1.568$, $e = 1.399$, $f = 1.396$, $g = 1.394$; $\alpha = 15.8$, $\beta = 9.5$, $\gamma = 110.5$, $\delta = 4.1$, $\lambda = 118.3$; $q = 2.633$
			20°C: $a = 5.842(1)$, $b = 8.243(2)$, $c = 12.174(2)$, $\beta = 94.55$	584.5(2)	1.182	2	—	

Comments: The C-4—H bond and the C-7—H bond make angles of 8.0 and 3.1°, respectively, with the C-3,C-4,C-5 plane and the C-6,C-7,C-8 plane.

(*R*)-1-Hydroxy[2.2]metacyclophane (*m*-bromobenzoate) (67)[60]

Orthorhombic (ether/hexane)	$P2_12_12_1$	CuKα	$a = 28.756(3)$ $b = 11.067(6)$ $c = 5.821(3)$	—	1.44 (1.40)	4	0.064	—

4,12-Dimethyl[2.2]metacyclophane (68)[61]

Monoclinic	$C2/c$	CuKα; MoKα	$a = 12.62(1)$ $b = 9.25(1)$ $c = 10.97(1)$ $\beta = 91.56(2)$	1,280.1	1.226(3) [1.22(1)]	4	0.061	$a = 1.517$ $\alpha^* = 0.415(16.7°)$ $b = 1.573$ $\beta^* = 0.187(15.4°)$ $e = 1.411$ $\gamma = 111.5$ $f = 1.397$ $\delta^* = 0.088(7.3°)$ $g = 1.389$ $\lambda = 118.9$ $u = 1.507$ $q = 2.819$

Comments: The molecule has an anti-step structure, with the aromatic rings adopting a boat conformation to reduce transannular interaction. The distortions are similar to, but greater than, those in [2.2]metacyclophane because of the substituent methyl groups. The methyl C atom lies in the C3,C-4,C-5 plane, but this atoms is raised 0.585 Å from the C-2,C-3,C-5,C-6 plane. The methyl groups appear to be stationary, and one H atom is positioned in the center of the proximate aromatic π cloud, with an average nonbonded contact to the C atoms of this ring of about 2.7 Å.

4,12-Bisbromomethyl[2.2]metacyclophane (69)[62]

Monoclinic	$P2_1/c$	CuKα	$a = 7.84(1)$ $b = 14.01(2)$ $c = 7.50(1)$ $\beta = 114.71(5)$	748.4	1.749(4) [1.76(1)]	2	0.086	$a = 1.519$ $\alpha^* = 0.424(16.2°)$ $b = 1.568$ $\beta^* = 0.178(14.7°)$ $e = 1.407$ $\gamma = 112.5$ $f = 1.387$ $\delta^* = 0.080(6.7°)$ $g = 1.381$ $\lambda = 118.5$ $u = 1.492$ $q = 2.77$

Comments: The molecule has an anti-step structure. The aromatic rings are boat-shaped, as in [2.2]metacyclophane and 4,12-dimethyl[2.2]metacyclophane, and the overall structure is similar to these compounds. The C—Br bond length is 1.981(5), slightly larger than the mean for normal aliphatic C—Br bonds, and may be due to crowding. The Br atoms are positioned away from the opposing aromatic ring, whereas the H atoms on the bromomethyl C atom are positioned (theoretically) an average of 2.7 Å from that ring (although not centrally located in the π cloud as in 4,12-dimethyl [2.2]metacyclophane). The bromomethyl C atom is positioned 0.494 Å above the C-3,C-5,C-6,C-8 plane.

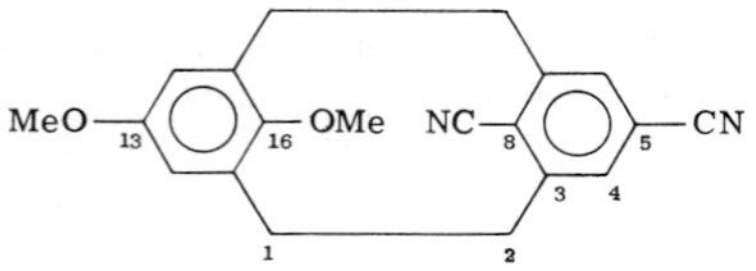

70 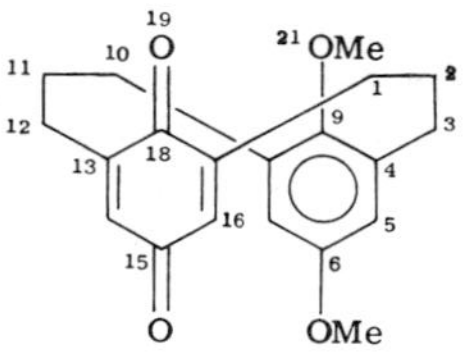 **71** 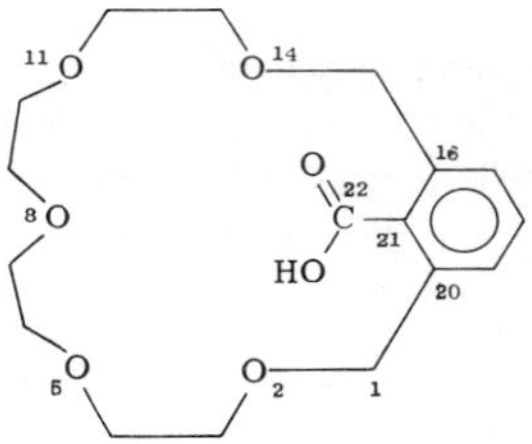 **72**

Crystal system	Space group	Radiation	Cell dimensions (Å)	Cell volume (Å^3)	Density (g/cm^3), calculated (measured)	Z	R	Significant bond lengths and nonbonded distances (Å), and bond and interplanar angles (deg)
5,8-Dicyano-13,16-dimethoxy[2.2]metacyclophane (70)[11]								
Monoclinic	$P2_1/c$	—	$a = 7.591$ $b = 15.466$ $c = 14.747$ $\beta = 108.34$	—	—	4	0.05	$q = 2.715$

Comments: The aromatic rings are distorted into boat shapes. Whereas the outer methoxyl group is nearly coplanar with the aromatic ring to which it is attached (torsional angle, 0.4°), the inner methoxyl group is nearly perpendicular to that ring (torsional angle, 80.6°). The outer C—C≡N bond angle is 178.8°, and the inner C—C≡N bond angle is 174.1°.

B. Structures containing three or more atoms in the bridge

Crystal system	Space group	Radiation	Cell dimensions (Å)	Cell volume (Å^3)	Density (g/cm^3), calculated (measured)	Z	R	Meta ring	Quinoid ring
6,9-Dimethoxy[3.3](2,6)*p*-benzoquinonometacyclophane (71)[63]									
Monoclinic (MeOH/acetone, 1 : 1)	$P2_1$	MoKα	$a = 7.067(1)$ $b = 9.838(1)$ $c = 11.881(1)$ $\beta = 91.68(1)$	—	1.31	2	0.044	$a = 1.516(4)$ $b = 1.514(4)$	$a = 1.506(5)$ $b = 1.524(5)$

$e = 1.396(4)$		$e = 1.489(5)$
$f = 1.388(4)$		$f = 1.336(4)$
$g = 1.391(4)$		$g = 1.459(5)$
$v = 1.379(3)$		$v = 1.223(3)$
$\beta = 2.9$		$\beta = 13.1$
	$p = 3.03$	
	$q = 3.05$	
	$\theta = 14.9$	

Comments: Whereas the meta ring is only slightly deformed, the quinoid ring is severely distorted. In a projection of the quinoid ring onto the meta ring plane the C-15,C-18 axis is almost (2.2°) parallel to the C-6,C-9 axis, and the centers of the two rings are only slightly displaced with respect to one another. There is a disorder in the central C atom of one of the bridges, and the C-9 methoxyl group is almost perpendicular (87.6°) to the aromatic plane in order to reduce the nonbonded interaction between the methoxyl group and the central C atom of the bridge. The methoxyl group at the C-6 atom is twisted 17.6° out of the aromatic plane. The nonbonded distances between O-19 and O-21 is 3.04 Å, and the nonbonded distance between C-5 and C-16 and C-6 and C-15 is 3.39 and 3.58 Å, respectively.

21-Carboxy-2,5,8,11,14-pentaoxa[15]metacyclophane (72)[64]

Monoclinic	$P2_1/c$	MoKα	$-160°C$	1,703.3	1.327	4	0.055	$a = 1.509(4)$	$u = 1.506$
			$a = 9.809(3)$					$b = 1.424(3)$	C-22=O = 1.216
			$b = 12.848(4)$					$c = 1.419(4)$	C-22—O = 1.321
			$c = 13.888(3)$					$d = 1.509(5)$	$\alpha = 0.0; 0.07$
			$\beta = 103.31(2)$					$e = 1.401(3)$	$\gamma = 108.2$
								$f = 1.392(4)$	$\lambda = 119.5(2)$
								$g = 1.394(4)$	

Comments: The aromatic ring is planar. Only the C-15 bridging atom is displaced from the aromatic plane due to the proximate carbonyl group O atom. The carboxyl group plane forms an angle of 59° with the aromatic ring plane, so that reduced nonbonded interactions with the interannular atoms can be achieved and intramolecular H bonding of the hydroxyl H atom with the macrocyclic ether O atoms can be maximized. Thus, the OH · · · O-8 nonbonded distance is 1.8 Å, whereas the O-2 · · · C-22 and O-14 · · · C-22 distances are 2.746 and 2.767 Å, respectively.

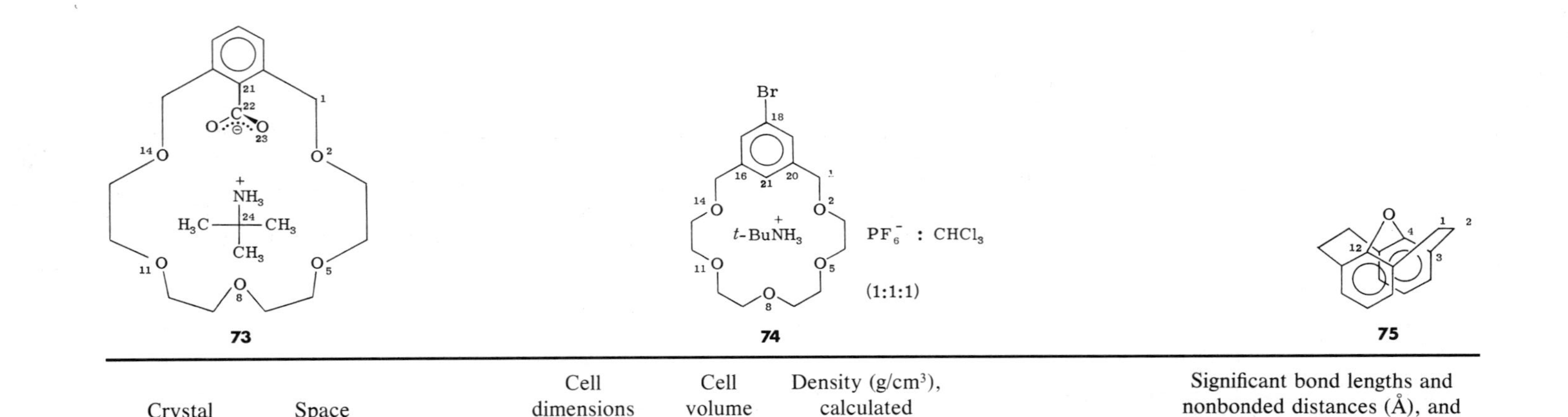

Crystal system	Space group	Radiation	Cell dimensions (Å)	Cell volume (Å³)	Density (g/cm³), calculated (measured)	Z	R	Significant bond lengths and nonbonded distances (Å), and bond and interplanar angles (deg)	
***tert*-Butylammonium 2,5,8,11,14-pentaoxa[15]metacyclophane-21-carboxylate (73)**[65]									
Orthorhombic	*Pnma*	CuKα	120K	2,172.6	1.264	4	0.038	a = 1.507(2)	u = 1.522(3)
			a = 17.811(3)					b = 1.437(2)	C-22—O = 1.548(3)
			b = 13.920(2)					c = 1.432(2)	α^* = 0.06
			c = 8.763(2)					d = 1.503(2)	γ = 111.5(1)
								e = 1.403(2)	λ = 119.9(1)
								f = 1.394(2)	
								g = 1.393(2)	

Comments: The aromatic ring is planar and perpendicular to the carboxylate plane. The five ether O atoms of the macroring are arranged alternately above and below their least-squares plane. This arrangement, with O-23, allows for a symmetric hexagonal electronegative cavity in the host for binding the guest ammonium ion. The C-24—N axis is almost perpendicular (3.4°) to the least-squares plane of the five ether O atoms and one of the carboxylate O atoms (O-23). The H bonding of the ammonium group is to the O-5 and O-11 atoms and O-23 atom of the carboxylate group, the distances being 2.21, 2.21, and 1.70 Å, respectively. The aromatic ring plane and the plane formed by the five ether O atoms and O-23 forms an angle of 66°. The polyether ring conformation is significantly distorted from the "crown" model because of *m*-xylyl unit and guest complexation. The ethylenoxy units maintain a gauche conformation, the torsion angle around the corresponding C—C units being 72°.

18-Bromo-2,5,6,11,14-pentaoxa[15]metacyclophane : *tert*-butylammonium hexafluorophosphate : $CHCl_3$ (1 : 1 : 1) (74)[66]

Triclinic ($CHCl_3$/ethyl ether)	$P\bar{1}$	MoKα	a = 12.416(6) b = 13.571(4) c = 9,731(4) α = 104.64(3) β = 91.03(4) γ = 89.21(4)	—	1.494	2	0.112		—

Comments: The complex is of the "nesting" type in which the *tert*-butylammonium cation and the metacyclo ring are on the same face of the macroring and the ammonium group is H-bonded to O-2, O-8, and O-14. Chloroform sits on the opposite face of the macroring.

C. STRUCTURES CONTAINING HETEROATOMS IN THE BRIDGE

4,12-Oxido-*syn*-[2.2]metacyclophane (75)[67]

Orthorhombic	$Cmc2_1$	CuKα	a = 8.78(1) b = 16.04(2) c = 8.42(1)	1,185.8	1.245(5) [1.24(2)]	4	0.057	a = 1.516(7) b = 1.540(10) e = 1.375(5) f = 1.365 g = 1.370 v = 1.403(7)	α^* = 0.18 β^* = 0.10 γ = 119.0 δ^* = 0.02 λ = 116.3 θ = 99.6

Comments: The molecule is wedge-shaped. The apexes of the wedges all point in approximately the same direction in the crystal. The aromatic rings are slightly distorted into boat shapes. The C—O—C angle is 101.4° and is quite different from that found in normal ethers. The lower than normal value is probably the result of the angle of inclination (θ = 99.6°) of the two aromatic rings.

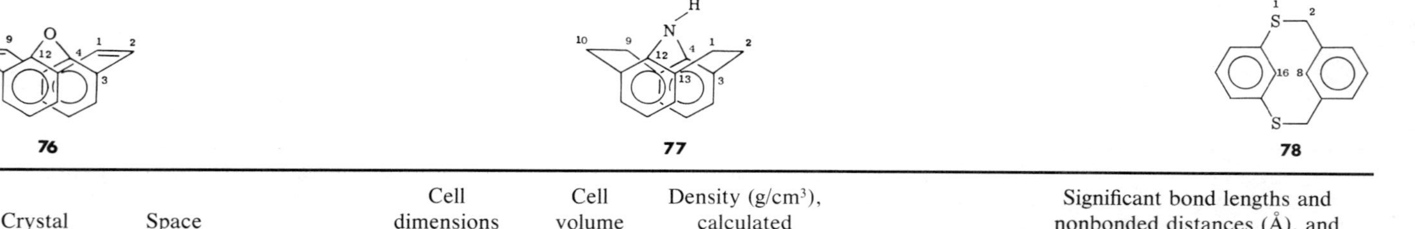

Crystal system	Space group	Radiation	Cell dimensions (Å)	Cell volume (Å³)	Density (g/cm³), calculated (measured)	Z	R	Significant bond lengths and nonbonded distances (Å), and bond and interplanar angles (deg)	
			4,12-Oxido-*syn*-[2.2]metacyclophane-1,9-diene (76)[68]						
Orthorhombic	*Pbca*	CuKα	−160°C a = 11.96(1) b = 17.13(2) c = 10.67(1)	2,186	1.326 [1.30(1)]	8	0.069	a = 1.459 b = 1.359 e = 1.396 f = 1.403 g = 1.385 v = 1.385 q = 2.12	α^* = 0.291 β^* = 0.161 γ = 125.3 δ^* = 0.049 θ = 123.6 λ = 115.7

Comments: The molecule is similar in wedge-shape structure to 4,12-oxido-*syn*-[2.2]metacyclophane. The wedge shape (which prevents the rings from adopting a trans orientation) is maintained by the compression of the ether linkage and the tension in the bridging olefin units. The C—O—C angle is 99.7°, 1.7° smaller than the corresponding angle in 4,12-oxido-*syn*-[2.2]metacyclophane and 11.8° smaller than normal ether C—O—C angles. The aromatic rings assume a boat-shape distortion (characteristic of all [2.2]metacyclophanes), but the angle (θ) that the least-squares aromatic planes make with one another is 123.6°, 24° greater than in 4,12-oxido-*syn*-[2.2]metacyclophane.

Crystal system	Space group	Radiation	Cell dimensions (Å)	Cell volume (Å³)	Density (g/cm³), calculated (measured)	Z	R	Significant bond lengths and nonbonded distances (Å), and bond and interplanar angles (deg)	
			4,12-Imino-*syn*-[2.2]metacyclophane (77)[69]						
Monoclinic	*Ia*	CuKα	−160 (10)°C a = 8.756(5) b = 41.64(2) c = 9.952(2)	3,396	1.30	12	0.042	a = 1.516 b = 1.571 e = 1.396 f = 1.393 g = 1.390	β^* = 0.103 γ = 117.0 δ^* = 0.039 θ = 94 λ = 117.8

Monoclinic	Ia	CuKα	Ambient temperature	1,747	1.26 (1.24)	6	—	—
			$a = 8.86(1)$					
			$b = 42.20(4)$					
			$c = 4.98(1)$					
			$\beta = 110.2(1)$					

Comments: The structure is quite similar to that of 4,12-oxido-*syn*-[2.2]metacyclophane. The aromatic rings are inclined at an angle (θ) of 94° with respect to one another, and the C—N—C angle is 99.4°, much smaller than the corresponding angle in secondary amines, which is close to tetrahedral. Unlike the 4,17-oxido-*syn*-[2.2]metacyclophane this molecule does not have even approximate *mm*2 symmetry. Each phenyl ring is twisted in opposite directions about its C—N bond, from 2.5 to 4.8°, which gives rise to a twisting of the C-1—C-2 bond of 14° (dihedral angle C-13—C-1—C-2—C-3), minimizing the eclipsing interactions of the H atoms on the bridge (the C-1—C-2 and C-9—C-10 bonds make an angle of between 10.4 and 17.2° with respect to one another). The amino H atom lies well off the pseudo-twofold axis passing through the N atom because of the pyramidal configuration at this atom. This results in steric repulsion between the amino H atom and the H atom at C-1 of the bridge (nonbonded distance, 2.28 Å) and displaces the N atom 2.3–3.4° from the approximate twofold axis in the direction of the opposite bridge.

1,10-Dithia[2.2]metacyclophane (78)[70]

Monoclinic (benzene)	$P2_1/n$	CuKα	$a = 7.868(1)$	—	1.365 (1.36)	4	0.036	S-Bridged ring	CH_2-Bridged ring
			$b = 27.269(1)$					$a = 1.785(3)$	$a = 1.500(4)$
			$c = 5.600(1)$					$b = 1.855(3)$	
			$\beta = 98.17(1)$					$e = 1.390(4)$	$e = 1.395(3)$
								$f = 1.386(3)$	$f = 1.389(3)$
								$g = 1.386(4)$	$g = 1.388(4)$
								$\alpha^* = 0.337$	$\alpha^* = 0.323$
								$\beta^* = 0.086$	$\beta^* = 0.079$
								$\gamma = 99.4(1)$	$\gamma = 111.1(2)$
								$\delta^* = 0.050$	$\delta^* = 0.034$
								$\lambda = 119.8(3)$	$\lambda = 118.8$
								$q = 2.712$	
								$\theta = 5$	

Comments: Both rings are distorted into boat shapes, but because of the S atoms the distortions are not as great as those found in [2.2]metacyclophane. The two aromatic rings are inclined 5° with respect to one another (θ), unlike the rings in [2.2]metacyclophane. The H atoms on the aromatic ring are slightly displaced out of their respective aromatic planes toward the opposite aromatic ring, and the H atoms bound to the C-8 and C-16 atoms are 2.48 Å from the opposite aromatic plane. The aromatic rings are oriented anti with respect to one another.

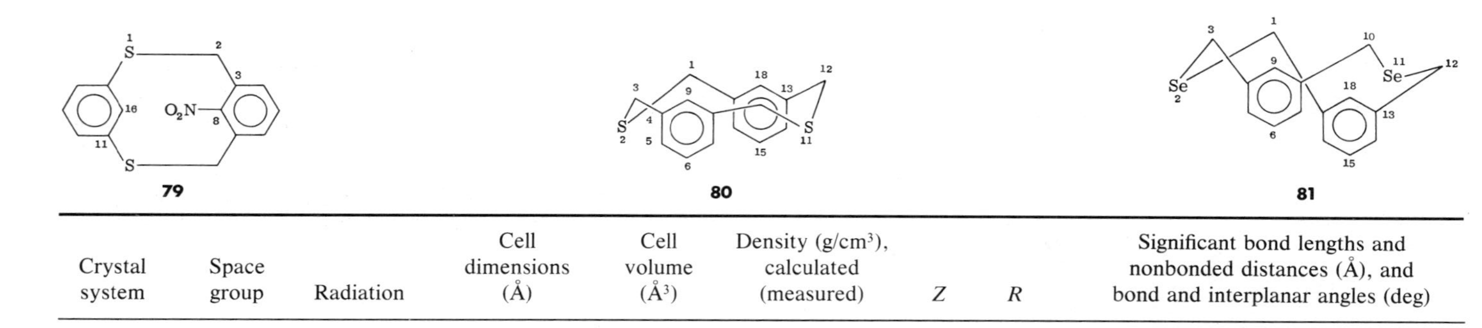

Crystal system	Space group	Radiation	Cell dimensions (Å)	Cell volume (Å³)	Density (g/cm³), calculated (measured)	Z	R	Significant bond lengths and nonbonded distances (Å), and bond and interplanar angles (deg)	
					8-Nitro-1,10-dithia[2.2]metacyclophane (79)[71]				
Triclinic (benzene)	$P\bar{1}$	Ni-Filtered CuKα	$a = 13.865(1)$	—	1.512 (1.51)	2	0.053	S-Bridged ring	CH_2-Bridged ring
			$b = 7.626(1)$					$a = 1.778(2)$	$a = 1.500(3)$
			$c = 6.988(1)$					$b = 1.857(2)$	
			$\alpha = 120.07(1)$					$e = 1.392(3)$	$e = 1.399(2)$
			$\beta = 84.13(1)$					$f = 1.400(3)$	$f = 1.398(3)$
			$\gamma = 94.44(1)$					$g = 1.383(3)$	$g = 1.378(3)$
									$w = 1.483(2)$
								$\alpha^* = 0.362; 0.199$	$\alpha^* = 0.375$
								$\beta^* = 0.099$	$\beta^* = 0.132$
								$\gamma = 101.1(2)$	$\gamma = 110.7(1)$
								$\delta^* = 0.044$	$\delta^* = 0.055$
								$\lambda = 119.7(2)$	$\lambda = 116.4(2)$
								$q = 2.781$	
								$\theta = 13$	

Comments: Both aromatic rings are bent into boat shapes, but the nitro-substituted ring is more severely bent. The plane of the nitro group makes an angle of 47° with the adjacent phenyl ring, placing one of the nitro group O atoms closer to the opposite aromatic ring π cloud. This O atom is 2.995(3) Å from the C-11 atom and is 2.84 Å from the opposite aromatic ring plane. The H atom at C-16 is bent toward the opposite aromatic ring. The aromatic rings are oriented anti with respect to one another.

syn-2,11-Dithia[3.3]metacyclophane (80)[72]

Monoclinic	$P2_1/n$	Zr-Filtered MoKα	$a = 18.982(5)$ $b = 7.957(3)$ $c = 9.241(4)$ $\beta = 100.03(4)$	—	1.317 (1.34)	4	0.049	$a = 1.510(9)$ $b = 1.811(7)$ $e = 1.384(7)$ $f = 1.383(9)$ $g = 1.389(9)$ $q = 3.05$ $\theta = 20.6$	$\alpha^* = 0.08$ $\beta^* = 0.010$ $\gamma = 115.4(5)$ $\lambda = 118.4(6)$

Comments: The aromatic rings are planar and are oriented syn to one another. The S atoms of the bridges are also oriented syn with respect to one another and with respect to the rings. The H atoms at C-9 and C-18 are directed toward one another and are 2.73 Å apart. The C-6 and C-15 atoms are 3.92 Å apart. The atoms of each aromatic ring do not eclipse the atoms of the other ring. The two rings are twisted 7° about an axis that passes through the C-9 and C-18 atoms.

syn-2,11-Diselena[3.3]metacyclophane (81)[73]

Monoclinic	$P2_1/n$	MoKα	$a = 19.26(1)$ $b = 8.014(3)$ $c = 9.24(1)$ $\beta = 98.9(1)$	1.409(2)	1.726 (1.733)	4	0.039	$a = 1.502(8)$ $b = 1.964(6)$ $e = 1.386(7)$ $f = 1.391(8)$ $g = 1.378(8)$ $q = 3.15$	$\alpha^* = 0.045(5)$ $\gamma = 115.2(4)$ $\lambda = 119.2(4)$ $\theta = 19.1$

Comments: The molecule has an approximate twofold axis that bisects the lines joining the C-6,C-15 atoms and C-9,C-18 atoms. The aromatic rings are planar but form a dihedral angle of 19.1° (θ). The Se atoms are syn with respect to one another, and so are the aromatic rings. The aromatic ring atoms do not eclipse the corresponding atoms on the second aromatic ring. The rings are twisted with respect to one another to avoid this eclipsing. The distance between the ring centers is 3.55 Å, and the H-9 and H-18 atoms are 2.85 Å from each other.

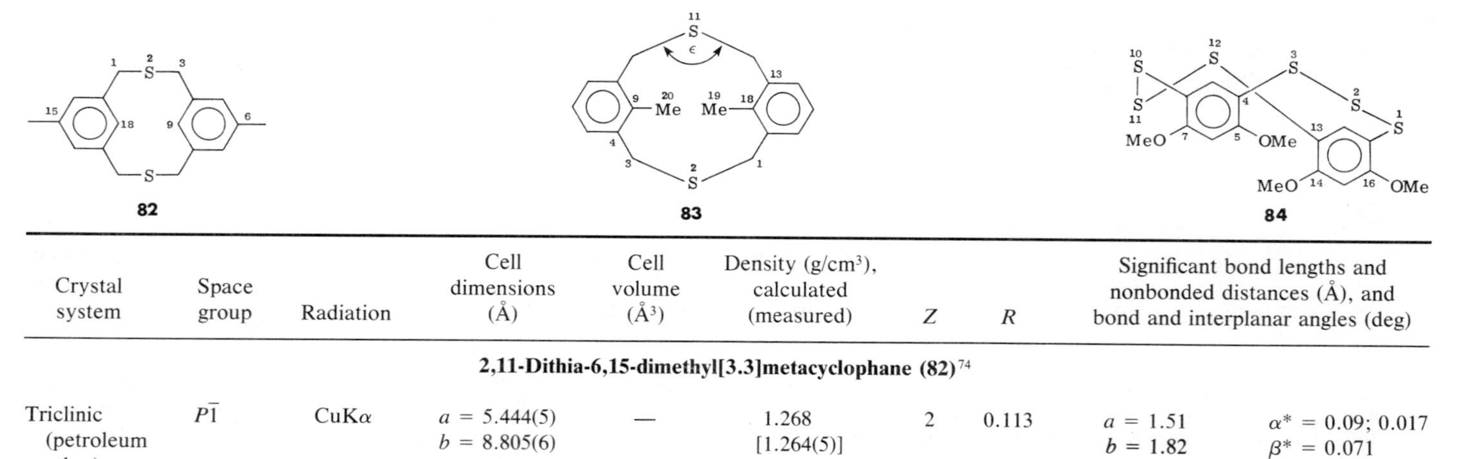

Crystal system	Space group	Radiation	Cell dimensions (Å)	Cell volume (Å³)	Density (g/cm³), calculated (measured)	Z	R	Significant bond lengths and nonbonded distances (Å), and bond and interplanar angles (deg)	
			2,11-Dithia-6,15-dimethyl[3.3]metacyclophane (82)[74]						
Triclinic (petroleum ether)	$P\bar{1}$	CuKα	$a = 5.444(5)$ $b = 8.805(6)$ $c = 16.93(1)$ $\alpha = 101.19(8)$ $\beta = 90.58(8)$ $\gamma = 98.47(8)$	—	1.268 [1.264(5)]	2	0.113	$a = 1.51$ $b = 1.82$ $e = 1.39$ $f = 1.41$ $g = 1.38$ $u = 1.53$	$\alpha^* = 0.09; 0.017$ $\beta^* = 0.071$ $\gamma = 113.2; 117$ $\delta = \sim 0$ $\lambda = 119.2$ $\theta = 6.6$ $q = 2.98$

Comments: The molecule exists in an anti-stepped orientation, with the methyl substituents pointing in opposite directions. Except for the C-9 and C-18 atoms, the aromatic rings are nearly planar and are inclined (θ) 6.6° with respect to one another. The methyl substituents lie in the aromatic plane. Although the two CH_2—S—CH_2 bridges adopt nonequivalent conformations, their dimensions are nearly equivalent and there is no evidence of conformational disorder. The S atoms seem to be syn with respect to one another.

syn-2,11-Dithia-9,18-dimethyl[3.3]metacyclophane (83)[75]

Monoclinic	$P2_1/n$	MoKα	$a = 15.514(5)$	—	1.30	4	0.071	$a = 1.51(1)$	$\alpha^* = 0.19(6.4°)$
			$b = 12.141(6)$		[1.25(2)]			$b = 1.78(2)$	$\beta = \sim 0$
			$c = 8.649(4)$					$e = 1.38(1)$	$\gamma = 117.5(9)$
			$\beta = 109.88(3)$					$f = 1.40(1)$	$\delta = \sim 0$
								$g = 1.38(1)$	$\varepsilon = 106.8(6)$
								$u_{9,20} = 1.54(1)$	$\lambda = 119.2(9)$
									$\theta = 11.5$
									$p = 3.30(1)$
									$q = 3.32(1)$
									$CH_3 \cdots CH_3 = 3.52(1)$

Comments: The aromatic rings are syn with respect to one another but do not lie directly above one another. There is a partial disorder (20%) associated with one of the bridge S atoms, and therefore it is not entirely clear whether the S atoms are also syn with respect to one another. However, the principal species is one in which the methyl groups point in one direction and the S atoms and the aromatic rings point in the opposite direction. The aromatic rings are nearly planar, and the C-19 and C-20 atoms are raised out of their mean aromatic plane by 0.28 Å to minimize $CH_3 \cdots CH_3$ interactions.

5,7,14,16-Tetramethoxy-1,2,3,10,11,12-hexathia[3.3]metacyclophane (84)[76]

Triclinic	$P\bar{1}$	MoKα	$a = 13.86(1)$	997.7	1.55	2	0.047	$a = 1.772(6)$	$\alpha^* = 0.093$
			$b = 10.86(1)$		(1.56)			$b = 2.058(3)$	$\gamma = 103.8(2)$
			$c = 8.155(7)$					$e = 1.382(9)$	$\lambda = 118.3(5)$
			$\alpha = 93.83(8)$					$f = 1.423(8)$	$\theta = 125.3$
			$\beta = 96.12(8)$					$g = 1.386(9)$	
			$\gamma = 124.18(8)$					$v = 1.350(7)$	

Comments: There is no significant deviation from planarity of the aromatic rings. The aromatic rings are syn with respect to one another, as are the S bridges, and they are both oriented in the same direction, giving the molecule an M shape when viewed down the S-2,S-11 axis. The aromatic rings are not parallel. The 1,2,4,5 substituents in each ring are nearly coplanar with the ring (including the methoxyl methyl C atom).

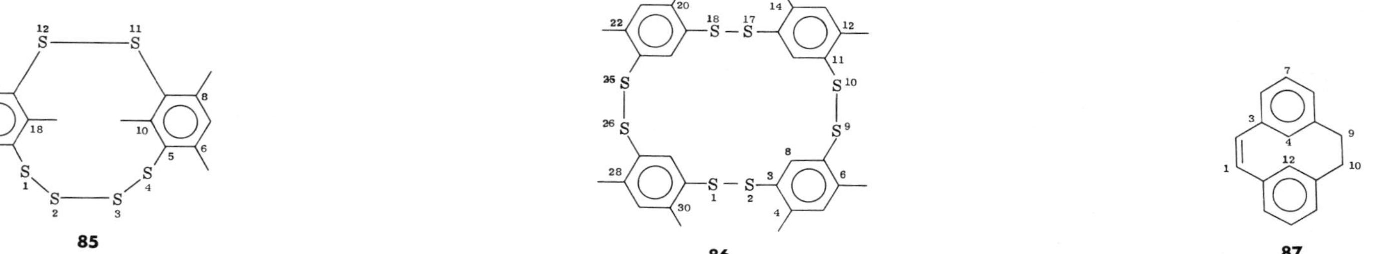

Crystal system	Space group	Radiation	Cell dimensions (Å)	Cell volume (Å³)	Density (g/cm³), calculated (measured)	Z	R	Significant bond lengths and nonbonded distances (Å), and bond and interplanar angles (deg)	
6,8,10,14,16,18-Hexamethyl-1,2,3,4,11,12-hexathia[4.2]metacyclophane (85)[76,77,78]									
Orthorhombic	$Pca2_1$	MoKα	a = 14.895(7) b = 17.154(7) c = 15.622(7)	3,991.6	1.43 (1.44)	8	0.033	S—S side	S—S—S—S side
								a = 1.786(4)	a = 1.784(4)
								b = 2.084(2)	b = 2.061(2)
									c = 2.074(2)
								e = 1.412(5)	e = 1.408(6)
								f = 1.403(7)	f = 1.400(7)
								g = 1.395(7)	g = 1.392(7)
								α^* = 0.019; 0.076	α^* = 0.043; 0.073
								γ = 102.0(2)	γ = 105.1(2)
								λ = 121.3(4)	λ = 122.2(4)

Comments: There is no significant deviation from planarity of the aromatic rings. The aromatic rings are anti with respect to one another. In the crystals there are two crystallographically independent molecules, A and B, which are very similar in shape. Both molecules exhibit the anti conformation of the aromatic rings. Because of the different sulfide bridges (disulfide and polysulfide) the benzenoid ring planes make a dihedral angle of 157° with respect to one another and are inclined toward the pseudo-C_2 axis passing through the midpoints of bonds S-2—S-3 and S-11—S-12. Whereas the C-10 and C-18 methyl groups are displaced an average of 0.030 Å from these respective aromatic planes away from the cyclophane cavity, the C-6, C-8, C-14, and C-16 methyl groups are displaced an average of 0.021 Å toward the cavity. Looking down the axis that bisects the S-2—S-3 and the S-11—S-12 bonds, it is observed that these two bond vectors make an angle of greater than 45° with respect to one another. The values in the last columns are given for molecule A.

4,6,12,14,20,22,28,30-Octamethyl-1,2,9,10,17,18,25,26-octathia[2^4]metacyclophane (86)[79]

Monoclinic	$P2_1/c$	MoKα	a = 18.189(8) b = 10.597(5) c = 17.604(8) β = 109.96(7)	3,189.3	1.40 (1.41)	4	0.032	a = 1.787(5) b = 2.023(2)	α^* = 0.067 (for three rings) α^* = 0.005 (for fourth ring)

Comments: The molecule assumes a saddle shape (approximate S_4 symmetry), with the adjacent aromatic rings being nearly perpendicular to one another (average 86°). Whereas the S atoms bound to three of the aromatic rings are displaced an average of 0.067 Å (α^*) from their respective aromatic planes, the S atoms bound to the fourth ring are nearly coplanar (α^* = 0.005 Å). The methyl substituents of three aromatic rings are displaced an average of 0.026 Å from their aromatic planes; those of the fourth ring are displaced 0.078 Å from their aromatic plane. (The unusual displacements of the S and methyl groups do not occur in the same ring.) The torsion angles around the S—S bonds range from 75.0 to 96.6°.

D. STRUCTURES WITH UNSATURATION IN THE BRIDGE

[2.2]Metacyclophane-1-ene (87)[80]

Orthorhombic (EtOH) (monoclinic form was also isolated, but the structure was not determined)	$Pbca$	CuKα	a = 10.932(1) b = 17.336(1) c = 12.089(1)	2,291.1	1.20 (1.18)	8	0.047	$a_{2,3}$ = 1.473 $a_{5,9}$ = 1.505 $b_{1,2}$ = 1.336 $b_{9,10}$ = 1.552 e = 1.391 f = 1.385 g = 1.383	α_2^* = 0.49 α_9^* = 0.42 β^* = 0.16 γ_2 = 124.9 γ_9 = 111.2 δ^* = 0.06 λ = 117.8 q = 2.592(4)

Comments: The molecular geometry is intermediate between [2.2]metacyclophane and [2.2]metacyclophane-1,9-diene. The rings are boat-shaped.

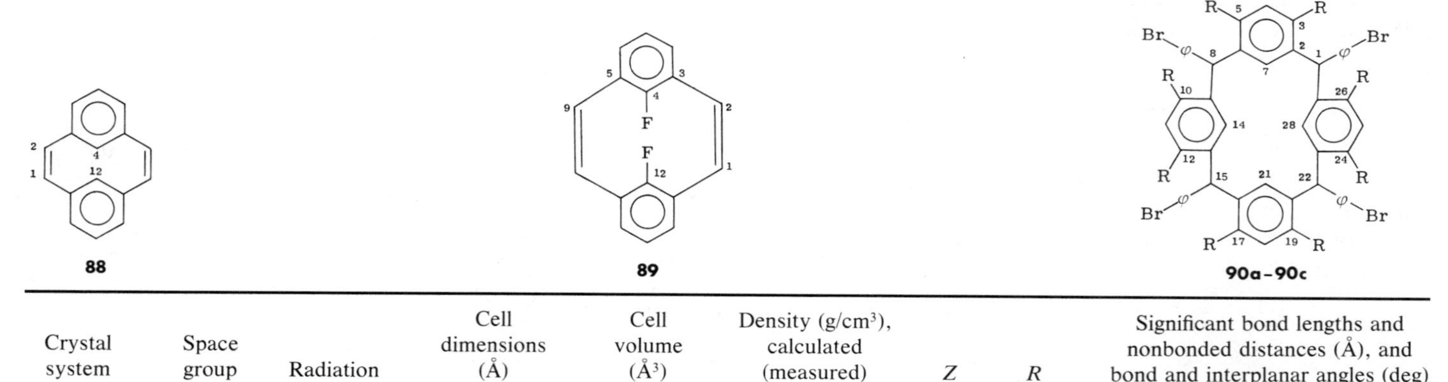

Crystal system	Space group	Radiation	Cell dimensions (Å)	Cell volume ($Å^3$)	Density (g/cm^3), calculated (measured)	Z	R	Significant bond lengths and nonbonded distances (Å), and bond and interplanar angles (deg)
			[2.2]Metacyclophane-1,9-diene (88)[81]					
Monoclinic	$P2_1/a$	CuKα	$a = 13.25(1)$ $b = 5.640(5)$ $c = 7.350(5)$ $\beta = 95.39(5)$	547	1.24 (1.24)	2	0.054	$a = 1.483$ $\alpha^* = 0.49(19.3°)$ $b = 1.341$ $\beta^* = 0.23(18.8°)$ $e = 1.404$ $\gamma = 126.0$ $f = 1.396$ $\delta^* = 0.08(6.7°)$ $g = 1.383$ $\lambda = 117.7$ $q = 2.567$

Comments: The aromatic rings are oriented anti to one another, their mean planes are parallel, and they suffer severe boat distortion. The aromatic C—H bonds are displaced toward the other aromatic ring as in other cyclophanes, the greatest displacement (0.2 Å) being for the C-4—H and C-12—H bonds. This brings the H-4 atom to within 2.46 Å of the C-12 atom.

anti-4,12-Difluoro[2.2]metacyclophane-1,9-diene (89)[82]

Monoclinic	$P2_1/c$	CuKα	20°C a = 14.22(1) b = 7.072(5) c = 12.05(1) β = 114.0(1)	—	1.44 (1.43)	4	0.054	a = 1.330(5) b = 1.474(4) e = 1.390(4) f = 1.398(5) g = 1.372(5) x = 1.360(3)	α^* = 0.51(20.2°) β^* = 0.23(19.3°) γ = 125.9 δ^* = 0.08(6.7°) λ = 115.7 q = 2.66

Comments: The molecule has the anti conformation, with dimensions similar to those of the parent hydrocarbon. Each F atom is displaced 0.18 Å from the plane of the nearest three ring C atoms (C-3, C-4, and C-5) toward the opposite ring.

E. MULTIBRIDGED STRUCTURES OF THE $[m^y]$ TYPE

anti,syn-1,8,15,22-Tetra-_p_-bromophenyl[1⁴]metacyclophane 3,5,10,12,17,19,24,26-octabutyrate (90a: R = —O—C(=O)—/\/)[83]

Triclinic	$P\bar{1}$	—	a = 13.99(5) b = 15.00(2) c = 20.23(6) α = 81.66(25) β = 75.26(30) γ = 89.92(35)	4,061(10)	1.365(4) [1.40(5)]	2	0.105	a = 1.528 e = 1.394 f = 1.382 g = 1.368	γ = 111 λ = 115

Comments: The four bridging C atoms (C-1, C-8, C-15, and C-22) are very nearly coplanar (plane p). Two opposing metacyclo rings are oriented syn with respect to one another, whereas the other two are anti to one another. The pair that are anti to one another are nearly coplanar (their planes make angles of 8 and 13° with plane p), whereas the pair that are syn to one another face each other, are inclined toward one another, and their planes are inclined 74 and 78° with respect to plane p. The four p-bromophenyl groups appended to the bridge methine C atoms are all syn with respect to one another, are found on one side of plane p, and are directed anti to the metacyclo ring pair, which is syn-related. The appended p-bromophenyl group planes make angles of between 75 and 85° with respect to plane p and therefore face one another around the periphery of the macrocycle.

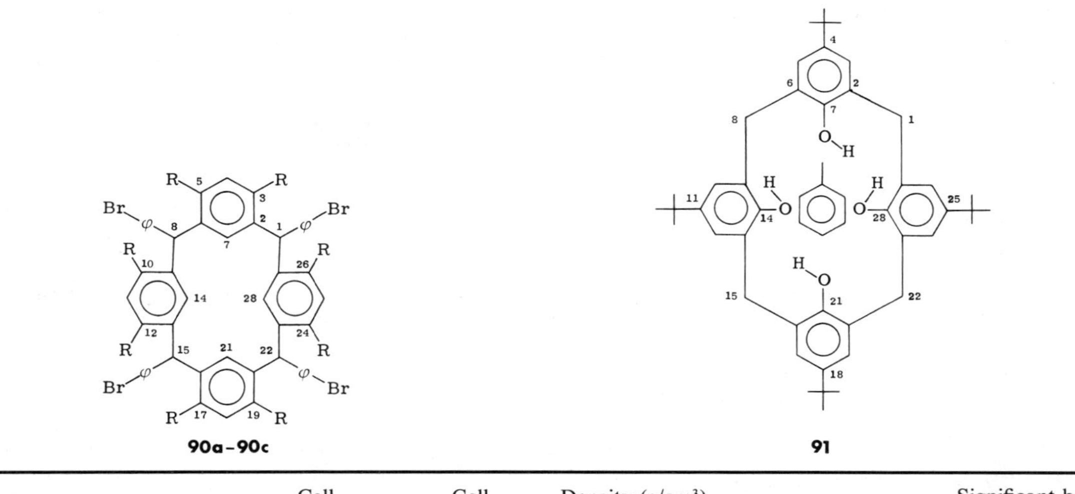

Crystal system	Space group	Radiation	Cell dimensions (Å)	Cell volume (Å³)	Density (g/cm³), calculated (measured)	Z	R	Significant bond lengths and nonbonded distances (Å), and bond and interplanar angles (deg)
***anti,syn*-1,8,15,22-Tetra-*p*-bromophenyl[1⁴]metacyclophane 3,5,10,12,17,19,24,26-octaacetate (90b: R = —OAc)** [84]								
Monoclinic ($CHCl_3$)	$C2/c$	CuKα	a = 24.229(8) b = 11.959(5) c = 24.334(8) β = 110.24(1)	—	1.450 (1.47)	4	0.117	a = 1.53(2) e = 1.40(2) f = 1.40(2) g = 1.38(2)

Comments: The structure is similar to that found for the *anti,syn*-octabutyrate (see previous entry), with all four *p*-bromophenyl rings syn to one another and syn to a pair of opposite metacyclo rings. The second pair of metacyclo rings are anti and very nearly coplanar. All the aromatic rings are planar (the maximum deviation from planarity is 0.011 Å), and the average angle between any two of the three aromatic planes bound to the methine centers is 82.7°. The molecule has a twofold axis of symmetry. The syn-oriented rings are inclined ~84.5° to the C-1,C-8,C-15,C-22 plane (plane *p*), whereas the anti-oriented rings are inclined ~8.8° to plane *p*. The planes of the *p*-bromophenyl units are nearly parallel to one another (average 9.3°) and are approximately perpendicular (average 84°) to plane *p*. The dihedral angle between the 1,4 axes of these *p*-bromophenyl rings is 12°.

anti,anti-1,8,15,22-Tetra-_p_-bromophenyl[1⁴]metacyclophane 3,5,10,12,17,19,24,26-octaacetate (90c: R = —OAc)[84]

Monoclinic ($CHCl_3$)	$P2_1/n$	CuKα	$a = 16.894(5)$ $b = 14.051(5)$ $c = 13.348(4)$ $\beta = 92.38(1)$	—	1.515 (1.50)	2	0.043	$a = 1.519(6)$ $e = 1.386(6)$ $f = 1.382(6)$ $g = 1.366(7)$

Comments: The molecule has a center of symmetry. One pair of opposite metacyclo rings are anti to one another and are very nearly coplanar. They are inclined ~14° with respect to the C-1,C-8,C-15,C-22 plane (plane *p*). The second pair of metacyclo rings are anti to one another in a step fashion and nearly parallel to one another. These metacyclo rings are almost perpendicular to plane *p* (angle of inclincation 87°). A pair each of the four *p*-bromophenyl units are also oriented anti to one another on opposite faces of plane *p*. The dihedral angle that their 1,4 axes make with one another is 25°. The angle between any of the three aromatic rings appended to the methine bridge C atoms is 87°. All aromatic rings are planar within 0.011 Å.

4,11,18,25-Tetra-_tert_-butyl-7,14,21,28-tetrahydroxy[1⁴]metacyclophane : toluene (91)[85]

Tetragonal (toluene)	$P4/n$	—	$a = 12.756(2)$ $c = 13.793(3)$	—	—	2	0.092	—

Comments: Both host and guest show disorder. The host, which has fourfold symmetry, is calix-shaped and all the aromatic ring hydroxyl groups point toward the apex of a cone. This shape is determined mainly by four intramolecular hydrogen bonds [O · · · O = 2.670(9) Å]. The C-4,C-7 (and related) vectors are inclined 123° to the fourfold axis. The toluene guest is held in the cavity of the calix mainly through interactions with the *tert*-butyl groups and is centered along the fourfold axis, with the methyl group of the toluene pointing toward the apex of the cone.

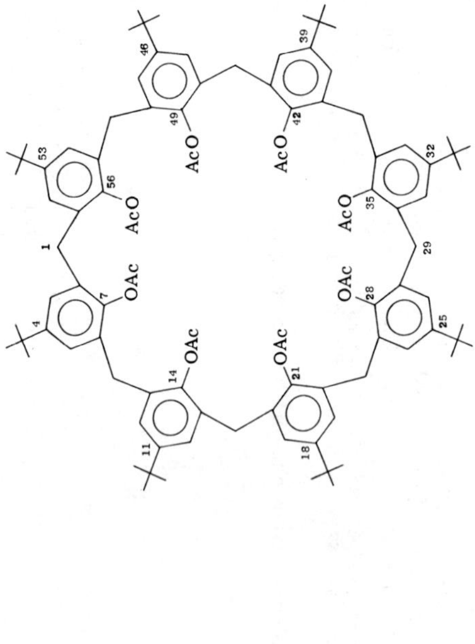

92

Crystal system	Space group	Radiation	Cell dimensions (Å)	Cell volume (Å^3)	Density (g/cm^3), calculated (measured)	Z	R	Significant bond lengths (Å), bond interplanar angles (deg), and nonbonded distances (Å)
4,11,18,25,32,39,46,53-Octa-*tert*-butyl[1^8]metacyclophane 7,14,21,28,35,42,49,56-octaacetate (92)[86]								
Triclinic (HOAc)	$P\bar{1}$	—	a = 11.518(3) b = 14.880(7) c = 16.356(6) α = 106.49(3) β = 77.67(2) γ = 100.00(1)	—	—	1	0.087	—

Comments: All the *tert*-butyl groups show disorder by 60° rotation about the bond to the aromatic rings. The molecule lies on a center of symmetry and has an ellipsoidal shape, with the polar acetyl units oriented toward the cavity center and the *tert*-butyl groups toward the outside. The major axis of the ellipse passes through the C-1,C-29 atoms, and the molecule has a step-like geometry about this axis (two sets of four phenyl units each forming the step). The benzenoid rings adopt a helical arrangement around the macroring, with average dihedral angles between the planes of the adjacent rings of 84.5(5)°. The cavity is not occupied in this crystal and has an approximate size of (1.51 × 3.73 × 4.56 Å) 107 Å^3, which is about twice the size of the cavity in α-cyclodextrin–H_2O_2 complex.

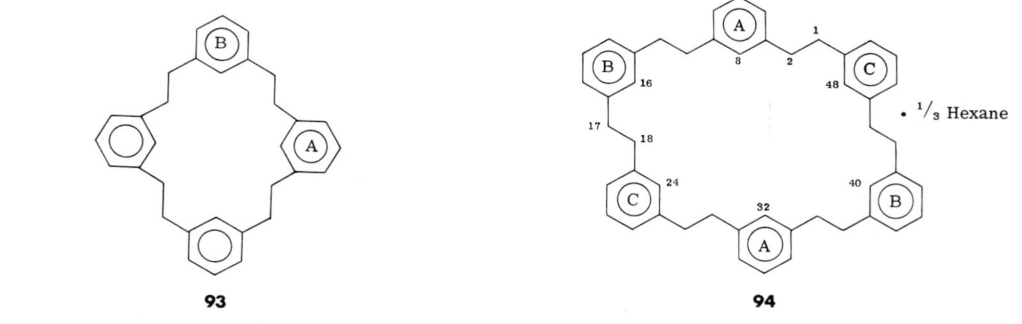

Crystal system	Space group	Radiation	Cell dimensions (Å)	Cell volume (Å^3)	Density (g/cm^3), calculated (measured)	Z	R	Significant bond lengths and nonbonded distances (Å), and bond and interplanar angles (deg)	
				[2^4]Metacyclophane (93)[87]					
Monoclinic	$P2_1/c$	CuKα	$a = 10.180(1)$ $b = 23.375(2)$ $c = 5.074(1)$ $\beta = 105.52(3)$	1,172.4	1.189 (1.14)	2	0.101	Gauche bridge‡ $a = 1.516$ $b = 1.538$	Anti bridge‡ $a = 1.523$ $b = 1.376$‡
(A second modification, whose structure was not determined, was found along with the above modification)			$a = 16.74$ $b = 8.97$ $c = 16.82$ $\beta = 98.9$					A ring‡ $e = 1.402$ $f = 1.387$ $g = 1.396$ $\gamma = 116$ $\lambda = 118.5$	B ring‡ $e = 1.392$ $f = 1.385$ $g = 1.403$ $\alpha^* = 0.12$ $\gamma = 118.4$ $\lambda = 119$

Comments: The atoms of the molecule are oriented (around a center of symmetry) in a zigzag fashion around the periphery of a rough parallelogram. The aromatic rings are planar but alternate in a parallel (A) and perpendicular (B) fashion to one another around the macroring (i.e., opposite rings are parallel and adjacent rings are perpendicular to one another). The angle that the normals to the adjacent aromatic rings make with one another is 97°. The aliphatic bridges exist in two conformations, anti (torsion angle, 68°) and gauche (torsion angle, 177°). The four aromatic rings are connected to one another alternatively by these two bridges, i.e., ring A, aliphatic bridge anti; ring B, aliphatic bridge gauche; etc. One of the C atoms adjacent to the B ring in the anti bridge has a fairly large displacement from the aromatic ring plane (as in more strained cyclophanes) due to different modes of thermal vibration of the adjacent aromatic ring and the aliphatic bridge. There are some small transannular intramolecular C · · · H nonbonded interactions.

[2^6]Metacyclophane · $\frac{1}{3}$ *n*-hexane (94)[88]

								A ring	B ring
Monoclinic (*n*-hexane)	*C*2/*c*	Ni-Fil-tered CuKα	$a = 34.608(4)$	4,124.85	1.006	4	0.10	$a = 1.522; 1.536$	$a = 1.528; 1.543$
			$b = 5.334(1)$		(1.078)			$b = 1.510; 1.539$	$b = 1.539; 1.438$
			$c = 26.992(3)$					$e = 1.393$	$e = 1.393$
			$\beta = 124.1(0)$					$f = 1.382$	$f = 1.355$
								$g = 1.379$	$g = 1.374$
								$\gamma = 110.9; 112.2$	$\gamma = 109.4; 113.5$
								$\lambda = 118.2$	$\lambda = 118.2$

C ring
$a = 1.532; 1.506$
$b = 1.438; 1.510$
$e = 1.381$
$f = 1.377$
$g = 1.369$
$\gamma = 115.7; 112.3$
$\lambda = 120.9$

Comments: The overall molecule is very approximately planar, with an open central cavity that encapsulates one-third of an *n*-hexane molecule at the center of the large macroring (30 atoms). The macroring structure approximates the configuration of a C_{30} cycloalkane. There are three distinct aromatic rings (A, B, C), each of which is planar and tilted to the mean molecular plane by 166.4, 160.0, and 6.4° respectively. One of the bridging C—C bonds (C-17—C-18 between rings B and C) is unusually small for a σ bond and is ascribed to heavy thermal motion that has a predominant optical mode between these atoms. A similar situation is observed in [2^4]metacyclophane.

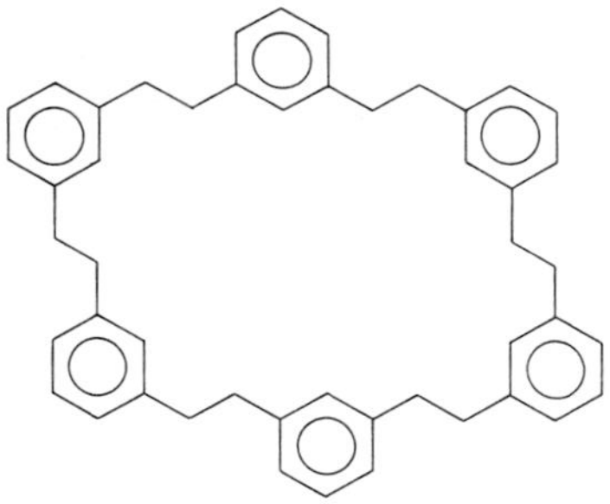

95

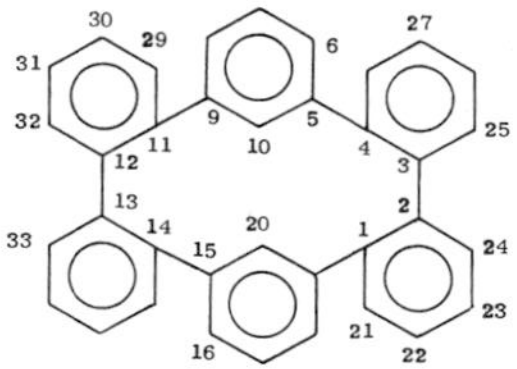

96

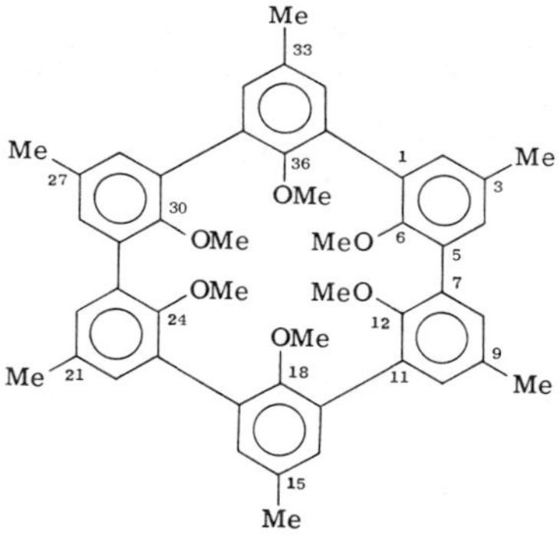

97

Crystal system	Space group	Radiation	Cell dimensions (Å)	Cell volume (Å^3)	Density (g/cm^3), calculated (measured)	Z	R	Significant bond lengths and nonbonded distances (Å), and bond and interplanar angles (deg)
				[2^6]Metacyclophane (95)[89]				
—	—	—	—	—	—	—	—	—

Comments: A preliminary report describes a variety of host–guest complexes using [2^6]metacyclophane as the host. Guests include (*a*) *p*-xylene, *o*-xylene (space group *Pnmn*; $Z = 2$); (*b*) benzene, *m*-xylene, *n*-heptane, cyclohexane, geraniol (space group $I2/a$; $Z = 4$); (*c*) squalene (space group $P2_1/n$; $Z = 2$). In the uncomplexed host no channels exist and the [2^6]metacyclophane structure is squashed. In the complexes the host forms long channels by stacking doughnut-shaped [2^6]metacyclophane molecules in which guests are included. The host is flexible and its conformation varies, giving rise to two different types of inclusion modes depending on the guest types. Those in group (*a*) have their aromatic planes parallel to the host plane, whereas those in group (*b*) are not parallel and penetrate the host cavity (see Chapter 11, Fig. 5).

1,2;3,4;11,12;13,14-Tetrabenzo[4.4]metacyclophane (96)[90]

Monoclinic	$P2_1/c$	MoKα	$a = 7.535(0)$ $b = 14.920(2)$ $c = 10.508(1)$ $\beta = 94.55(1)$	1.177(6)	1.29	2	0.042	$a = 1.488$ $b = 1.404$ $c = 1.496$ $e = 1.392$ $f = 1.392$ $g = 1.387$ $p = 3.039$ $q = 3.289$	$\beta^* = 0.034$ $\gamma = 123.8$ $\delta^* = 0.020$ $\lambda = 118.4$

Comments: The molecule can be looked on as a 14-membered ring with four ortho-fused and two meta-fused benzo rings attached to the macrocycle. Two ortho-fused rings (forming a biphenyloid unit) are separated from the other two ortho-fused rings on two sides by an intervening meta-fused ring, which forms the [4.4]metacyclophane framework. The aromatic rings are not all planar, but the mean planes of each adjacent aromatic ring form angles of about 53° with respect to one another. The meta rings are oriented anti with respect to one another and are parallel to one another with a transannular distance of 3.053 Å. These meta rings are slightly distorted into boat shapes. The aromatic rings of the biphenyloid units are planar (with nonbonded contacts between C-1 · · · C-4 and C-24 · · · C-25 of 3.288 and 2.960 Å, respectively), but the units are bent (axes C-3,C-27 and C-2,C-22 form an angle of 190.8° with respect to one another). The H atoms at C-10 and C-20 lie directly over the opposite aromatic ring and have a nonbonded contact of 2.54 Å with the H atoms at C-24, C-25, C-32, and C-33.

3,9,15,21,27,33-Hexamethyl-6,12,18,24,30,36-hexamethoxy[0^6]metacyclophane (spherand) (97)[91]

Trigonal	$R\bar{3}$	—	22°C $a = 11.697(3)$ $\alpha = 114.25(2)$	—	—	1	0.053		$\alpha^* = 0.16$

Comments: The molecule possesses $\bar{3}$ symmetry. The octahedral cavity defined by the six methoxyl O atoms, three above and three below the mean molecular plane, is 1.62 Å in diameter. The nonbonded O · · · O distances in this cavity are 2.92, 3.32, and 4.42 Å, respectively, for the pseudo-ortho, pseudo-meta, and pseudo-para interactions. These O atoms are displaced 0.20 Å from the best plane of the aromatic ring to which they are bound, and the methoxyl group plane (C—O—CH_3) makes an angle of 62° with the same plane. The distorted aromatic rings, which are alternately twisted above and below the mean molecular plane, form dihedral angles of 52° with respect to one another around the biphenyl-linking bonds. The aromatic rings are also folded 6.3° about the O—Ar—CH_3 axis.

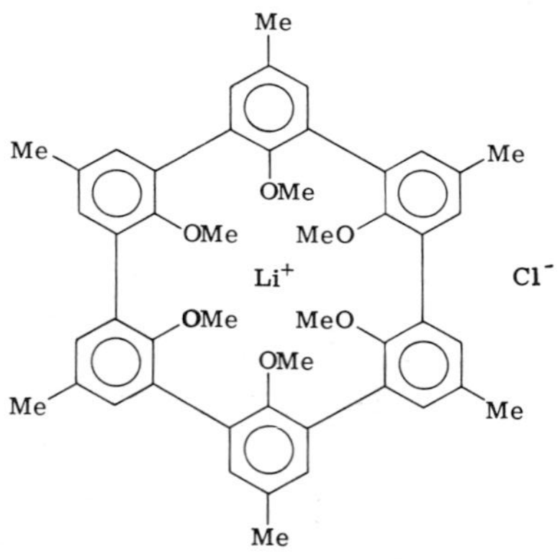

98

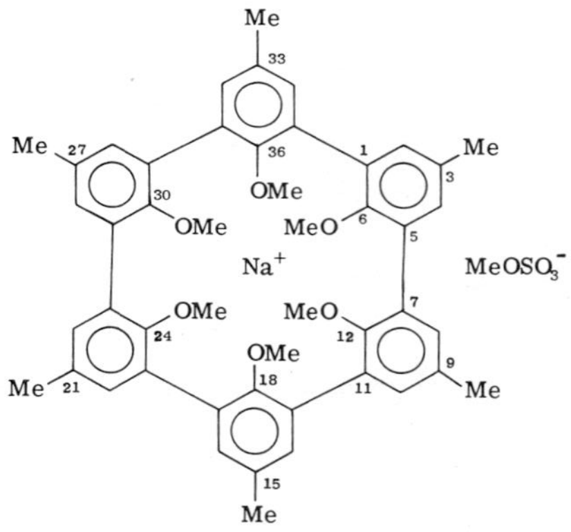

99

Crystal system	Space group	Radiation	Cell dimensions (Å)	Cell volume ($Å^3$)	Density (g/cm^3), calculated (measured)	Z	R	Significant bond lengths and nonbonded distances (Å), and bond and interplanar angles (deg)
			Lithiospherium chloride [spherand : lithium chloride complex (1 : 1)] (98)[91]					
Trigonal	$R\bar{3}$	—	22°C $a = 11.152(1)$ $\alpha = 110.60(1)$	—	—	1	0.047	$\alpha^* = 0.16$

Comments: The conformational organization of the host is similar to that of the uncomplexed spherand. It possesses a snowflakelike shape with a nearly spherical hole lined with 24 unshared electrons. The Li^+ occupies the spherand cavity and the Cl^- occupies a cavity defined by 12 methyl groups (six methoxyl methyl groups, three each above and below the ion, and six aryl methyl groups surrounding it in the median plane). The $Li^+ \cdots Cl^-$ distance is 5.2 Å along the $\bar{3}$ axis. The introduction of Li^+ into the spherand host causes a decrease in the diameter of the spherand cavity to 1.48 Å and a comparable reduction of the nonbonded O · · · O distances in the cavity to 2.78, 3.24, and 4.28 Å, respectively, for the pseudo-ortho, pseudo-meta, and pseudo-para interactions. In a similar fashion the methoxyl O atom moves closer to the mean aromatic plane to which it is attached (the out-of-plane distance being 0.07 Å as compared with 0.20 Å in the uncomplexed spherand). The dihedral angle between the methoxyl C—O—CH_3 group and the mean aromatic plane increases to 85° (from 62° in the uncomplexed spherand), whereas the dihedral angle between adjacent aromatic ring planes about the biphenyl-linking bond increases slightly to 56° (52° in the uncomplexed spherand). The folding axes about the O—Ar—CH_3 is reduced to 2.6° (compared with 6.3° in the uncomplexed spherand).

Sodiospherium methyl sulfate [spherand : sodium methyl sulfate complex (1 : 1)] (99)[91]

Monoclinic	$P2/c$	—	$a = 11.572(5)$ $b = 10.467(5)$ $c = 22.072(7)$ $\beta = 108.97(3)$	—	—	2	0.15	$\alpha^* = 0.16$

Comments: The structure is not fully refined. There is some disorder of the methyl sulfate ion and of a molecule of solvation. The general structural features are nonetheless evident. The host has the same conformational organization as that of the uncomplexed spherand, and the complex is similar in geometry to the LiCl complex (see previous entry): cavity diameter, 1.75 Å; nonbonded O · · · O distances, pseudo-ortho 3.00 Å, pseudo-meta 3.43 Å, pseudo-para 4.55 Å; displacement of O atoms from mean aromatic plane, 0.12 Å; dihedral angle between mean aromatic plane and the methoxyl group plane, 84°; twist angle between adjacent aromatic ring planes about the biphenyl link, 61°; folding angle about the O—Ar—CH_3 axes, 4.8°.

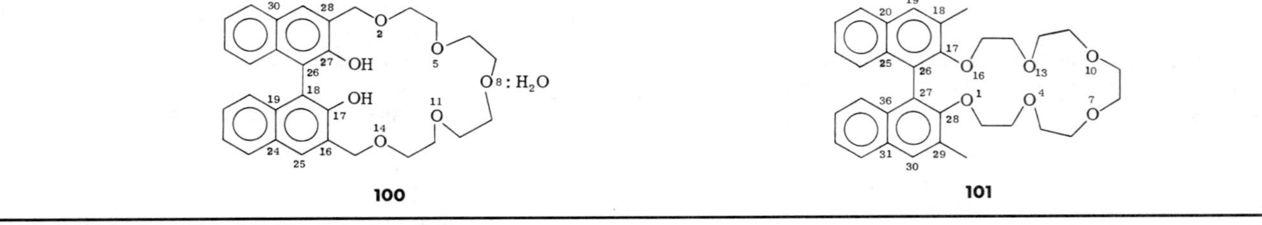

Crystal system	Space group	Radiation	Cell dimensions (Å)	Cell volume (Å^3)	Density (g/cm^3), calculated (measured)	Z	R	Significant bond lengths and nonbonded distances (Å), and bond and interplanar angles (deg)
			III. CONDENSED AROMATIC AND NONBENZENOID RING PHANES					
			17,27-Dihydroxy-2,5,8,11,14[15](3,3′)α-binaphthylophane : H_2O (100)[92]					
Monoclinic	$P2_1/c$	CuKα	$a = 12.334(3)$ $b = 10.638(2)$ $c = 21.704(4)$ $\beta = 110.81(2)$	2,662.0	1.304	4	0.065	$a = 1.504(6)$ $e_{16,17} = 1.427(5)$ $b = 1.434(6)$ $e_{16,25} = 1.374(6)$ $c = 1.425(6)$ $f_{17,18} = 1.381(5)$ $d = 1.494(7)$ $f_{24,25} = 1.411(6)$ $\gamma = 107.8(3)$ $g_{18,19} = 1.427(4)$ $\lambda = 118.8(4)$ $g_{19,24} = 1.428(5)$

Comments: The component species of the crystal associate by H bonding. The H_2O molecule is too small to fit into the macrocyclic cavity and is thus preferentially coordinated to O-18 and O-14 of the polyether bridge. This fixes atoms C-8 through C-15 into an energetically favored conformation that corresponds to synclinal torsion angles around the C—C bonds (69–71°) and antiplanar torsion angles around the C—O bonds (166–179°). The atoms in the bridge not coordinated to the H_2O exhibit an irregular conformation (probably disordered). The naphthalenoid rings are planar and are twisted ~78° with respect to one another about the binaphthyl link.

18,29-Dimethyl-1,4,7,10,13,16-hexaoxa[16](2,2′)α-binaphthylophane (101)[93,94]

Triclinic	$P\bar{1}$	MoKα	193(3)K	1,363.5	1.258	2	0.099	$a = 1.398(10)$	$e_{17,26} = 1.383(9)$
								$b = 1.447(10)$	$e_{17,18} = 1.419(12)$
			$a = 8.637(5)$					$c = 1.504(12)$	$f_{18,19} = 1.430(11)$
			$b = 11.974(7)$					$d = 1.416(15)$	$f_{25,26} = 1.374(11)$
			$c = 13.740(7)$					$\gamma = 114.2(6)$	$g_{19,20} = 1.432(11)$
			$\alpha = 104.32(4)$					$\lambda = 118.4(7)$	$g_{20,25} = 1.408(11)$
			$\beta = 85.20(4)$						
			$\gamma = 97.36(5)$						
			300K	1,392.4	1.232	2	—		
			$a = 8.738(3)$						
			$b = 12.037(5)$						
			$c = 13.771(4)$						
			$\alpha = 104.11(3)$						
			$\beta = 84.57(3)$						
			$\gamma = 96.46(3)$						

Comments: The molecule is conformationally disordered. The mean plane of the macroring forms an angle of 40° with the C-26—C-27 bond, which places one of the methyl groups on one face of the cavity.

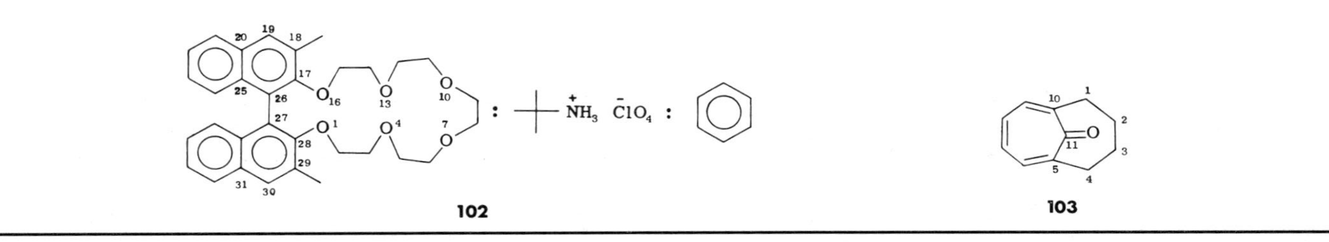

Crystal system	Space group	Radiation	Cell dimensions (Å)	Cell volume (Å³)	Density (g/cm³), calculated (measured)	Z	R	Significant bond lengths and non-bonded distances (Å), and bond and interplanar angles (deg)
18,29-Dimethyl-1,4,7,10,13,16-hexaoxa[16](2,2′)α-binaphthylophane : *tert*-butylammonium perchlorate : benzene (1 : 1 : 1) complex (102)[93,94]								
Triclinic	$P\bar{1}$	MoKα	133K	2,012.3	1.268	2	0.043	a = 1.391(5) $e_{17,26}$ = 1.372(5)
			a = 8.848(3)					b = 1.450(4) $e_{17,18}$ = 1.429(5)
			b = 11.023(5)					c = 1.500(6) $f_{18,19}$ = 1.423(5)
			c = 20.750(10)					d = 1.429(5) $f_{25,26}$ = 1.372(6)
			α = 91.80(4)					γ = 115.3(3) $g_{19,20}$ = 1.434(5)
			β = 91.41(4)					λ = 123.1(3) $g_{20,25}$ = 1.415(5)
			γ = 95.63(3)					
			300K	2,053.2	1.243	2	—	
			a = 8.902(5)					
			b = 11.117(5)					
			c = 20.885(12)					
			α = 91.69(4)					
			β = 91.07(5)					
			γ = 96.30(4)					

Comments: The molecule is ordered, and the polyether chain has an unstrained conformation with synclinal and antiplanar conformations around the C—C and C—O bonds. The O atoms point toward the center of the cavity (alternately above and below the mean O plane), and the cavity diameter is 2.8 Å. The ammonium ion, being too large to fit into the cavity, is displaced 1.0 Å from the mean O atom plane and H-bonds to three ether O atoms. The dihedral angle around the C-26—C-27 bond (i.e., the naphthyl planes) is 81.9°.

[4](2,7)Troponophane (103)[95]

Orthorhombic	$P2_12_12_1$	CuKα	a = 10.705(2)	—	—	4	0.0458	a = 1.509	α = 39.5
			b = 12.174(2)					b = 1.551	β = 61.5
			c = 6.800(1)					c = 1.527	γ = 111.4
								e = 1.479	δ = 28.9
								f = 1.348	λ = 118.0
								g = 1.447	
								i = 1.367	
								v = 1.214	

Comments: The troponoid moiety is bent into a tub conformation similar to that found in 4,5-benzo[4](2,7)troponophane. The C═O vector is bent 4.4° out of the C-5,C-10,C-11 plane toward the aliphatic bridge. The aliphatic bridge is twisted, with the C-2 atom 14.4° above and the C-3 atom 28.8° below the C-1,C-4,C-5,C-10 plane (C-2 is directed toward the O atom; see Chapter 8, Fig. 1). Preliminary X-ray data indicate that C-2 and C-3 in the dehydro derivative ([4](2,7)troponophane-2-ene) are nearly coplanar with the C-1,C-4,C-5,C-10 plane.

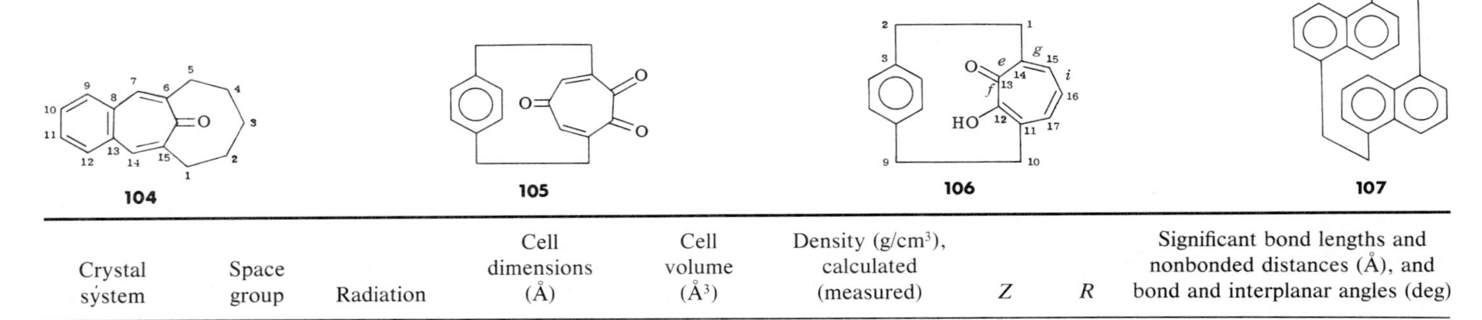

Crystal system	Space group	Radiation	Cell dimensions (Å)	Cell volume (Å³)	Density (g/cm³), calculated (measured)	Z	R	Significant bond lengths and nonbonded distances (Å), and bond and interplanar angles (deg)
			4,5-Benzo[5](2,7)troponophane (104)[96]					
Monoclinic (EtOH)	$P2_1/n$	CuKα	a = 9.803(2) b = 16.184(7) c = 7.802(2) β = 90.39(2)	1,237.7(7)	1.203 (1.200)	4	0.058	Tropono ring (meta type) a = 1.498 α = 3.70 b = 1.539 β = 51.3 c = 1.543 γ = 111.9 e = 1.476 δ = 24.3 f = 1.346 λ = 120.8 g = 1.462 i = 1.411 v = 1.207

Comments: The seven-membered ring adopts a boat form (see β and δ) because of the pentamethylene chain bridging the 2,7 positions on the ring. The C=O vector is bent out of the C-6,C-16,C-15 plane by 6.5°. The benzo moiety is bent slightly out of planarity (the interplanar angles defined by C-9, C-10,C-11,C-12 and C-12,C-13,C-8,C-9 is 2.29°), but its least-squares plane is inclined 5.8° to the C-6,C-7,C-14,C-15 plane in the direction opposite to the C=O vector. The pentamethylene bridge together with the C-6, C-15, and C-16 atoms forms an eight-membered ring, which adopts a stretched crown, chair–chair conformation.

[2.2]Paracyclo(3,7)tropoquinonophane (105)[97,98]

—	—	—	—	—	—	—	—	—

Comments: See Chapter 8, Fig. 4, for relevant angles in the tropoquinone ring. Both rings are bent out of planarity.

[2.2]Paracyclo(3,7)tropolonophane (106)[99]

Orthorhombic	*Pmnb*	CuKα	$a = 9.501(2)$ $b = 15.586(4)$ $c = 8.539(2)$	—	—	—	0.054	Para ring: $a = 1.506$, $e = 1.384$, $f = 1.380$, $\lambda = 117.2$ Tropolono ring: $a = 1.517$, $e = 1.401$, $f = 1.469$, $g = 1.388$, $i = 1.371$, $v = 1.294$, $\lambda = 130.0$ $b = 1.536$ $p = 2.772$ $q_{C \cdots C—O} = 3.190$ $q_{C \cdots C} = 3.245$

Comments: Both rings are distorted from planarity, although the tropolone ring is more distorted. Unlike tropolone, which as intermolecular H bonding in the crystalline state, this compound has a monomeric structure. The C—O bonds of the tropolone ring and the C—H bonds of the benzenoid ring are bent out of their respective aromatic planes and toward the other ring by 11.4 and 3.5°, respectively. The C-15 and C-17 atoms are bent out of the C-11,C-12,C-13,C-14 plane and away from the benzenoid ring by 25.2° and the C-16 atom is further bent in the same direction by 11.0° out of the C-11,C-14,C-15,C-17 plane (see Chapter 8, Fig. 3).

[2.2](1,5)Naphthalenophane (107)[100]

—	—	—	—	—	—	—	—	—

Comments: The naphthalenoid rings are severely distorted, with the nonbonded distance between the ring atoms varying from 2.80 to 3.57 Å. The molecule is not chiral.

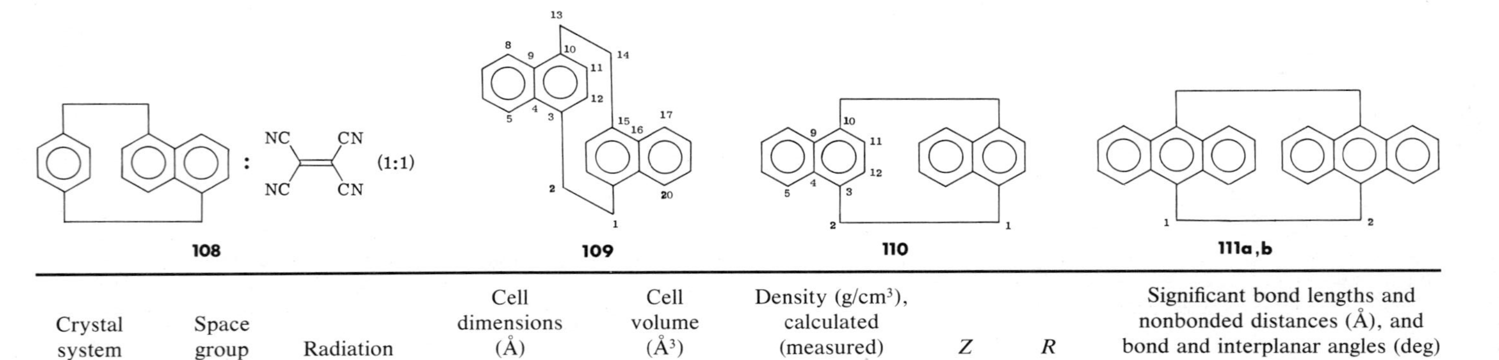

Crystal system	Space group	Radiation	Cell dimensions (Å)	Cell volume ($Å^3$)	Density (g/cm^3), calculated (measured)	Z	R	Significant bond lengths and nonbonded distances (Å), and bond and interplanar angles (deg)	
[2.2](1,5)Naphthalenoparacyclophane : TCNE (1 : 1) (108)[101]									
—	—	—	—	—	—	—	—	—	

Comments: The donor and acceptor molecules appear in the crystal in a stacked order with alternate molecules of donor and acceptor. Within the stack the TCNE molecule is practically equidistant from the neighboring naphthalenoid ring (3.464 Å) and benzenoid ring (3.457 Å).

Crystal system	Space group	Radiation	Cell dimensions (Å)	Cell volume ($Å^3$)	Density (g/cm^3), calculated (measured)	Z	R	Significant bond lengths and nonbonded distances (Å), and bond and interplanar angles (deg)	
***anti*-[2.2](1.4)Naphthalenophane (109)**[102]									
Monoclinic	$P2_1/c$	CuKα	$a = 14.63(2)$	1,606.1	1.275	4	0.155	$a = 1.508$	$\alpha = 14.4; 10.1$
			$b = 13.45(2)$		(1.277)			$b = 1.577$	$\beta = 12.2; 16.7$
			$c = 8.17(2)$					$e_{10,11} = 1.402$	$\gamma = 113.1$
			$\beta = 91.7(2)$					$e_{9,10} = 1.453$	$\lambda = 115.7$
								$f_{11,12} = 1.403$	$\varphi = 167$
								$f_{4,9} = 1.419$	
								$p = 2.79$	
								$q = 3.10$	

Comments: The bridged portion of the naphthalenoid ring is boat-shaped. Atoms C-3 and C-10 are about 0.107 Å out of the C-4,C-9,C-11,C-12 plane. There is a translation of about 0.3 Å along the long axis of the naphthalenoid rings such that the two rings approach one another and overlap more. This movement tends to remove eclipsing interactions of the central ring atoms and nonbonded interaction between the H atoms at C-5, C-8, C-17, and C-20 with those of the bridge on C-1, C-2, C-13, and C-14. No twisting of the naphthalenoid rings with respect to one another is observed

syn-[2.2](1,4)Naphthalenophane (110)[102]

Monoclinic	$P2_1/c$	—	$a = 8.37(6)$ $b = 11.86(9)$ $c = 16.85(9)$ $\beta = 105.39$	1,613.01	—	—	—	$a = 1.514$ $b = 1.570$ $e_{10,11} = 1.367$ $e_{3,12} = 1.429$ $f_{11,12} = 1.404$ $f_{9,4} = 1.472$ $p = 2.77$ $q = 3.09$	$\gamma = 113.1$ $\lambda = 117.2$ $\varphi = 163$

Comments: The bridged portion of the naphthalenoid ring is distorted from planarity into a boat shape. The outer portion of the naphthalenoid ring is about 18° from being coplanar with the least-squares plane of the bridged portion. The rings are twisted with respect to one another by approximately 15° (about the normal to the mean planes of the bridged portion of the naphthalenoid ring and centered in those rings), which allows for the removal of transannular eclipsing of the naphthalenoid ring atoms and removal of eclipsing of the bridge H atoms.

[2.2](9,10)Anthracenophane (α form) (111a)[103]

Monoclinic ($CHCl_3$)	$P2_1/c$	CuKα	$a = 8.28$ $b = 13.26$ $c = 21.55$ $\beta = 116$	—	1.314	4	—	—	

[2.2](9,10)Anthracenophane (β form) (111b)[104]

Monoclinic	$P2_1/c$	CuKα	$a = 10.277(1)$ $b = 12.794(2)$ $c = 8.466(1)$ $\beta = 112.89(1)$	1,025.4	1.322	2	0.040	$a = 1.520$ $b = 1.565$ $e = 1.406$ $f = 1.438$ $p = 2.772$ $q = 3.169$	$\gamma = 112.4$ $\lambda = 118.0$ $\varphi = 160.9$ $\omega \cong 78$

Comments: The anthracene rings are laterally displaced with respect to their long axes by 0.6 Å to minimize transannular π–π contact. As a result the C-1—C-2 bond is twisted, giving rise to a torsional angle (C_{ring}—C-1—C-2—C_{ring}) of 16°. This twist also reduces eclipsing interactions of the H atoms on the bridge C atoms.

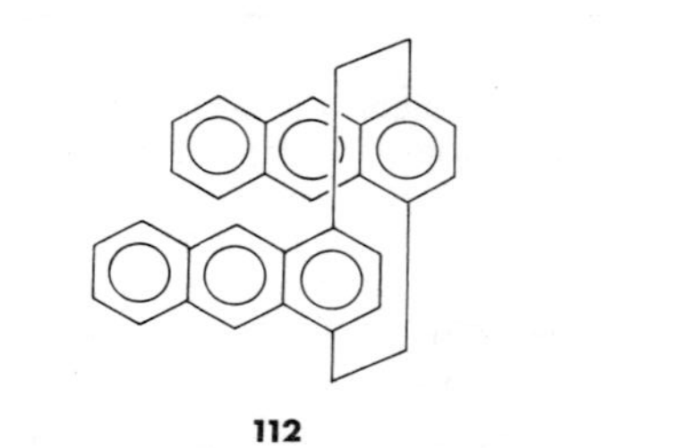

112

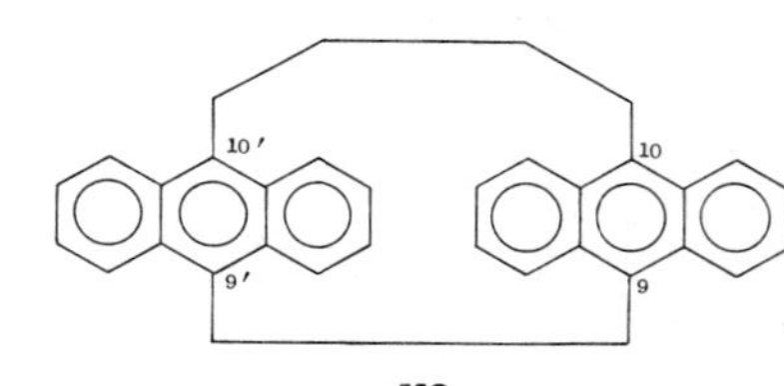

113

Crystal system	Space group	Radiation	Cell dimensions (Å)	Cell volume (Å^3)	Density (g/cm^3), calculated (measured)	Z	R	Significant bond lengths and nonbonded distances (Å), and bond and interplanar angles (deg)	
[2.2](1,4)Anthracenophane (112) [105,106]									
Monoclinic	$A2$	MoKα	−149°C	—	—	8	0.055	Molecule A	Molecule B
			$a = 25.539(2)$					$a = 1.517$	$a = 1.512$
			$b = 8.152(1)$					$b = 1.593$	$b = 1.594$
			$c = 20.561(3)$					$e_{outer} = 1.365$	$e_{outer} = 1.370$
			$\beta = 106.03(1)$					$e_{inner} = 1.442$	$e_{inner} = 1.452$
								$f_{outer} = 1.420$	$f_{outer} = 1.418$
								$f_{inner} = 1.447$	$f_{inner} = 1.440$
								$\alpha = 13.2$	$\alpha = 13.4$
								$\beta = 9.8$	$\beta = 10.3$
								$\gamma = 112.8$	$\gamma = 112.8$
								$\omega = \sim 84$	$\omega = \sim 78$
								$\lambda = 117.8$	$\lambda = 117.3$
								$p = 2.769$	$p = 2.785$
								$q_{outer} = 3.099$	$q_{outer} = 3.092$
								$q_{inner} = 3.096$	$q_{inner} = 3.147$

Comments: The analysis showed two distinct anthracenophane molecules (A and B) in each asymmetric unit of the crystal lattice. The least-squares planes of the distant six-membered rings form angles of inclination (θ) of 9.1 and 5.1° in A and B, respectively. The bridged six-membered rings form angles of inclination (θ) of 0.2 and 1.4° in A and B, respectively. The nonbonded aromatic C · · · C distances increase toward the nonbridged end of the anthracene nuclei. The molecule undergoes photoisomerization where the 9,10 positions of the two anthracene rings form bonds. The following crystal information is available for the photoproduct, which is also a cyclophane:

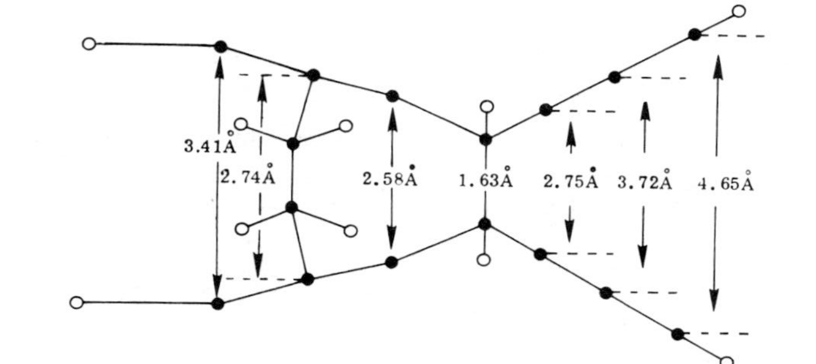

[2.4](9,10)Anthraceneophane (113)[107]

Monoclinic (chlorobenzene/ petroleum ether)	$P2_1/c$	—	a = 10.551(1) b = 26.512(3) c = 8.952(1) β = 114.735(8)	—	—	4	0.041	φ = ~176.9

Comments: The molecule shows a disordered structure resulting from the superposition of two molecular conformations that differ in the ethane bridge conformation. Nonbonded distances between C-9 · · · C-9′ and C-10 · · · C-10′ are 2.797(3) and 3.874(3) Å, respectively. The anthrocenoid moieties are distorted by folding (2.4 and 3.8° about the C-9 · · · C-10 axis) and twisting (about an axis normal to the central rings) of the middle rings in both conformers (as is found in other cyclophane systems). The torsional angle in the four-carbon bridge is <4°.

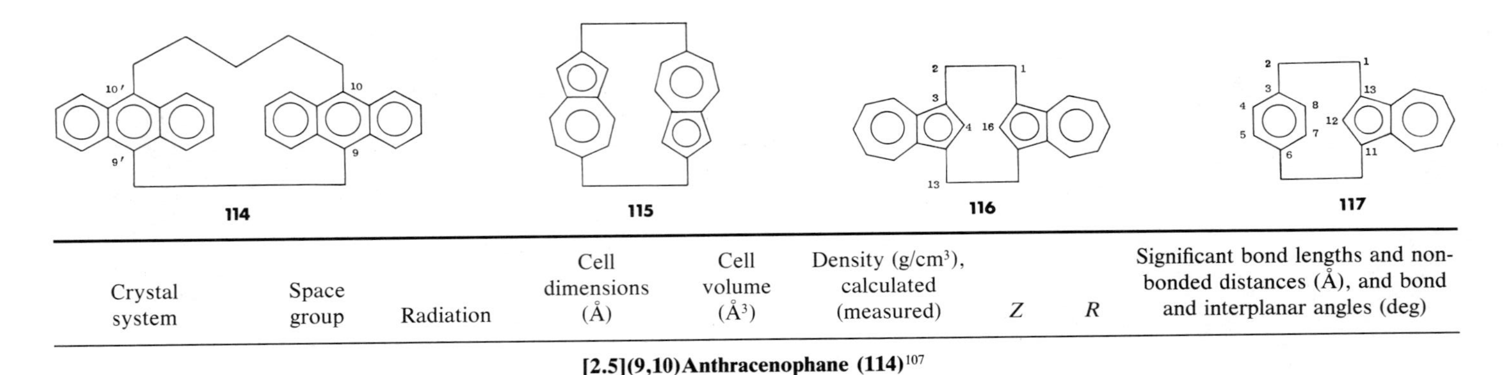

Crystal system	Space group	Radiation	Cell dimensions (Å)	Cell volume (Å^3)	Density (g/cm^3), calculated (measured)	Z	R	Significant bond lengths and nonbonded distances (Å), and bond and interplanar angles (deg)
[2.5](9,10)Anthracenophane (114)[107]								
Monoclinic (chlorobenzene/ petroleum ether)	$P2_1/c$	—	$a = 10.607(3)$ $b = 26.881(8)$ $c = 9.499(2)$ $\beta = 115.87(2)$	—	—	4	0.052	—

Comments: The molecule shows a disordered structure resulting from the superposition of two molecular conformations that differ in the ethane bridge conformation. Nonbonded distances between C-9 · · · C-9′ and C-10 · · · C-10′ are 2.900(5) and 4.647(5) Å, respectively. The anthracene moieties are distorted by folding about the C-9 · · · C-10 axis and by twisting about an axis normal to the central rings of the anthracenoid units in both conformers (as is found in other cyclophane systems).

Crystal system	Space group	Radiation	Cell dimensions (Å)	Cell volume (Å^3)	Density (g/cm^3), calculated (measured)	Z	R	Significant bond lengths and nonbonded distances (Å), and bond and interplanar angles (deg)
[2.2](2,6)Azulenophane (115)[108]								
—	$P4_2/mnm$	—	—	—	—	2	—	$p = 2.8$ $q = 3.2$ $r = 3.7$

Comments: Crystals of syn or anti isomers or the mixture were disordered. The structure was solved using a crystal of the mixture of syn and anti isomers assuming an anti orientation of the rings. It clearly demonstrated that the azuleno rings were bent into a bow shape, which is common among cyclophanes.

[2.2](1,3)Azulenophane (116)

Monoclinic	$P2_1/a$	CuKα	$a = 7.714(1)$ $b = 12.612(1)$ $c = 8.96(1)$ $\beta = 109.93(1)$	—	—	2	0.071	$a = 1.514$ $b = 1.571$ $e = 1.398$ $f = 1.408$ $g = 1.510$ $p = 2.839$ $q = 2.744$	$\alpha = 19.3$ $\beta = 5.3$ $\gamma = 110.6$ $\delta = 3.6$ $\varphi = 174.7$ $\lambda = 107.7$

Comments: The azulene rings are slightly distorted from planarity. The C-4—H and C-16—H bonds are bent out of the planes of the nearest three C atoms toward the opposite ring by 9° (see Chapter 8, Fig. 5).

[2.2](1,3)Azulenoparacyclophane (117)[110]

								Para ring	1,3-Azuleno ring (meta type)
Monoclinic	$P2_1/a$	CuKα	$a = 9.023(1)$ $b = 19.980(2)$ $c = 8.259(1)$ $\beta = 109.02(1)$	—	—	4	0.068	$a = 1.517$ $e = 1.403$ $f = 1.386$ $\alpha = 13.8$ $\beta = 14.4$ $\gamma = 109.2$ $\varphi = 164.1$ $\lambda = 116.5$	$a = 1.511$ $e = 1.395$ $f = 1.410$ $g = 1.507$ $\alpha = 13.3$ $\beta = 6.3$ $\gamma = 112.5$ $\delta = 1.2$ $\varphi = 173.7$ $\lambda = 107.3$

Common to both rings: $b = 1.579$; $\theta = \sim 8$; $p = 2.736$; $q_{\text{C-12}\cdots\text{C-7—C-8}} = 3.014$; $q_{\text{C-12}\cdots\text{C-4—C-5}} = 3.312$

Comments: Both rings are nonplanar, although the benzenoid ring is more so than the azuleno ring. The C—H bonds on the benzenoid ring are bent toward the azulenoid ring, although the inner C—H bonds are less bent (7°) out of the C-3,C-8,C-7,C-6 plane than the outer C—H bonds (18°) out of the C-3,C-4,C-5,C-6 (see Chapter 8, Fig. 5).

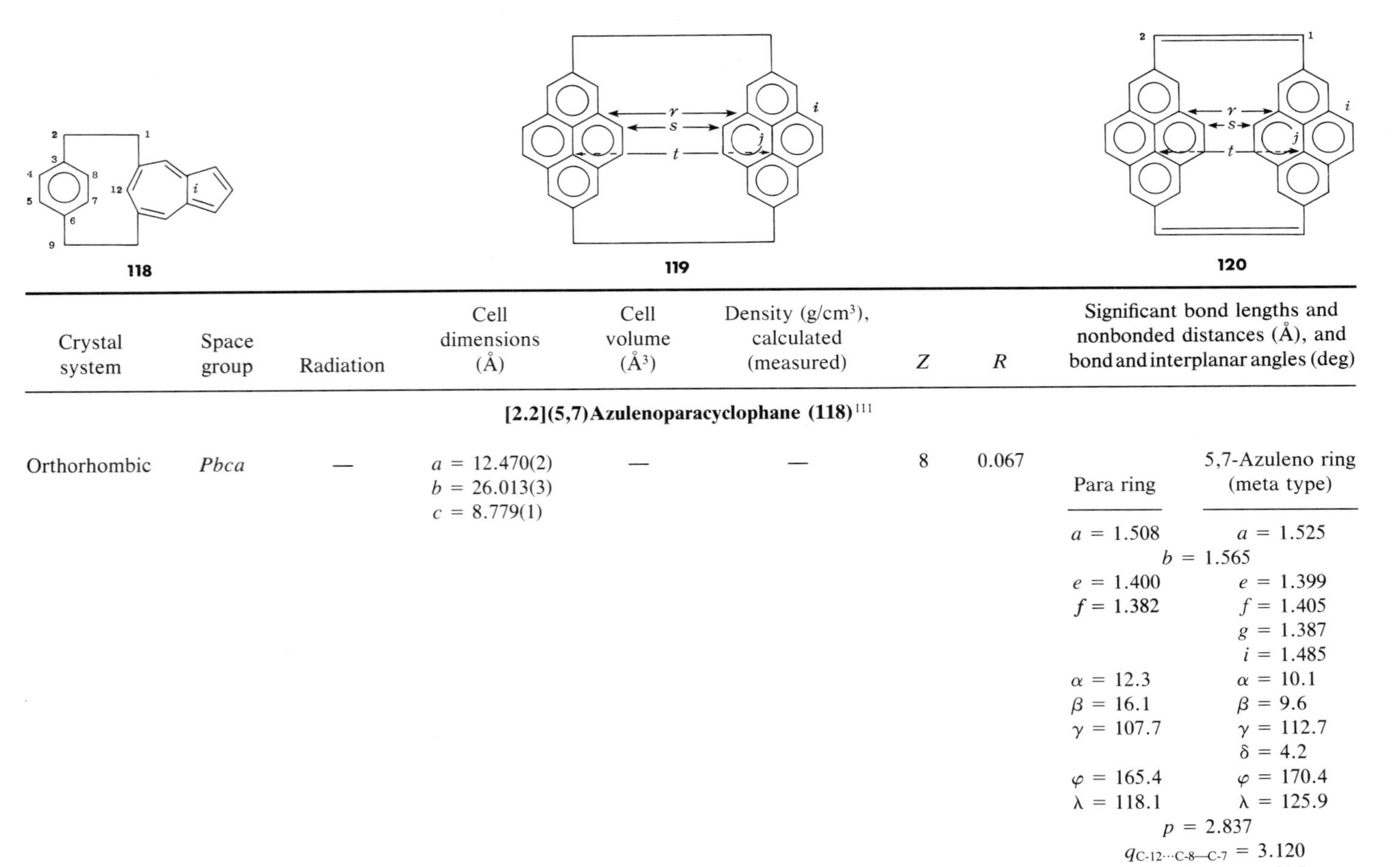

Crystal system	Space group	Radiation	Cell dimensions (Å)	Cell volume (Å^3)	Density (g/cm^3), calculated (measured)	Z	R	Significant bond lengths and nonbonded distances (Å), and bond and interplanar angles (deg)	
[2.2](5,7)Azulenoparacyclophane (118)[111]									
Orthorhombic	*Pbca*	—	$a = 12.470(2)$ $b = 26.013(3)$ $c = 8.779(1)$	—	—	8	0.067	Para ring	5,7-Azuleno ring (meta type)
								$a = 1.508$	$a = 1.525$
								$b = 1.565$	
								$e = 1.400$	$e = 1.399$
								$f = 1.382$	$f = 1.405$
									$g = 1.387$
									$i = 1.485$
								$\alpha = 12.3$	$\alpha = 10.1$
								$\beta = 16.1$	$\beta = 9.6$
								$\gamma = 107.7$	$\gamma = 112.7$
									$\delta = 4.2$
								$\varphi = 165.4$	$\varphi = 170.4$
								$\lambda = 118.1$	$\lambda = 125.9$
								$p = 2.837$	
								$q_{C\text{-}12\cdots C\text{-}8—C\text{-}7} = 3.120$	
								$q_{C\text{-}12\cdots C\text{-}4—C\text{-}5} = 3.225$	

are bent toward the inside of the cyclophane cavity from their respective aromatic planes by 4, 11, and 7°, respectively (see Chapter 8, Fig. 5).

[2.2](2,7)Pyrenophane (119)[112,113]

Monoclinic (toluene)	$P2_1/c$	—	a = 9.038(1) b = 13.141(1) c = 9.734(2) β = 97.93(1)	—	—	2	0.048	a = 1.514 b = 1.568 e = 1.390 f = 1.395 g = 1.419 i = 1.434 j = 1.419	p = 2.79 q = 3.19 r = 3.65 s = 3.80 t = 3.76

Comments: The pyrenoid portion of the cyclophane is bow-shaped, and the pyrenoid units are not planar.

[2.2](2.7)Pyrenophane-1,19-diene (120)[114]

Monoclinic	$P2_1/c$	—	−160°C a = 9.829(2) b = 12.631(2) c = 10.001(2) β = 113.82(1)	1,135.9(4)	1.323	2	0.064	a = 1.497 b = 1.344 e = 1.397 f = 1.402 g = 1.420 i = 1.440 j = 1.427	α = 7.1 β = 10.8 γ = 118.9 λ = 119.2
			20°C a = 9.889(2) b = 12.865(3) c = 9.965(2) β = 113.96(1)	1,158.6(1)	1.297 (1.298)	2	—		p = 2.790 q = 3.230 r = 3.763 s = 3.881 t = 3.903

Comments: The pyrenoid units are similar to those found in [2.2](2,7)pyrenophane and have a long shallow-boat shape. The two pyrenoid moieties do not eclipse one another and are slightly shifted from an exactly overlapping position. The torsional angle around the C-1=C-2 bond is 4.2°.

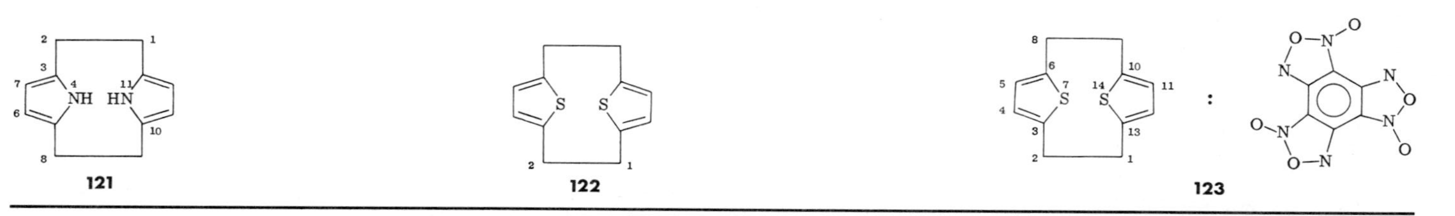

Crystal system	Space group	Radiation	Cell dimensions (Å)	Cell volume ($Å^3$)	Density (g/cm^3), calculated (measured)	Z	R	Significant bond lengths and nonbonded distances (Å), and bond and interplanar angles (deg)

IV. HETEROPHANES

[2.2](2,5)Pyrrolophane (121)[115]

Crystal system	Space group	Radiation	Cell dimensions (Å)	Cell volume ($Å^3$)	Density (g/cm^3), calculated (measured)	Z	R	Significant bond lengths and nonbonded distances (Å), and bond and interplanar angles (deg)
Monoclinic (sublimed)	$P2_1/c$	MoKα	80K $a = 7.492(1)$ $b = 8.492(1)$ $c = 8.192(1)$ $\beta = 116.34(1)$	—	—	2	0.0397	$a = 1.501(2)$ $\alpha = 18$ $b = 1.572(2)$ $\beta = 6$ $e = 1.374(2)$ $\gamma = 110.0$ $f = 1.376(2)$ $\lambda = 106.9(1)$ $g = 1.423(2)$ $p = 2.55$

Comments: The pyrrole rings are nearly planar and are anti-parallel to one another. The H atoms on the pyrrole ring N atom are bent toward the opposite aromatic ring. This H atom is 2.43 Å from the least-squares plane of the proximate ring, whereas the N atom to which the H atom is bound is 2.62 Å from that plane. The C-6 and C-7 atoms are 2.58 Å from that plane.

[2.2](2,5)Thiophenophane (122)[116]

Monoclinic	$P2_1/c$	MoKα	$a = 6.052(6)$	532.9	1.37	2	0.030	$a = 1.512$	$\alpha^* = 0.52$
			$b = 12.63(1)$		(1.38)			$b = 1.592$	$\beta^* = 0.20$
			$c = 7.111(2)$					$e = 1.728$	$\gamma = 112.9$
			$\beta = 101.34(7)$					$f = 1.369$	$\lambda = 109.53$
								$g = 1.435$	$\omega = 65$
								$p = 2.63$	
								$q = 3.225$	

Comments: The molecule has a stepped-anti orientation of the heteroaromatic rings. The thiophenoid rings are envelope-shaped, with the S atoms being displaced 0.196 Å out of the plane of the other four atoms of the aromatic ring. The latter planes are parallel to one another. The conformation around the bridging C-1—C-2 bond is intermediate between gauche and eclipsed, the torsion angle being 34.4°.

[2.2](2,5)Thiophenophane : benzotrifuroxan (1 : 1) (123)[117]

Triclinic	$P\bar{1}$	CuKα	$a = 9.71(3)$	996.7	1.574	2	0.137	Ring A[‡]	Ring B[‡]
			$b = 8.01(3)$		(1.566)				
			$c = 15.28(3)$					$a = 1.46$	$a = 1.47$
			$\alpha = 102.2(2)$					$b = 1.61$	
			$\beta = 96.2(2)$					$e = 1.78$	$e = 1.74$
			$\gamma = 117.9(2)$					$f = 1.44$	$f = 1.37$
								$g = 1.42$	$g = 1.40$
								$\lambda = 1.03.9$	$\lambda = 110.0$
								$\gamma = 109.4$	$\gamma = 113.5$
								$p = 2.81$	
								$q = 3.19$	

Comments: The thiophenophane portion of the complex has an anti-stepped structure with the least-squares plane of the two rings being parallel. The thiophenoid rings are slightly envelope-shaped, with C-3 and C-6 being about 0.08–0.1 Å from the plane of the C-4, C-5, and S-7 atoms. The two thiophenoid rings are not crystallographically equivalent. One is more closely associated with the benzo portion of the benzotrifuroxanoid moiety, and the other is more closely associated with the furoxanoid portion of the benzotrifuroxanoid moiety. The thiophenoid rings and the benzotrifuroxanoid rings are inclined 9.5 ° with respect to one another, and the complex is made up of plane-to-plane stacks in which the thiophenophane and benzotrifuroxan alternate with one another in the stacks.

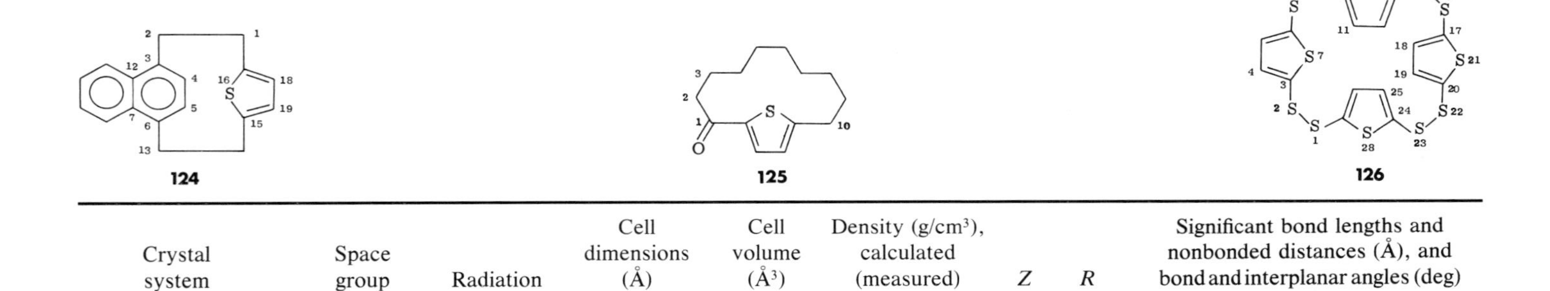

Crystal system	Space group	Radiation	Cell dimensions (Å)	Cell volume (Å³)	Density (g/cm³), calculated (measured)	Z	R	Significant bond lengths and nonbonded distances (Å), and bond and interplanar angles (deg)
			[2.2](2,5)Thiopheno(1,4)naphthalenophane (124)[118]					
Monoclinic (EtOH/H_2O)	$P2_1/c$	MoKα	$a = 9.677(3)$ $b = 11.233(3)$ $c = 13.117(4)$ $\beta = 111.73(3)$	1,342.4	1.30 (1.30)	4	0.067	see below

Para ring	Thiophenoid ring (meta type)
$a = 1.505(11)$	$a = 1.505(5)$
$b = 1.571(6)$	
$e_{3,4} = 1.366(5)$	$e = 1.721(4)$
$e_{3,12} = 1.437(4)$	
$f_{4,5} = 1.403(3)$	$f = 1.362(5)$
$f_{7,12} = 1.435(3)$	
	$g = 1.417(5)$
$\alpha = 15.6$	$\alpha = 2.9$
$\beta = 11.3$	$\beta = 0$
$\gamma = 112.2(3)$	$\gamma = 113.4(3)$
$\lambda = 117.2(3)$	$\lambda = 109.6(2)$
$\varphi = 19.2$	
$p = 2.86$	
$\theta = 0$	

Comments: The thiophenoid ring is planar, whereas the naphthalene ring is bent about the C-3—C-6 axis. The thiophenoid ring is parallel to the least-squares plane of the naphthalenoid ring. The nonbonded distances between the S atom and C-18 and C-19 atoms and the least-squares plane of the bridged portion of the naphthalenoid ring are 2.93 and 2.49 Å, respectively. The H atoms on C-4 and C-5 are bent toward the thiophenoid ring.

[10](2,5)Thiophenophan-1-one

Tetragonal	$P4_2/b$	MoKα	$a = 17.226(7)$ $c = 9.079$	—	1.15 (1.16)	8	**0.141**	Keto bridge	CH_2 bridge
								$a = 1.46(2)$ $b = 1.53(2)$ $c = 1.53(2)$ $d = 1.57(2)$ $e = 1.73(1)$ $f = 1.38(2)$	$a = 1.54(2)$ $b = 1.55(2)$ $c = 1.56(2)$ $d = 1.56(2)$ $e = 1.73(1)$ $f = 1.39(2)$
								$g = 1.44(2)$	
								$\alpha^* = 0.25$ $\gamma = 110(1)$ $\lambda = 111(1)$	$\alpha^* = 0.15$ $\gamma = 125(1)$ $\lambda = 112(1)$

Comments: The angle between the carbonyl group plane and the thiophenoid ring plane is 18.4°.

1,2,8,9,15,16,22,23-Octathia[2^4](2,5)thiophenophane (126)[120]

Orthorhombic ($CHCl_3$/toluene); a second crystal type was isolated but was twinned and was subjected to preliminary studies only	*Pnma*	—	$a = 10.642(3)$ $b = 19.409(4)$ $c = 11.074(3)$	—	1.71	4	0.104	$a = 1.70$ $b = 2.08$ $e = 1.73$ $f = 1.37$ $g = 1.43$	$\lambda = 110.8$
Monoclinic ($CHCl_3$/toluene)	($P21/c$; $P2/c$; Pc)	—	$a = 11.78$ $b = 10.72$ $c = 9.78$ $\beta = 109.0$	—	1.69 (1.63)	2	—		—

Comments: Three of the four thiophenoid rings are rotated so that their S atoms are pointing toward the outer periphery of the macrocycle, whereas the fourth thiophenoid ring is rotated so that its S atom is pointing toward the inside of the macrocycle. The angles that the thiophenoid ring planes make with a plane crossing the midpoints of the disulfide bridge bonds range from 8.8 to 31.8°. The torsional angles around the four disulfide bridge bonds range from a strained value of 60.6° to a favorable value of 89.2°. Nonbonded distances between C-20 · · · C-24, C-19 · · · C-24, and C-19 · · · C-25 are 3.33, 343, and 3.43 Å, respectively. The C-4 · · · C-20 nonbonded distance is rather short (3.36 Å), despite the transannular nature of this interaction over a rather large macrocycle.

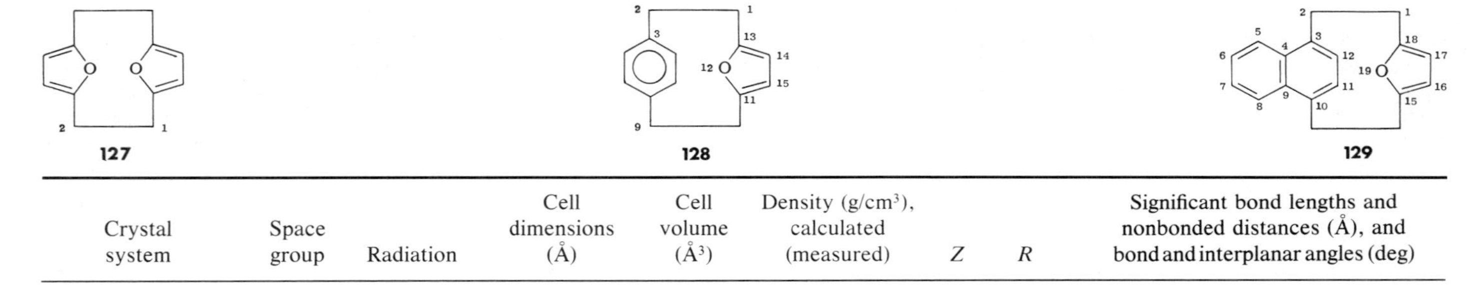

Crystal system	Space group	Radiation	Cell dimensions (Å)	Cell volume ($Å^3$)	Density (g/cm^3), calculated (measured)	Z	R	Significant bond lengths and nonbonded distances (Å), and bond and interplanar angles (deg)
[2.2](2,5)Furanophane (127)[116]								
Orthorhombic	*Pbca*	MoKα	$a = 13.60(1)$ $b = 7.507(7)$ $c = 9.725(7)$	992.9	1.26 (1.28)	4	**0.041**	$a = 1.505$, $\alpha^* = 0.40$ $b = 1.569$, $\beta^* = 0.08$ $e = 1.390$, $\gamma = 111.4$ $f = 1.351$, $\lambda = 109.09$ $g = 1.451$, $\omega = 53$ $p = 2.34$ $q = 2.511$

Comments: The molecule has a stepped-anti orientation of the heteroaromatic rings. The furanoid rings are envelope-shaped with the O atoms displaced 0.085 Å from the plane of the other four atoms of the aromatic ring. The latter planes are parallel to one another. The conformation around the C-1—C-2 bond is nearly gauche, the torsion angle being 50.7°.•

Crystal system	Space group	Radiation	Cell dimensions (Å)	Cell volume ($Å^3$)	Density (g/cm^3), calculated (measured)	Z	R	Para ring	Furanoid ring (meta type)
[2.2](2,5)Furanoparacyclophane (128)[118]									
Tetragonal	$P\bar{4}2_1c$	MoKα	$a = 15.990(5)$ $c = 8.373(2)$	2,140.3	1.227 (1.23)	8	0.077	$a = 1.507(12)$ $b = 1.572(14)$ $e = 1.392(13)$ $f = 1.373(13)$	$a = 1.479(13)$ $e = 1.366(10)$ $f = 1.342(13)$ $g = 1.440(13)$

$\alpha = 13.2$	$\alpha = 7.6$
$\beta = 16.8$	$\beta = 0$
$\gamma = 108.8(8)$	$\gamma = 115.0(8)$
$\lambda = 117.6(8)$	$\lambda = 108.6(8)$
$\varphi = 15.7$	
$p = 2.83$	
$\theta = 19.3$	

Comments: The furanoid ring is planar, whereas the benzenoid ring is boat-shaped. The nonbonded distances between the O atom and C-14 and C-15 and the least-squares plane of the aromatic ring are 2.58 and 3.25 Å, respectively.

[2.2](2,5)Furano(1,4)naphthalenophane (129)[121]

								Naphthalenoid ring	Furanoid ring (meta type)
Orthorhombic (benzene–hexane, 3 : 2)	*Pbca*	Ni-Filtered CuKα	$a = 7.859(2)$ $b = 11.482(3)$ $c = 28.818(8)$	—	1.269 (1.26)	8	0.101	$a = 1.494(15)$	$a = 1.490(13)$
								$b = 1.574(16)$	
								$e_{3,4} = 1.431(11)$	$e = 1.382(13)$
								$e_{3,12} = 1.367(13)$	$f = 1.366(13)$
								$f_{11,12} = 1.394(14)$	$g = 1.392(14)$
								$f_{4,9} = 1.430(10)$	
								$\alpha = 14$	$\alpha = 7$
								$\beta = 16$	$\beta = \sim 0$
								$\gamma = 108.9$	$\gamma = 113.5$
								$\lambda = 117.4$	$\lambda = 108.6$
								$\theta = 23$	

Comments: The furanoid ring is planar and oriented anti to the naphthalenoid ring. The nonbridged portion of the naphthalenoid ring is planar, but the bridged portion is bent into a boat shape. The aromatic rings are inclined 23° (θ) with respect to one another, which staggers the atoms of the aromatic rings and prevents them from being directly above one another. Whereas the O atom is 2.60 Å from the least-squares plane of the bridged portion of the naphthalenoid ring, C-16 and C-17 are 3.43 Å from this plane. Atoms C-15 and C-18 are 2.91 Å from the same plane. The C-11 and C-12 H atoms are displaced out of the C-10,C-11,C-12,C-3 plane and are bent toward the furanoid ring. The nonbridged portion of the naphthalenoid ring (C-4, C-5, C-6, C-7, C-8, C-9) is inclined slightly toward the inner part of the cyclophane macrocycle.

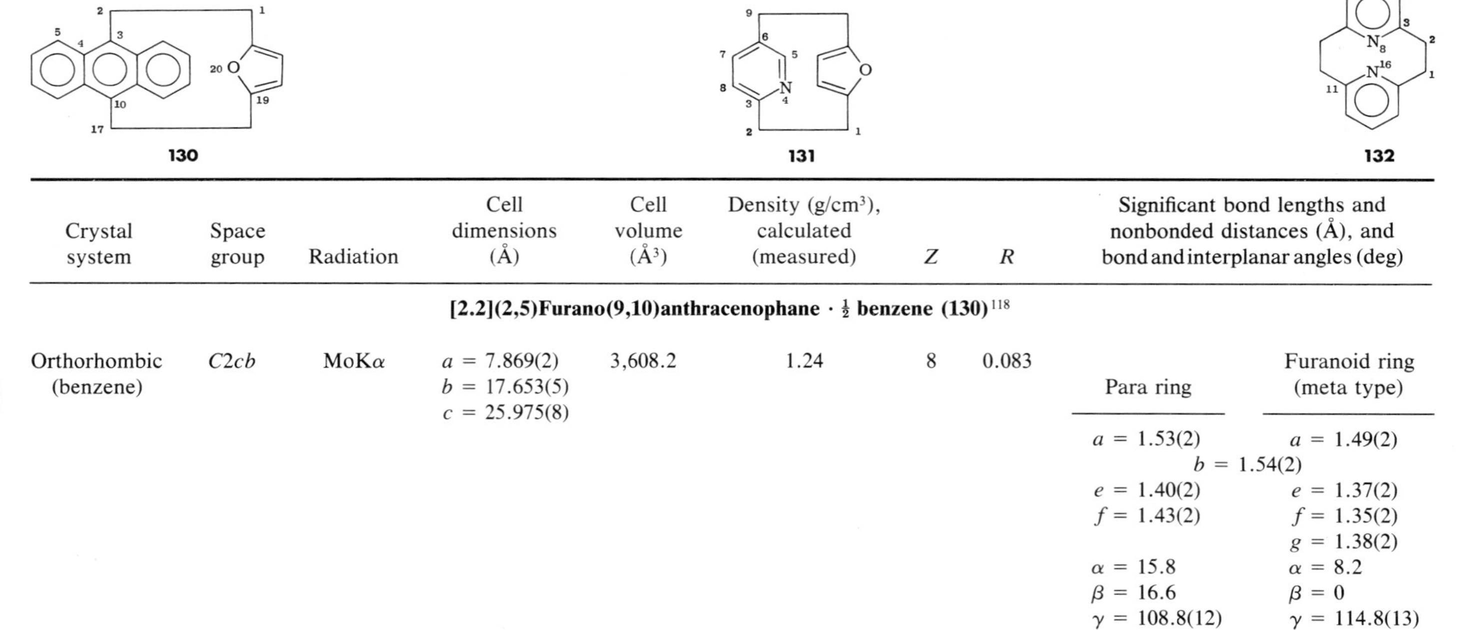

Crystal system	Space group	Radiation	Cell dimensions (Å)	Cell volume ($Å^3$)	Density (g/cm^3), calculated (measured)	Z	R	Significant bond lengths and nonbonded distances (Å), and bond and interplanar angles (deg)
[2.2](2,5)Furano(9,10)anthracenophane · ½ benzene (130)[118]								
Orthorhombic (benzene)	$C2cb$	MoKα	$a = 7.869(2)$ $b = 17.653(5)$ $c = 25.975(8)$	3,608.2	1.24	8	0.083	Para ring: $a = 1.53(2)$; $b = 1.54(2)$; $e = 1.40(2)$; $f = 1.43(2)$; $\alpha = 15.8$; $\beta = 16.6$; $\gamma = 108.8(12)$; $\lambda = 118.3(11)$; $\varphi = 18.4$ Furanoid ring (meta type): $a = 1.49(2)$; $b = 1.54(2)$; $e = 1.37(2)$; $f = 1.35(2)$; $g = 1.38(2)$; $\alpha = 8.2$; $\beta = 0$; $\gamma = 114.8(13)$; $\lambda = 105.4$ $p = 3.10$ $\theta = 23.3$

Comments: The furanoid ring is planar, whereas the anthracenoid ring is butterfly-shaped and bent about the C-3—C-10 axis ($\varphi = 18.4°$). The nonbonded distances between the O atom and the C-22,C-23 atoms, and the least-squares plane of the central anthracene ring are 2.72 and 3.55 Å, respectively. The molecule crystallizes with four solvent molecules (benzene) per unit cell.

[2.2](2,5)Furano(2,5)pyridinophane (131)[122]

								Pyridino ring (para type)	Furano ring (meta type)
Monoclinic	*Cc*	MoKα	23(1)°C	1,108.07	1.30	4	0.032		
			$a = 10.187(3)$						
			$b = 12.618(3)$					$a_{2,3} = 1.505(4)$	$a = 1.496(4)$
			$c = 8.311(2)$					$a_{6,9} = 1.408(4)$	
			$\beta = 107.64(1)$					$b = 1.575(5)$	
								$e_{3,4} = 1.346(4)$	$e = 1.381(3)$
								$e_{3,8} = 1.385(4)$	
								$f_{4,5} = 1.336(4)$	$f = 1.344(4)$
								$f_{7,8} = 1.380(4)$	
								$g = 1.389(4)$	$g = 1.434(4)$
								$\alpha^* = 0.172$	
									$\beta^* = 0.02$
								$\gamma = 108.5(2)$	$\gamma = 114.1(2)$
								$\lambda = 118.7(3)$; 116.7(3)	$\lambda = 108.9(2)$
								$\theta = 23$	

Comments: Neither the pyridino ring nor the furano ring is planar, but their least-squares planes are inclined 23° (θ) with respect to one another. The O and N atoms are syn with respect to one another. The O atom is positioned 2.611 Å from the N-4,C-5,C-7,C-8 plane. The N · · · O nonbonded distance is 2.843 Å, whereas the O · · · C-5 nonbonded distance is 2.812 Å.

[2.2](2,6)Pyridinophane (132)[123]

Orthorhombic	*Pbca*	MoKα	$a = 13.49(8)$	1,141.0	1.22	4	0.047	$a = 1.510$	$\alpha^* = 0.34$
			$b = 7.525(7)$		(1.23)			$b = 1.581$	$\beta^* = 0.11$
			$c = 11.24(8)$					$e = 1.350$	$\gamma = 110.5$
								$f = 1.395$	$\delta^* = 0.05$
								$g = 1.384$	$\lambda = 121.6$
									$\omega = 46$
								$p = 2.22$	
								$q = 2.545$	

Comments: The pyridinoid rings are arranged in an anti-stepped geometry, and are slightly boat-shaped.

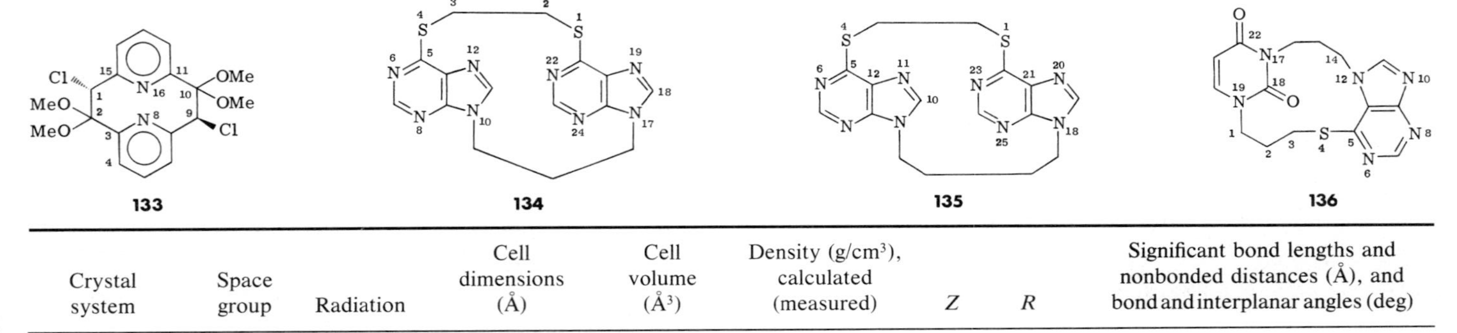

Crystal system	Space group	Radiation	Cell dimensions (Å)	Cell volume (Å³)	Density (g/cm³), calculated (measured)	Z	R	Significant bond lengths and nonbonded distances (Å), and bond and interplanar angles (deg)	
1α,9β-Dichloro-2,2,10,10-tetramethoxy[2.2](2,6)pyridinophane (133)[124]									
Monoclinic (sublimation)	$P2_1/c$	CuKα	$a = 14.706(1)$	1,833.1	1.45	4	0.037	$a_{1,15} = 1.514(4)$	$\alpha^* = 0.29; 0.36$
			$b = 17.168(1)$		[1.44(1)]			$a_{2,3} = 1.530(4)$	$\beta^* = 0.10$
			$c = 7.269(1)$					$b = 1.575(4)$	$\gamma = 112.8(3); 110.7(3)$
			$\beta = 92.72(1)$					$e = 1.337(4)$	$\delta^* = 0.05$
								$f = 1.383(4)$	$\lambda = 122.7(3)$
								$g = 1.383(4)$	
								$v = 1.403(4)$	
								$z = 1.790(3)$	
								$q = 2.475(5)$	

Comments: There are two independent centrosymmetric molecules, type 1 and type 2, two each located around two inversion centers in the unit cell. The pyridino rings are bent into boat shapes and are anti to one another, with the Cl atoms on the bridge facing in the same direction as the pyridino ring N atom. The N · · · N nonbonded contact is about 0.07 Å shorter than in [2.2](2,6)pyridinophane. The aromatic rings appear to be parallel to one another.

1,4-Dithia[4.3](6,9)purinophane (134)[125]

Monoclinic	$P2_1/n$	—	$a = 10.561(1)$ $b = 16.674(3)$ $c = 9.136(1)$ $\beta = 90.29(1)$	1,608.8(4)	—	4	0.048	$\theta = 5.5$

Comments: The general structural features of the purinoid rings are similar to those of 6-methylthiopurine. The purinoid rings are planar and are stacked nearly parallel to one another ($\theta = 5.5°$), with the smallest transannular distances being 3.174 (between N-17 and C-9) and 3.245 Å (between C-18 and N-8). The purinoid rings crisscross one another (when one is looking down an axis normal to both ring planes), and the bridges are oriented syn with respect to one another.

1,4-Dithia[4.4](6,9)purinophane (135)[125]

Monoclinic	$P2_1/n$	—	$a = 10.870(2)$ $b = 17.058(2)$ $c = 9.289(1)$ $\beta = 94.15(1)$	1,717.9(4)	—	4	0.052	$\theta = 6.6$

Comments: The general structural features of the purinoid rings are not very different from those of 6-methylthiopurine. The purinoid rings are planar and are stacked almost parallel to one another ($\theta = 6.6°$), with the smallest transannular nonbonded distances being 3.523 (between N-11 and N-23) and 3.539 Å (between C-12 and C-21). The purinoid rings crisscross one another (when one is looking down an axis normal to both ring planes), and the bridges are oriented syn with respect to one another.

4-Thia[4.3](1,3)pyrimidino(6,7)purinophane (136)[126]

Monoclinic	$P2_1/c$	CuKα	$a = 10.861(1)$ $b = 14.924(1)$ $c = 10.861(1)$ $\beta = 118.63(1)$	1,545.2	—	4	0.047	$\theta = 50.4$

Comments: The aromatic rings are not stacked above one another as in 1,4-dithia[4.3](6,9)purinophane or 1,4-dithia[4.4](6,9)purinophane. The angle of inclination (θ) of the two rings is 50.4°. The aromatic rings are planar.

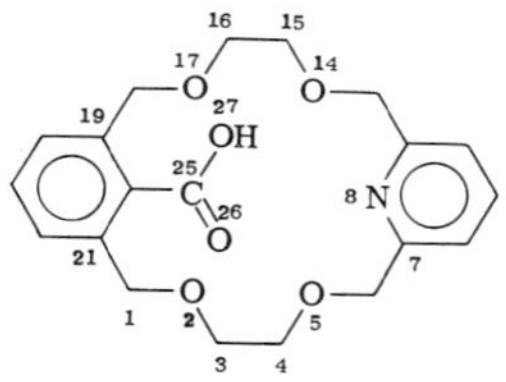

137

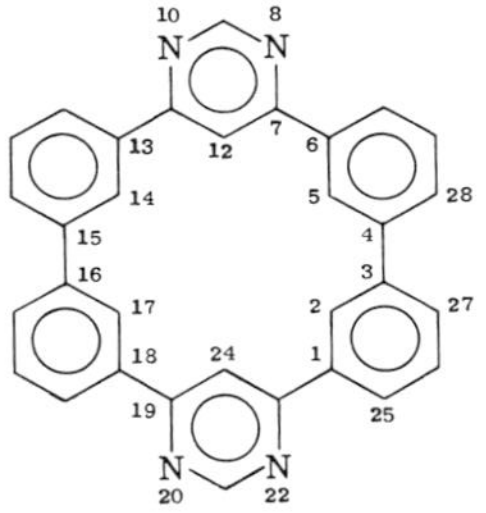

138

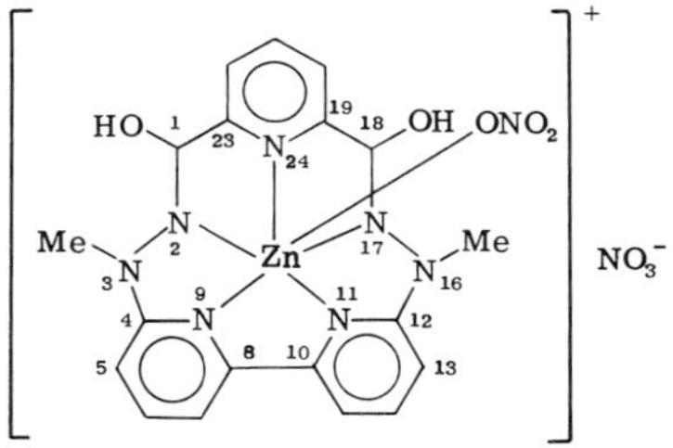

139

Crystal system	Space group	Radiation	Cell dimensions (Å)	Cell volume ($Å^3$)	Density (g/cm^3), calculated (measured)	Z	R	Significant bond lengths and nonbonded distances (Å), and bond and interplanar angles (deg)	
2,5,14,17-Tetraoxa-20-carboxy[6.6](2,6)pyridinometacyclophane (137)[127]									
Orthorhombic	*Pbca*	MoKα	−160°C	3,527.7	1.406	8	0.062	Meta ring	Pyridino ring
			a = 8.752(3)					a = 1.509(7)	a = 1.507(7)
			b = 20.498(6)					b = 1.432(6)	b = 1.419(6)
			c = 19.664(5)					c = 1.437(6)	c = 1.430(5)
								d = 1.511(7)	
								e = 1.389(6)	e = 1.358(6)
			Ambient temperature	3,596.4	1.379	—	—	f = 1.402(7)	f = 1.383(6)
								g = 1.385(7)	g = 1.389(7)
								u = 1.530(7)	
			a = 8.843(2)					C=O = 1.213(6)	
			b = 20.558(4)					C—O = 1.369(6)	
			c = 19.783(6)					γ = 108.0(4)	γ = 109.5(4)
								λ = 119.4(4)	λ = 121.0(4)

Comments: The molecular framework approximates C_2 symmetry, with the ether O atoms of the bridges pointing toward the macroring cavity. The C-3—C-4 and C-15—C-16 bonds adopt gauche conformations. The carboxyl group points toward the cavity of the macroring and is inclined 57° to the aromatic plane to which it is bound. The OH · · · N distance is 1.8 Å, and the O-27 · · · N distance is 2.66 Å, indicating strong H bonding. The O-2 · · · C-25 and O-17 · · · C-25 distances are 2.84 and 2.87 Å, respectively, indicating the dipole interaction between these centers. The benzenoid and pyridinoid rings are planar and nearly parallel to one another. The cavity of the marcrocycle is covered by a lipophilic skin of C—H bonds.

1,3;4,6;13,15;16,18-Tetra-1,3-benzo[6.6](4,6)pyrimidenophane (138)[128]

Monoclinic	$P2_1/n$	CuKα	a = 17.890(4)	—	1.402	2	0.047	a = 1.489(7)	e = 1.377(7)
			b = 5.110(1)					b = 1.382(1)	f = 1.329(7)
			c = 12.457(3)					c = 1.393(1)	g = 1.331(9)
			β = 106.71(2)					d = 1.485(6)	γ = 120.5
									λ = 118.2

Comments: The macrostructure is essentially flat, with nearly perfect D_{2h} symmetry. The normals to neighboring six-membered rings are inclined from 2 to 5° with respect to one another so that the C and N atoms are removed from the molecular plane by no more than 0.05 Å. This causes an unusually close contact between all the inner H atoms at C-2, C-5, C-12, C-14, C-17, and C-24 and a close contact between the C-27 and C-28 H atoms, the mean value for these contacts being 2.02 Å. This flat structure does not seem to be due to crystal packing forces but seems to be the result of the best orientation for maximum π interaction between the neighboring aromatic rings. Thus, the energy gained by maintaining a planar structure more than outweighs the nonbonded interactions between the H atoms.

{*cis*-1,18-Dihydroxy-3,16-dimethyl-2,3,16,17-tetraaza[3](6,6′)α-bipyridylo[3](2,6)pyridinophano-N^2,N^9,N^{11},N^{17},N^{24}}nitratozinc nitrate (139)[129]

Triclinic	$P\bar{1}$	MoKα	a = 12.015(3)	1,163	1.62	2	0.0580	—
			b = 10.646(2)		(1.60)			
			c = 10.434(2)					
			α = 108.23(2)					
			β = 113.04(2)					
			γ = 78.92(2)					

Comments: The molecule is folded about an axis passing through the N-2,N-17 atoms so that the N-2,N-17,N-24 plane is inclined 64° to the bipyridylo plane. Because of the coordination to Zn, the N atoms in the bipyridylo unit are syn-oriented. The Zn atom sits in the center of the quinquedentate macrocyclic cavity with average Zn—N-9,Zn—N-11, Zn—N-2,Zn—N-17, and Zn—N-24 bond distances of 2.101, 2.366, and 2.052 Å, respectively.

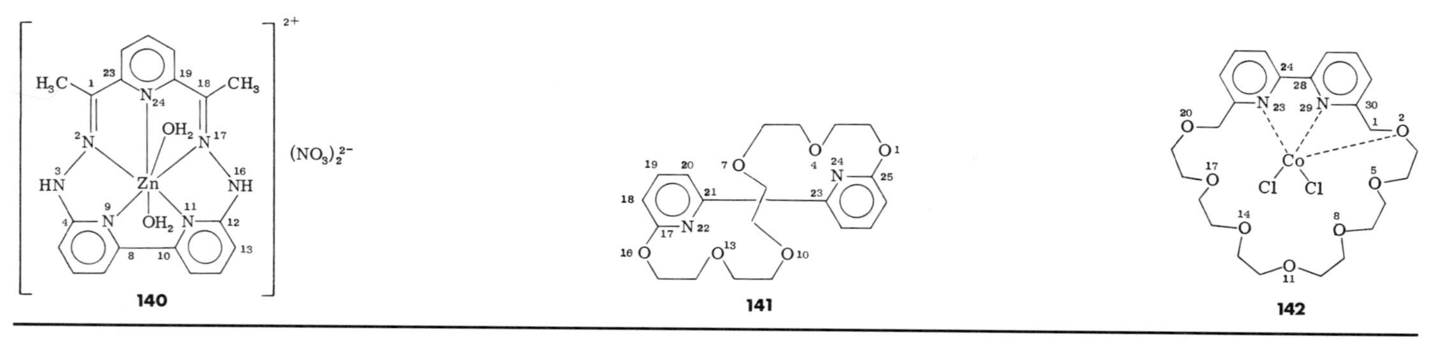

Crystal system	Space group	Radiation	Cell dimensions (Å)	Cell volume (Å³)	Density (g/cm³), calculated (measured)	Z	R	Significant bond lengths and nonbonded distances (Å), and bond and interplanar angles (deg)
{1,18-Dimethyl-2,3,16,17-tetraaza[3](6,6′)α-bipyridylo[3](2,6)pyridinophane-1,17-dieno-N^2,N^9,N^{11},N^{17},N^{24}}diaquozinc dinitrate (140)[130]								
Monoclinic (EtOH)	*Pn*	MoKα	$a = 14.132(4)$ $b = 9.226(4)$ $c = 8.739(2)$ $\beta = 94.50(1)$	1,139.41	1.65 (1.62)	2	0.097	—

Comments: The complex cation has a pentagonal–bipyramidal coordination geometry, with the Zn atom occupying the center of the quinquedentate cyclophane cavity and the five coordinating N atoms (N-2, N-9, N-11, N-17, and N-24) very nearly planar (deviation from the best least-squares plane of the five N atoms <0.01 Å). The Zn—N bond lengths to the above N atoms are 2.318, 2.045, 2.041, 2.292, and 2.283 Å, respectively, those to the N-9 and N-11 atoms being shortest because of the constraints of the cyclophane cavity. The two H_2O molecules occupy axial coordination positions and are situated above and below the N atom plane by 2.250 and 2.155 Å, respectively.

1,4,7,10,13,16-Hexaoxa[16](6,6′)α-bipyridylophane (141)[131]

Monoclinic (cyclohexane)	$P2_1/a$	MoKα	24°C	1,995	1.30	4	0.059	$a = 1.359$	$e_{17,18} = 1.391$
								$b = 1.468$	$e_{17,22} = 1.314$
			$a = 8.712(3)$					$c = 1.457$	$f_{18,19} = 1.372$
			$b = 14.205(5)$					$d = 1.411$	$f_{21,22} = 1.354$
			$c = 16.458(5)$					$\gamma = 118.9$	$g_{19,20} = 1.388$
			$\beta = 101.60(2)$					$\lambda = 124.2$	$g_{20,21} = 1.377$

Comments: The molecule has approximate C_2 symmetry. The torsion angle around the bipyridyl linkage (C-21—C-23) is 171°, and the N atoms of the rings are anti to one another. The polyether chain crosses the C-21—C-23 vector at the C-8—C-9 bond, and the O links to the aromatic rings are nearly coplanar with the rings and syn to the N atoms. The atoms near the center of the polyether chain are affected by large thermal motion.

{2,5,8,11,14,17,20-Heptaoxa[21](6,6′)α-bipyridylophano}cobalt dichloride (142)[132]

Monoclinic ($CHCl_3$ or benzene)	$P2_1/c$	—	$a = 9.692(4)$	—	1.452	4	0.064	—	
			$b = 28.692(10)$						
			$c = 9.835(3)$						
			$\beta = 97.89(3)$						

Comments: The bipyridylo unit is planar with a syn orientation of the N atoms. The geometry of the coordination sphere is a distorted trigonal bipyramid. The Co is pentacoordinate rather than the usual tetra- or hexacoordinate, with one of the coordination sites occupied by the O-2 atom. This Co—O distance is 2.319(5) Å, which is considerably longer than typical Co(II)—O distances of 1.93 Å.

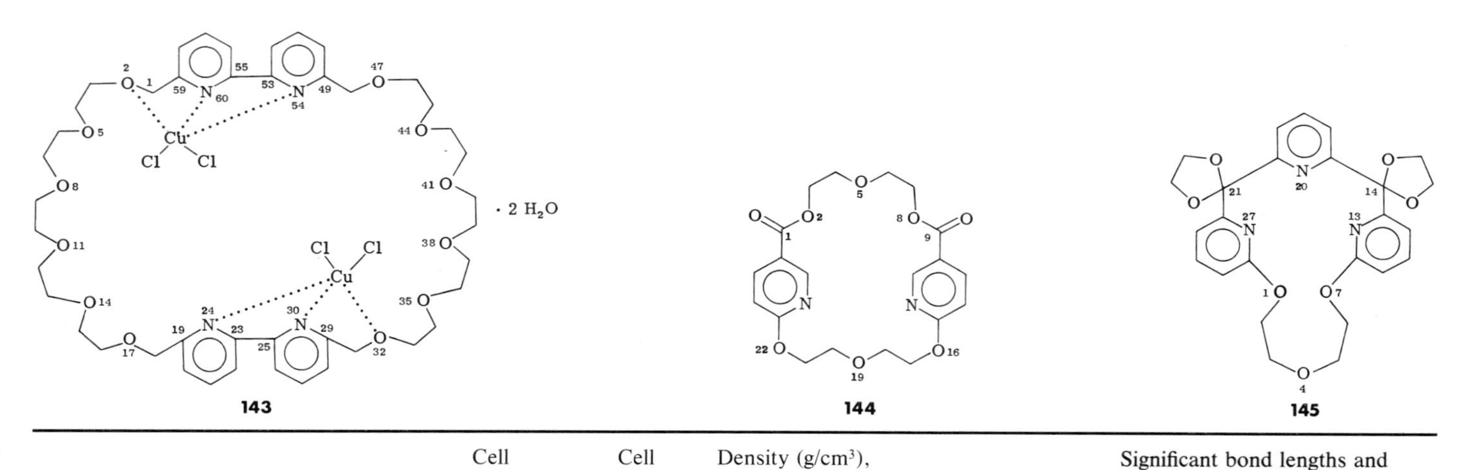

Crystal system	Space group	Radiation	Cell dimensions (Å)	Cell volume (Å^3)	Density (g/cm^3), calculated (measured)	Z	R	Significant bond lengths and nonbonded distances (Å), and bond and interplanar angles (deg)
{2,5,8,11,14,17,32,35,38,41,44,47-Dodecaoxa[18.18](6,6′)α-bipyridylophano}(copper dichloride)$_2$ · 2 H_2O (143) [133]								
Triclinic	$P\bar{1}$	—	a = 12.918(5) b = 12.896(5) c = 18.576(7) α = 102.08(3) β = 99.12(3) γ = 118.14(2)	—	1.489	2	0.073	—

Comments: Two independent but similar molecules exist in the unit cell, which differs basically in the conformation of the polyether bridges. The Cu atoms are coordinated in a distorted trigonal bipyramid shape. The two Cu—N distances are 2.049(10) and 1.946(10) Å for the axial and equatorial bonds, respectively, and 2.123(9) Å for the Cu—O bond. The bipyridylo rings are planar and are stacked in a parallel fashion, with 3.66 Å between ring centers and 7.16 Å between Cu atoms. The Cu atoms and the bipyridylo liganded N atoms are oriented anti to one another.

2,5,8,16,19,22-Hexaoxa[9.7](2,5)pyridinophane-1,9-dione (144) [134]

Triclinic (CHCl$_3$)	$P\bar{1}$	MoKα	$a = 7.651(2)$ $b = 10.237(2)$ $c = 12.754(2)$ $\alpha = 101.51(1)$ $\beta = 95.26(2)$ $\gamma = 94.23(2)$	—	1.429	2	0.061	$e_{C,N} = 1.337$ $e_{C,C} = 1.387$ $C—C_{bridge} = 1.502$ $C—O_{ether} = 1.437$

Comments: The pyridinoid rings are nearly parallel to one another (7.4°) and are separated by 4.33 Å between ring centers. The N atoms are syn with respect to one another. The ester linkages to the pyridinoid rings are different, one being syn to its ring N atom, the other being anti to its ring N atom. The ether linkages to the pyridinoid rings are both syn with respect to the N atoms.

14,21-Bisethylenedioxy-1,4,7-trioxa[7.1.1](2,6)pyridinophane (145) [135]

Monoclinic (EtOH)	$P2_1/c$	MoKα	22°C $a = 10.454(4)$ $b = 27.144(7)$ $c = 9.168(3)$ $\beta = 115.09(2)$	2,356	1.352	4	0.054	Unsymmetrically substituted rings

Unsymmetrically substituted rings

Ether bridge side	Ketal bridge side
$a = 1.359(7)$	$a = 1.540(6)$
$b = 1.442(7)$	$e = 1.352(7)$
$c = 1.504(9)$	$f = 1.362(8)$
$d = 1.417(6)$	$g = 1.396(8)$
$e = 1.213(7)$	$v = 1.423(6)$
$f = 1.389(7)$	
$g = 1.354(9)$	
$\gamma = 118.4(4)$	$\gamma = 110.0(4)$
$\lambda = 123.4(5)$	$\lambda = 123.7(4)$

Symmetrically substituted ring

$a = 1.517(7)$
$e = 1.332(6)$
$f = 1.395(7)$
$g = 1.379(9)$
$\lambda = 123.4(5)$

Comments: The gross conformation is globular; the main backbone is wrapped around the central cavity somewhat like the seam of a tennis ball, with the two unsymmetrically substituted pyridinoid rings nearly parallel to one another and separated by ~4.5 Å. The N atom lone pairs of these two rings point toward the cavity and anti to the lone pair of the symmetrically substituted pyridinoid ring, which is directed outside the cavity. The main C backbone plane forms an angle of ~60° with the symmetrically substituted ring. The polyether O atoms are directed outward from the cavity, and the chain is bound to the two unsymmetrically substituted pyridinoid rings in a fashion that is syn to the ring N atoms. The bridging atoms to these rings, O-1 and O-7, are in the aromatic plane.

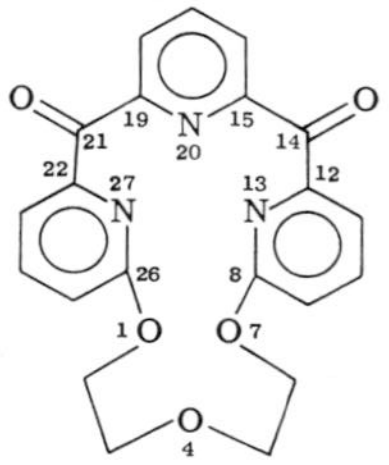

146

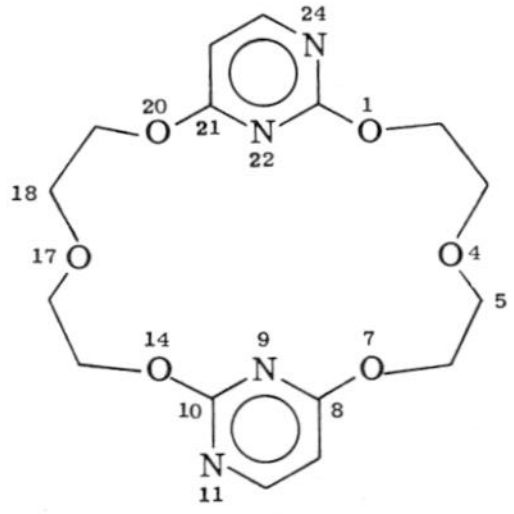

147

Crystal system	Space group	Radiation	Cell dimensions (Å)	Cell volume (Å^3)	Density (g/cm^3), calculated (measured)	Z	R	Significant bond lengths and nonbonded distances (Å), and bond and interplanar angles (deg)	
1,4,7-Trioxa[7.1.1](2,6)pyridinophane-14,21-dione (146)[135]									
Monoclinic (EtOH)	$A2/a$	MoKα	22°C	3,773	1.378	8	0.038	Unsymmetrically substituted rings	
			a = 20.124(7) b = 12.194(3) c = 15.385(2) β = 92.13(2)					Ether bridge side	Keto bridge side
								a = 1.361(3) b = 1.444(3) c = 1.502(4) d = 1.418(3) e = 1.323(3)	a = 1.504(3) e = 1.351(3)

$f = 1.381(3)$	$f = 1.370(4)$
$g = 1.367(4)$	$g = 1.396(4)$
	$v = 1.214(4)$
$\gamma = 117.7(2)$	$\gamma = 118.8(2)$
$\lambda = 124.2(2)$	$\lambda = 123.7(2)$

Symmetrically substituted ring
$a = 1.504(3)$
$e = 1.338(3)$
$f = 1.381(4)$
$g = 1.381(4)$
$\lambda = 123.3(2)$

Comments: The overall conformation is similar to a wheelchair, the chair back being the symmetrically substituted ring, the wheels being the unsymmetrically substituted rings, and the polyether chain being the foot rest. The "wheel" rings are approximately parallel (separation, ~4.4 Å) to one another, and the N atom lone pairs are directed inwardly. The lone pair on the symmetrically substituted pyridinoid ring is directed outward and deviates from antiparallelism with the other two lone pairs by 60°. The planes of the carbonyl groups are very nearly coplanar with the symmetrically substituted pyridinoid ring probably due to delocalization requirements. This causes the major structural difference between this molecule and its bisethylenedioxy derivative (previous entry), which is more globular because the carbonyl requirements open the "tennis ball" structure and render a "wheelchair" structure. The torsion angles around C-8—O-7 and C-26—O-1 are about 13.4° and render the polyether chain linkage syn to the N atoms of the rings containing C-8 and C-26. The O-4 atom is directed toward the middle of the molecule.

anti-1,4,7,14,17,20-Hexaoxa[7.7](2,6)pyrimidinophane (147)[136]

Monoclinic	$P2_1/c$	MoKα	$a = 10.688(2)$ $b = 4.952(3)$ $c = 16.315(4)$ $\beta = 99.06$	—	1.421	2	0.032	—

Comments: Because of the size of the macrocycle there are no unusual distortions in the pyrimidino rings or the polyether bridges as are found for the smaller [*m.n*]cyclophanes. The pyrimidino rings have an anti,anti orientation, and only a small cavity exists in the center of the macrocycle. Six atoms are directed toward the center of the macrocycle: N-9, N-22, O-1, O-14, and H at C-5 and H at C-18. The respective distances that these N, O, and H atoms make with the center of symmetry of the molecule are 2.72, 2.74, and 1.21 Å.

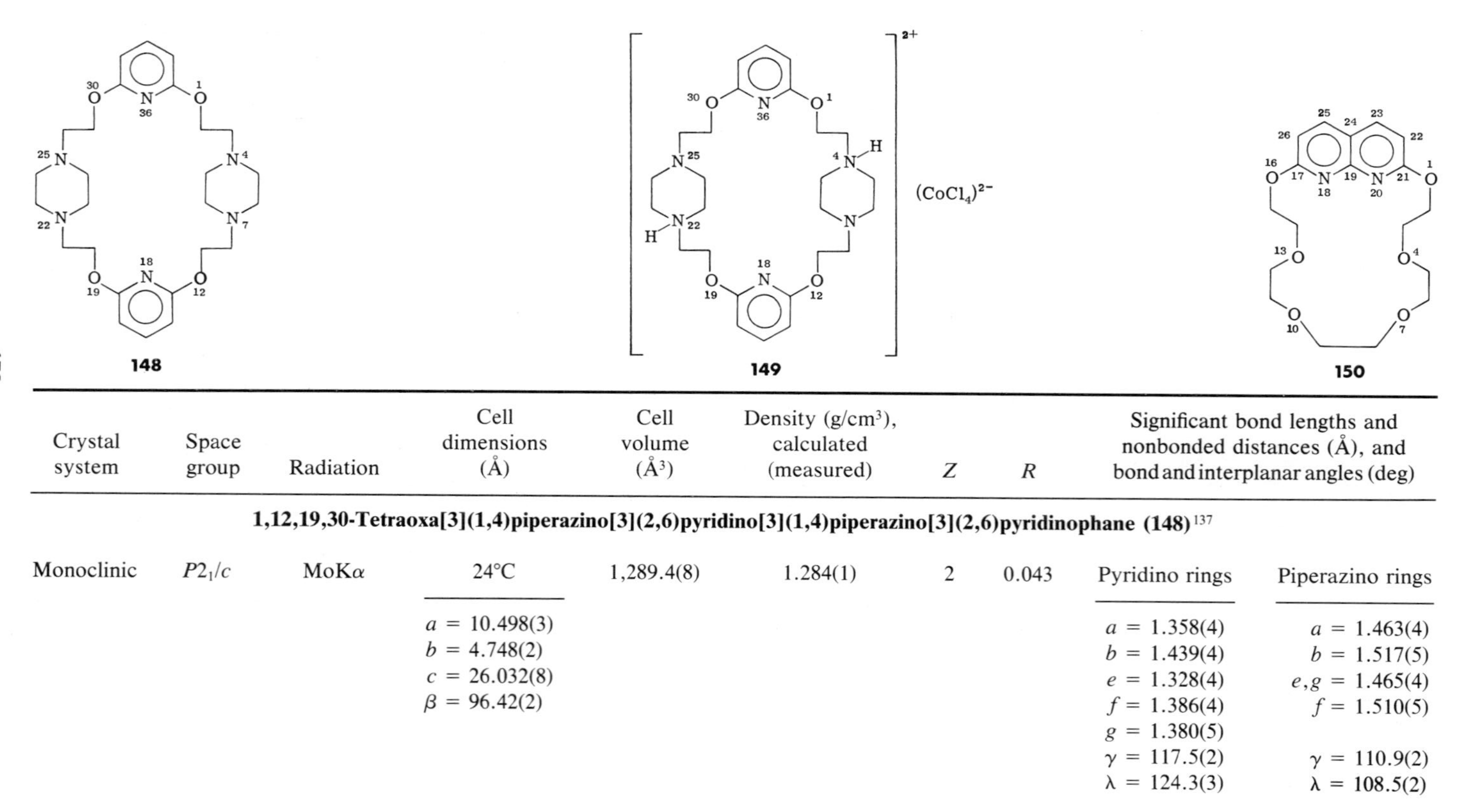

Crystal system	Space group	Radiation	Cell dimensions (Å)	Cell volume (Å³)	Density (g/cm³), calculated (measured)	Z	R	Significant bond lengths and nonbonded distances (Å), and bond and interplanar angles (deg)	
1,12,19,30-Tetraoxa[3](1,4)piperazino[3](2,6)pyridino[3](1,4)piperazino[3](2,6)pyridinophane (148) [137]									
Monoclinic	$P2_1/c$	MoKα	24°C	1,289.4(8)	1.284(1)	2	0.043	Pyridino rings	Piperazino rings
			a = 10.498(3)					a = 1.358(4)	a = 1.463(4)
			b = 4.748(2)					b = 1.439(4)	b = 1.517(5)
			c = 26.032(8)					e = 1.328(4)	e,g = 1.465(4)
			β = 96.42(2)					f = 1.386(4)	f = 1.510(5)
								g = 1.380(5)	
								γ = 117.5(2)	γ = 110.9(2)
								λ = 124.3(3)	λ = 108.5(2)

Comments: The piperazino rings are in a chair conformation and opposite to one another in the macrocycle. The pyridino rings are planar and nearly coplanar to one another on opposite sides of the macrocyclic ring, adopting an anti orientation so that the N atoms of the rings point toward the inside of the macrocyclic cavity. The piperazino ring planes (as part of the chain linking the two pyridino rings) are nearly perpendicular to the pyridino ring planes.

{4,22-Dihydro-1,12,19,30-tetraoxa[3](1,4)piperazino[3](2,6)pyridino[3](1,4)piperazino[3](2,6)pyridinophano}cobalt tetrachloride (149)[137]

Monoclinic	$P2_1/c$	MoKα	$a = 15.258(6)$ $b = 28.676(12)$ $c = 15.538(5)$ $\beta = 107.18(3)$	6,495(4)	1.432(1)	8	0.108	—

Comments: The precision of the determination does not allow for meaningful values of bond lengths and bond angles, but the determination does define the framework as being essentially the same as the uncomplexed structure (see previous entry). The two independent $[CoCl_4]^{2-}$ ions are essentially regular tetrahedra with Co—Cl distances averaging 2.262(8) and 2.258(9) Å. The N-4 and N-22 atoms are protonated, and crystal packing is determined by N—H · · · Cl H bonds. Each macrocyclic cation is surrounded by six $[CoCl_4]^{2-}$ anions and is H-bonded to one of them. The molecule as a whole is not very flexible.

1,4,7,10,13,16-Hexaoxa[16](2,7)naphthyridinophane (150)[138]

Triclinic	$P\bar{1}$	MoKα	$a = 8.969(3)$ $b = 10.325(3)$ $c = 20.107(5)$ $\alpha = 86.02(2)$ $\beta = 89.01(2)$ $\gamma = 79.15(2)$	—	1.327	4	0.055	$a = 1.368$ $b = 1.440$ $c = 1.509$ $d = 1.434$ $\gamma = 117.8$ $\lambda = 126.1$	$e_{21,22} = 1.385$ $e_{20,21} = 1.327$ $f_{22,23} = 1.370$ $f_{19,20} = 1.379$ $g_{23,24} = 1.433$ $g_{19,24} = 1.375$

Comments: Two distinct molecules exist in the asymmetric unit and lie in similar orientations with respect to the crystal axes. The only substantial difference between the two independent molecules is the conformation of the polyether bridge between the O-1 and O-7 atoms. The bridge is quite flexible and subject to subtle packing effects. The naphthyridino ring deviates only marginally from planarity, and the two N atoms are directed toward the inside of the macrocyclic cavity. Unlike 1,8-naphthyridine itself, in which the two N atoms lie on opposite sides of the aromatic plane, the N atoms in the phane are nearly coplanar (angle between planes C-17,N-18,C-19 and C-19,N-20,C-21 is ~0°).

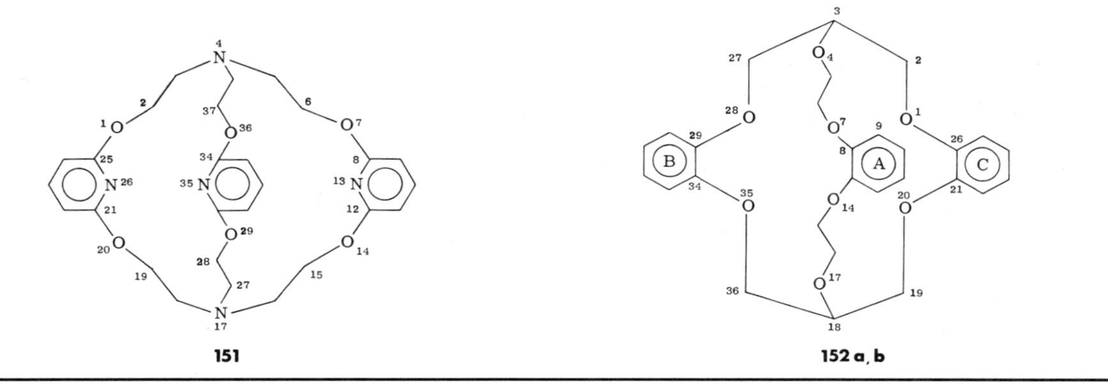

Crystal system	Space group	Radiation	Cell dimensions (Å)	Cell volume (Å³)	Density (g/cm³), calculated (measured)	Z	R	Significant bond lengths and nonbonded distances (Å), and bond and interplanar angles (deg)	

V. ORTHO RING PHANES (INCLUDING THYMOTIDES, CYCLOVERTRILS, AND RELATED STRUCTURES)

1,7,14,20,29,36-Hexaoxa-4,17-diaza[$3^{4,34}$ · $3^{17,30}$][7.7](2,6)pyridinophane (151)[139]

Crystal system	Space group	Radiation	Cell dimensions (Å)	Cell volume (Å³)	Density (g/cm³), calculated (measured)	Z	R	Significant bond lengths and nonbonded distances (Å), and bond and interplanar angles (deg)	
Trigonal	$R\bar{3}c$	—	$a = 10.937(2)$	—	1.355	2	0.060	$a = 1.333(9)$	$e = 1.348(7)$
			$\alpha = 83.33(2)$					$b = 1.449(9)$	$f = 1.388(10)$
								$c = 1.526(12)$	$g = 1.388(10)$
								$d = 1.459(9)$	
								$\gamma = 115.6(6)$	
								$\lambda = 122.7(7)$	

Comments: The macrobicyclic cryptand possess D_3 symmetry in the crystal, deviating from ideal D_{3h} symmetry by a slight twist about the C_3 axis passing through the bridgehead N atoms. The aromatic rings are planar and are inclined about 60° with respect to each other. The pyridino ring N atoms point toward the center of the crypt as well as the H atoms at C-2, C-6, C-15, C-19, C-28, and C-37. The most spectacular geometry in this molecule consists of the *trigonal planar bridgehead N atoms* (N-4 and N-17). All the bridgehead C—N—C angles are 120.0(5)°, indicating sp^2- instead of sp^3-hybridized N atoms.

1,4,7,14,17,20,28,35-Octaoxa[2^{3,29} · 2^{18,34}][7.7]orthocyclophane (152a)[140]

								Two-atom bridged rings	Four-atom bridged ring
Monoclinic	$P2_1/c$	—	$a = 11.539(2)$	2,849(1)	1.32	4	0.066		
			$b = 9.558(3)$		(1.32)				
			$c = 22.802(7)$						
			$\beta = 98.19(2)$					$a = 1.391(9)$	$a = 1.366(7)$
								$b = 1.417(9)$	$b = 1.432(8)$
								$c = 1.520(8)$	$c = 1.486(9)$
									$d = 1.424(7)$
								$e = 1.398(9)$	$e = 1.399(8)$
								$f = 1.376(9)$	$f = 1.390(9)$
								$g = 1.379(10)$	$g = 1.401(9)$
								$i = 1.357(10)$	$i = 1.337(10)$
								$\gamma = 117.7(6)$	$\gamma = 118.0(5)$

Comments: The bicyclophane has no discernible symmetry. The aromatic rings are planar, and the aromatic ring planes form angles of 51.0, 61.7, and 69.4° with respect to one another (AB, AC, BC, respectively). The O—C groups appended to the aromatic rings are significantly out of the aromatic planes to which they are attached. The O-28 atom is disordered.

1,4,7,14,17,20,28,35-Octaoxa[2^{3,29} · 2^{18,34}][7.7]orthocyclophane : potassium chloride : multihydrate (152b)[140,141]

								Two-atom bridged rings	Four-atom bridged ring
Monoclinic	$I2/c$	MoKα	$a = 22.578(6)$	—	1.349 ($n = 4.5$)	8	0.104		
			$b = 17.162(5)$		[1.35(1)]				
			$c = 16.742(5)$						
			$\beta = 99.48(2)$					$a = 1.38(1)$	$a = 1.38(1)$
								$b = 1.43(1)$	$b = 1.45(1)$
								$c = 1.50(1)$	$c = 1.49(1)$
									$d = 1.43(1)$
								$e = 1.37(2)$	$e = 1.38(1)$
								$f = 1.39(2)$	$f = 1.40(2)$
								$g = 1.38(2)$	$g = 1.39(2)$
								$i = 1.36(2)$	$i = 1.33(2)$
								$\gamma = 118.5$	$\gamma = 118$

Comments: The complex has a noncrystallographic plane of symmetry passing through the K atom and bisecting the bonds of the substituted aromatic ring C atoms. All eight O atoms are coordinated to the K atom (K · · · O distances range from 2.68 to 2.74 Å). All except O-28 and O-35 form a ring that is nearly coplanar with the K atom, whereas the latter two reside 2.4 Å from this plane on the same side of the cavity as the K atom. (Coordination about the K atom approximates an end-capped trigonal prism.) The torsion angles about the C—C bonds to the aromatic ring are 0°, whereas all other C—C torsion angles are about 60°. Most C—O torsion angles are 180°. The aromatic rings are planar, and two of the aromatic ring planes form angles of 34 and 40° with respect to the plane of the six O atoms. The aromatic ring planes form angles of 74.6, 36.5, and 68.9° with respect to one another (AB, AC, BC, respectively). The H_2O molecules and Cl^- ions are disordered.

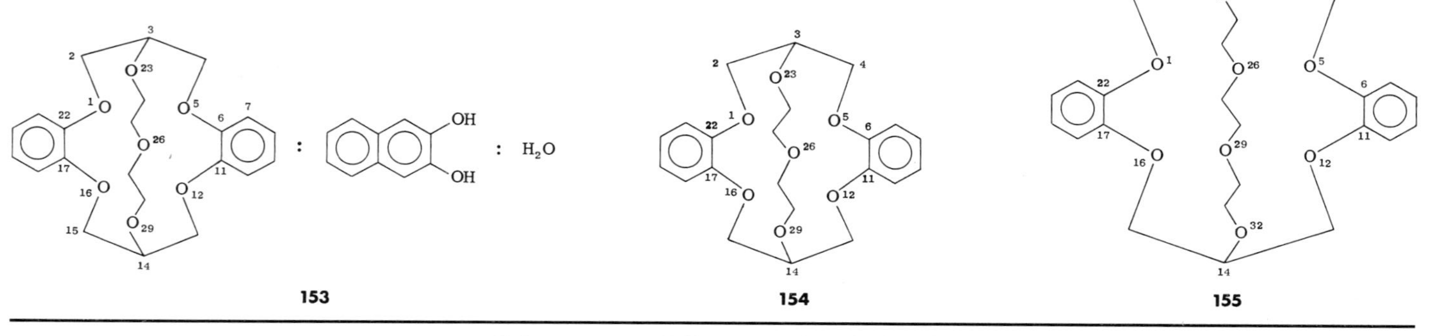

Crystal system	Space group	Radiation	Cell dimensions (Å)	Cell volume (Å^3)	Density (g/cm^3), calculated (measured)	Z	R	Significant bond lengths and nonbonded distances (Å), and bond and interplanar angles (deg)	
1,5,12,16,23,26,29-Heptaoxa[7^{3,14}][5.5]orthocyclophane : naphthalene-2,3-diol : H_2O (1 : 1 : 1) (153)[142]									
Monoclinc	$P2_1/a$	MoKα	a = 9.014(1) b = 33.302(7) c = 9.663(1) β = 95.42(1)	—	1.336	4	0.0685	a = 1.371(5) b = 1.432(5) c = 1.517(6) γ = 121.0(3); 116.3(3) λ = 119.6(4)	e = 1.395(6) f = 1.388(6) g = 1.392(7) i = 1.359(7)

Comments: The naphthalene-2,3-diol is linked to the bicyclophane through an H bond to the O of water, which is bound to the bicyclophane by H bonds to O-1, O-16, O-23, and O-29. The benzenoid rings are planar and, whereas one is nearly parallel to (coplanar), the other is perpendicular to the plane of the O-1, O-16, O-23, O-26, and O-29 atoms.

1,5,12,16,23,26,29-Heptaoxa[7^{3,14}][5.5]orthocyclophane (154)[143]

Rhombohedral	$R\bar{3}$	MoKα	a = 33.569(11) c = 9.182(5) Hexagonal axes corresponding to rhombohedral a = 19.621 α = 117.61	8,961	1.34 (1.34)	18	0.040	a = 1.367(9)‡ b = 1.424(2) c = 1.513(2) γ = 118.8(1) λ = 119.6(1)	e = 1.401(2) f = 1.379(2) g = 1.387(2) i = 1.369(3)

Comments: The molecule shows an open face containing four O atoms, which are available for cation complexation. The molecule is best considered a bridged analog of benzo-15-crown-5. The torsion angles about C—C and about C—O are about 60 and 180°, respectively, except for the C-6—O-5 angle, which is distorted and has a value of 56°. The C-6—O-5 bond is also longer than the other three similar bonds (1.379 Å). These deviations are an indication of the lack of conjugation of this O atom with the adjacent aromatic ring. The deviations of O-1, O-23, O-26, O-29, and O-16 from their mean plane (C) is similar to that found in benzo-15-crown-5. The angles between this plane and the two aromatic ring planes are 86 and 5°, whereas the angle between the aromatic ring planes is 85°. This molecule forms cation complexes easily because of the flexibility in the aliphatic chains.

1,5,12,16,23,26,29,32-Octaoxa[10^{3,14}][5.5]orthocyclophane (155)[143]

Rhombohedral	$R\bar{3}$	—	a = 34.772(7) c = 9.743(2) Hexagonal axes corresponding to rhombohedral a = 20.336 α = 117.5	10,202	1.31 (1.31)	18	0.0840	a = 1.364(7) b = 1.424(7) c = 1.510(9) γ = 118.2(5) λ = 120.0(7)	e = 1.385(8) f = 1.392(9) g = 1.399(10) i = 1.326(9)

Comments: The molecule shows an open face containing five O atoms, which are available for cation complexation. The molecule is best considered a bridged analog of benzo-18-crown-6. The torsion angle about C-27—C-28 is 177°, and the conformation is transoid. This causes the H atom at C-28 to point into the cavity and prevents O-29 from forming part of the open face of the molecule. The deviations of O-1, O-23, O-26, O-29, O-32, and O-16 from their mean plane (C) are much larger than those found for benzo-18-crown-6. The angles between this plane and the two aromatic ring planes are 92 and 19°, whereas that between the aromatic ring planes is 71°. This molecule forms cation complexes easily because of the flexibility of the aliphatic chains.

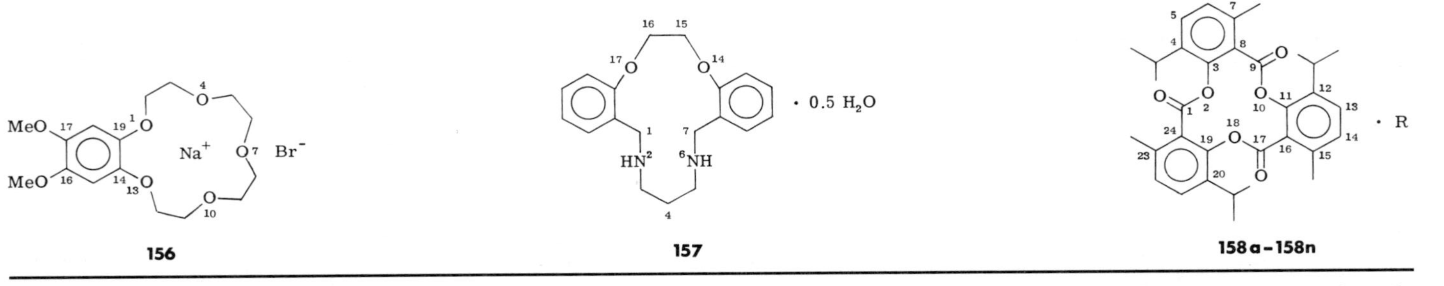

156 **157** **158a–158n**

Crystal system	Space group	Radiation	Cell dimensions (Å)	Cell volume (Å³)	Density (g/cm³), calculated (measured)	Z	R	Significant bond lengths and nonbonded distances (Å), and bond and interplanar angles (deg)	
16,17-Dimethoxy-1,4,7,10-13-pentaoxa[13]orthocyclophane : NaBr complex (156)[144]									
Monoclinic	$P2_1/c$	MoKα	a = 12.036(2) b = 8.222(6) c = 20.210(4) β = 102.21(4)	1,954.8	1.436 (1.432)	4	0.069	a = 1.39(9) b = 1.43(1) c = 1.48(1) d = 1.43(1) γ = 117.2(7) λ = 119.6(8)	e = 1.39(10 f = 1.39(1) g = 1.37(1) i = 1.40(1) v = 1.36

Comments: There are discrete molecules consisting of cyclophane $Na^+ \cdot Br^-$ ion pairs with the Na^+ coordinated to the five polyether O atoms (not to the methoxyl O atoms). The aromatic ring is planar. The Na · · · Br distance is 2.763(3) Å, and the Na · · · O lengths range from 2.37(1) to 2.45(1) Å.

Crystal system	Space group	Radiation	Cell dimensions (Å)	Cell volume (Å³)	Density (g/cm³), calculated (measured)	Z	R	Significant bond lengths and nonbonded distances (Å), and bond and interplanar angles (deg)
2,6-Diaza-14,17-dioxa[7.4]orthocyclophane · $\frac{1}{2}$ H_2O (157)[145]								
Monoclinic	$P2_1$	MoKα	a = 9.817(2) b = 19.016(4) c = 9.678(5) β = 95.33(2)	1,798.9	1.19	4	—	—

Comments: There are two independent molecules in the asymmetric unit that have slightly different configurations. The difference lies mainly in the conformation around the C-1—N-2 and C-7—N-6 bonds; one has a gauche and the other an anti arrangement. Both molecules have a pseudo-twofold axis passing through the C-4 atom and bisecting the C-15 —C-16 bond. The radius of the cavity defined by the bridging N and O atoms averages ~2.1 Å.

Tri-*o*-thymotide(TOT)-4,12,20-triisopropyl-7,15,23-trimethyl-2,10-18-trioxa[2.2.2]orthocyclophane-1,9,17-trione (158a: no R group)[146]

Orthorhombic	$Pna2_1$	CuKα	a = 16.049 b = 13.424 c = 13.969	—	—	4	—	—

Comments: Tri-*o*-thymotide has the propeller conformation with the carbonyl O atoms oriented in one direction and the aromatic rings in the opposite direction.

Tri-*o*-thymotide (158b: R = clathrates)[147]

—	—	—	—	—	—	—	—	—

Comments: Some crystallographic properties of 50 tri-*o*-thymotide clathrates have been described and discussed.

Tri-*o*-thymotide (158c: R = cetyl alcohol (6 : 1.3) clathrate)[146]

Hexagonal	$P6_1(P6_5)$	CuKα	a = 14.308 c = 29.016	—	—	—	—	—

Comments: The tri-*o*-thymotide has the propeller conformation with the carbonyl O atoms oriented in one direction and the aromatic rings in the opposite direction.

(*Continued*)

158a–158n

Crystal system	Space group	Radiation	Cell dimensions (Å)	Cell volume (Å^3)	Density (g/cm^3), calculated (measured)	Z	R	Significant bond lengths and nonbonded distances (Å), and bond and interplanar angles (deg)
Tri-*o*-thymotide (158d: R = ethanol (2:1) clathrate)[146]								
Trigonal	$P3_121$ ($P3_221$)	CuKα	$a = 13.443$ $c = 30.143$	—	—	6	—	—

Comments: The tri-*o*-thymotide has the propeller conformation with carbonyl O atoms oriented in one direction and the aromatic rings in the opposite direction.

Crystal system	Space group	Radiation	Cell dimensions (Å)	Cell volume (Å^3)	Density (g/cm^3), calculated (measured)	Z	R	Significant bond lengths and nonbonded distances (Å), and bond and interplanar angles (deg)
Tri-*o*-thymotide (158e: R = (*R*)-2-butanol (2 : 1) clathrate)[148]								
Tetragonal [(*R*)-2-butanol]	$P3_121$	—	298K $a = 13.598(3)$ $c = 30.447(8)$ 123K $a = 13.442(9)$ $c = 30.10(3)$	—	1.156 (1.149)	6	0.080	—

Comments: Six orthocyclophane units form a cage-type cavity encapsulating the guest molecule [(*R*)-2-butanol]. The orthocyclophane has a propeller-shaped structure.

(−)-Tri-*o*-thymotide (158f: R = (*R*)-2-butanol clathrate)[149]

—	—	MoKα	−150°C	—	—	—	0.069	—

Comments: (−)-Tri-*o*-thymotide has the *M* configuration and is left-handed. All the host molecules in a single crystal adopt the same chiral configuration about the cavity.

(+)-Tri-*o*-thymotide (158g: R = (*S*)(+)-2-bromobutane (2:1) clathrate)[150]

Tetragonal	$P3_221$	Ni-Filtered CuKα	−50°C a = 13.70(1) b = 30.25(1)	—	1.195 (1.191)	6	0.076	—

Comments: (+)-Tri-*o*-thymotide has the *P* (right-handed propeller) absolute configuration.

(−)-Tri-*o*-thymotide (158h: R = (*R,R*)(+)-2,3-Dimethyltriirane (2:1) clathrate)[150]

Tetragonal	$P3_121$	Ni-Filtered CuKα	−50°C a = 13.60(1) b = 30.28(1)	—	1.161 (1.158)	6	0.086	—

Comments: (+)-Tri-*o*-thymotide has the *P* (right-handed propeller) absolute configuration (see previous entry).

Tri-*o*-thymotide (158i: R = pyridine (2 : 1) clathrate)[151]

Trigonal	$P3_121$	—	a = 13.67(1) c = 29.90(1)	—	1.16	6	0.08	—

Comments: The tri-*o*-thymotide portion has a propeller-type conformation similar to that of uncomplexed tri-*o*-thymotide. All distances between tri-*o*-thymotide molecules are shorter than 3.90 Å, and those to the pyridine guest are less than 4.0 Å. The pyridine molecule is surrounded by eight tri-*o*-thymotide molecules. The angles between the normals to the least-squares planes of the three phenyl rings in the tri-*o*-thymotide portion are 27, 43, and 43° as compared with 39, 37, and 48° for uncomplexed tri-*o*-thymotide. The difference in the two cases reflects the flexibility in the host molecule and suggests the possibility of including larger molecular units. In fact, successful inclusion of *cis*-stilbene, *trans*-stilbene, and isopropyl 3-methyl-2-(*N*-2,4-dinitrophenylanalino)butyrate has been realized with a host–guest ratio of 2 : 1 (see next three entries).

(*Continued*)

158a–158n

Crystal system	Space group	Radiation	Cell dimensions (Å)	Cell volume (Å³)	Density (g/cm³), calculated (measured)	Z	R	Significant bond lengths and nonbonded distances (Å), and bond and interplanar angles (deg)
Tri-*o*-thymotide (158j: R = *cis*-stilbene (2 : 1) clathrate)[151,152]								
Triclinic	$P\bar{1}(P1)$	CuKα	$a = 11.323$ $b = 13.16$ $c = 24.862$ $\alpha = 95.52$ $\beta = 103.70$ $\gamma = 86.39$	3,580	1.150 (1.146)	2	—	—

Comments: The tri-*o*-thymotide is racemic. The packing exhibits three channels, two of which cross each other. The *cis*-stilbene molecules could not be observed in the refinement ($R = 0.21$).

Crystal system	Space group	Radiation	Cell dimensions (Å)	Cell volume (Å³)	Density (g/cm³), calculated (measured)	Z	R	Significant bond lengths and nonbonded distances (Å), and bond and interplanar angles (deg)
Tri-*o*-thymotide (158k: R = *trans*-stilbene (2 : 1) clathrate)[151,152]								
Triclinic	$P\bar{1}$	CuKα	$a = 11.639$ $b = 13.027$ $c = 24.409$ $\alpha = 96.05$ $\beta = 103.21$ $\gamma = 84.81$	3,574	1.152 (1.150)	2	0.131	—

Comments: In each unit cell there are two stilbene and four tri-*o*-thymotide units. The latter units have the expected propeller conformation, the aromatic rings being planar and the carbonyl groups pointing toward the C_3 axis in a direction opposite to the aromatic rings.

Tri-*o*-thymotide (158l: R = isopropyl 3-methyl-2-(*N*-2,4-dinitrophenylanalino)butyrate (2 : 1) clathrate)[151]

Triclinic	—	—	a = 11.36 b = 13.79 c = 23.36 α = 90.24 β = 90.02 γ = 89.18	—	—	—	—	—

Tri-*o*-thymotide (158m: R = methyl-*cis*-cinnamate (2 : 1) clathrate)[152]

Triclinic	$P\bar{1}(P1)$	—	a = 25.0 b = 11.3 c = 13.0 α = 92.0 β = 94.0 γ = 102	—	1.13 (1.08)	2	—	—

Tri-*o*-thymotide (158n: R = methyl-*trans*-cinnamate (2 : 1) clathrate)[152]

Triclinic	$P\bar{1}(P1)$	—	a = 24.19 b = 11.51 c = 13.04 α = 91.3 β = 96.6 γ = 101.8	—	1.14 (1.11)	2	—	—

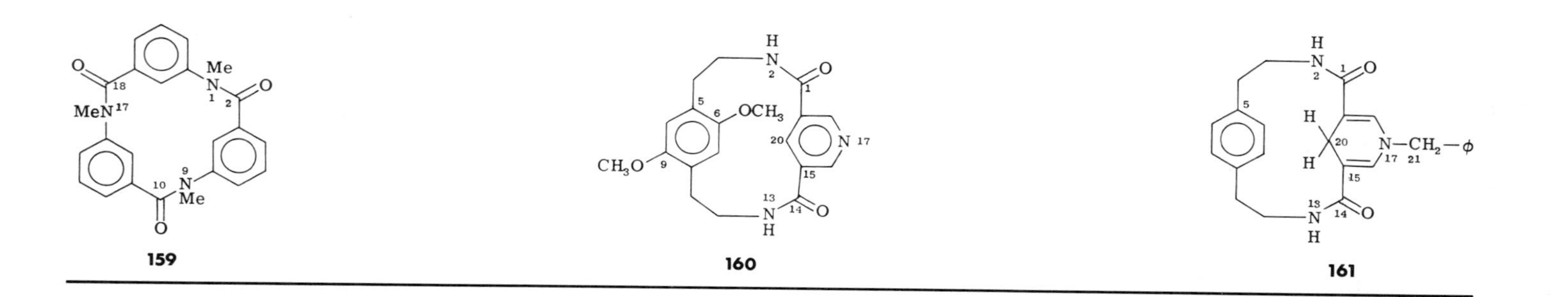

Crystal system	Space group	Radiation	Cell dimensions (Å)	Cell volume (Å^3)	Density (g/cm^3), calculated (measured)	Z	R	Significant bond lengths and nonbonded distances (Å), and bond and interplanar angles (deg)
			1,9,17-Trimethyl-1,9,17-triaza[2.2.2]metacyclophane-2,10,18-trione (159)[153]					
Triclinic (toluene)	$P\bar{1}$	CuKα	a = 13.546(1) b = 9.216(1) c = 9.469(1) α = 107.45(1) β = 109.49(1) γ = 72.43(1)	1.036	1.28	2	0.042	—

Comments: The molecule adopts a crown conformation in the solid state. The *N*-methylamide units adopt a cisoid conformation, and all the phenyl groups have a syn relationship. The C—CO—NMe—C units are anti to the phenyl rings.

Crystal system	Space group	Radiation	Cell dimensions (Å)	Cell volume (Å^3)	Density (g/cm^3), calculated (measured)	Z	R	Significant bond lengths and nonbonded distances (Å), and bond and interplanar angles (deg)	
			6,9-Dimethoxy-2,13-diaza[4]paracyclo[4](3,5)pyridinophane-1,14-dione (160)[154]						
Monoclinic	$P2_1/n$	CuKα	a = 10.2801(4) b = 9.4333(4) c = 18.7174(7) β = 101.42(1)	—	—	4	0.041	Para ring	Pyridino ring (meta type)
								a = 1.516(4)	a = 1.504(4)

								Para ring	Dihydropyridino ring (meta type)
								$b = 1.535(4)$ $c = 1.460(4)$	$b = 1.336(3)$
								$v_{C=O} = 1.233(3)$ $v_{C\text{-}6-O} = 1.382(4)$	
								$\alpha = 0$ $\beta^* = 0.08$ $\gamma = 114.4$	$\gamma = 114.1$
								$\theta = 77$	

Comments: The benzenoid ring is planar. The pyridino ring is slightly twisted about the N-17–C-20 axis. The H atom at C-20 points almost exactly into the center of the benzenoid π cloud and is ~2.80 Å from the nearest C atom of the benzenoid ring. There is considerable strain in the bridging chain, and the amido group deviate markedly from their ideal planar conformation.

17-Benzyl-17,20,dihydro-2,13-diaza[4]paracyclo[4](3,5)pyridinophane-1,14-dione (161)[154]

								Para ring	Dihydropyridino ring (meta type)
Monoclinic	—	CuKα	$a = 12.980(2)$ $b = 5.719(2)$ $c = 26.721(2)$ $\beta = 99.26(2)$	—	—	4	0.048	$a = 1.529(6)$ $b = 1.498(8)$ $c = 1.464(5)$	$a = 1.484(4)$ $b = 1.351(5)$
								$v_{C=O} = 1.231(5)$ $w_{N-C\text{-}21} = 1.462(5)$	
								$\alpha^* = 0.02$ $\beta^* = 0.2$ $\gamma = 113.7(4)$	$\beta^* = 0.21$ $\gamma = 114.4(3)$ $\delta^* = 0.12$

Comments: The benzenoid and dihydropyridinoid rings are both distorted from planarity into boat shapes. One of the H atoms at C-20 of the dihydropyridinoid ring is approximately above the center of the benzenoid ring with the shortest nonbonded H · · · C distance of 2.94 Å. There is considerable strain in the bridging chain, and the amido groups deviate markedly from their ideal planar conformation.

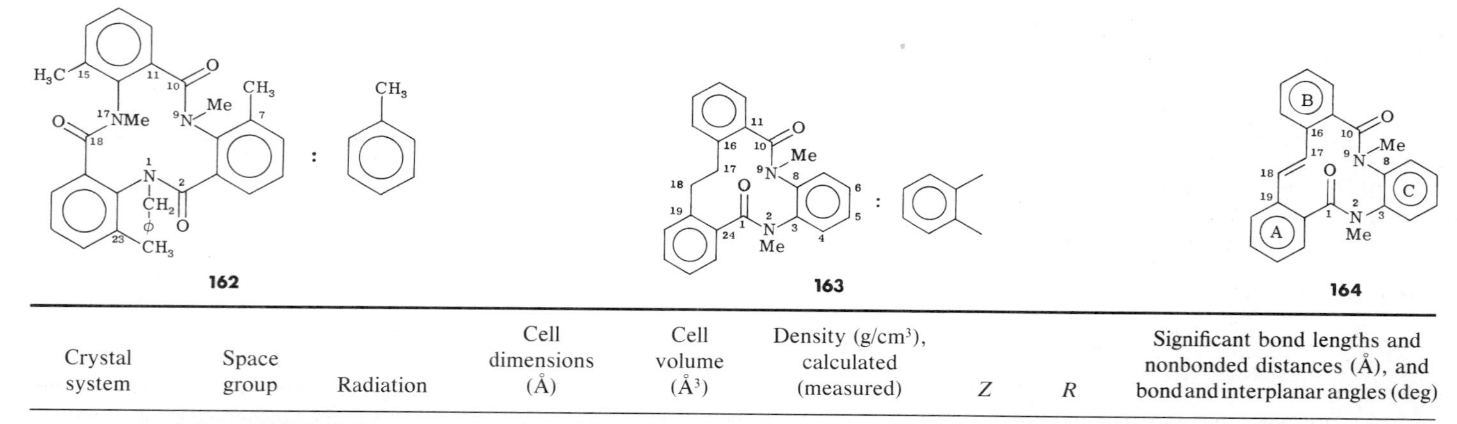

Crystal system	Space group	Radiation	Cell dimensions (Å)	Cell volume (Å^3)	Density (g/cm^3), calculated (measured)	Z	R	Significant bond lengths and nonbonded distances (Å), and bond and interplanar angles (deg)
1-Benzyl-7,9,15,17,23-pentamethyl-1,9,17-triaza[2.2.2]orthocyclophane-2,10-18-trione : toluene (1 : 1) clathrate (162)[155]								
Orthorhombic (toluene)	$P2_12_12_1$	CuKα	a = 17.358(1) b = 17.262(1) c = 10.993(1)	3,294	1.23	4	0.049	—

Comments: The host molecules adopt enantiomeric helical conformations within a lattice structure that contains chiral channels occupied by the toluene guest. The planes of the *trans*-amide bridges are nearly orthogonal to their adjacent aromatic rings and because of no *N*-hydrogen substituent (no H bonding) the molecule adopts a more nearly perfect helical geometry than other trianthranilides. The large continuous channels run parallel to the c axis and have an average minimum dimension of ~5.5 Å and an average maximum dimension of ~9.0 Å.

2,9-Diaza-2,9-dimethyl[2.2.2]orthocyclophane-1,10-dione : *o*-xylene (1 : 1) clathrate (163)[156]

Monoclinic	$C2/c$	CuKα	$a = 43.431(2)$ $b = 7.707(1)$ $c = 13.586(1)$ $\beta = 96.35(1)$	4,520	1.40	8	0.061	—

Comments: The molecule adopts an almost perfect *P* conformation (propeller) with the noncrystallographic twofold axis bisecting the C-17—C-18, C-3—C-8, and C-5—C-6 bonds. The O atom and methyl group of the amide bridges are trans-oriented, and the unit has a torsional angle of 180° around the C-1—C-2 bonds. The C-2—C-3 and C-1—C-24 torsional angles are 74 and 71°, respectively. The guest molecule (*o*-xylene) is disordered.

2,9-Diaza-2,9-dimethyl[2.2.2]orthocyclophan-17-ene-1,10-dione (164)[157]

Orthorhombic (MeOH)	$Pna2_1$	CuKα	$a = 10.708$ $b = 28.233$ $c = 6.197$	—	—	4	0.053	—

Comments: The molecule adopts a distorted propeller conformation, with a pseudo-C_2 axis bisecting the olefinic and C-3—C-8 bonds. The angles between the ring planes are 35, 33, and 13° for AB, AC and BC, respectively. The torsional angle around the amide linkages is 171°, with the methyl group and oxygen atom anti to one another and the planes of the amide groups being roughly perpendicular to the aromatic rings to which they are appended. The *N*-methyl groups are anti to one another and lie above and below the mean 12-membered-ring plane. The C-17 olefinic atom protrudes into the cavity, being nearly equidistant (3.16 and 3.24 Å) to the two N atoms and only 2.97 Å from the C-1 atom. The distortion in the macroring is due to the conjugational demands satisfied by the approach toward a planar stilbenoid group. [The torsional angles around C-16—C-17 and C-18—C-19 are 32 and 17°, respectively, and the bond angles at C-17 and C-18 (129 and 126°, respectively) are very near those found in *trans*-stilbene.]

165

166

167

Crystal system	Space group	Radiation	Cell dimensions (Å)	Cell volume (Å^3)	Density (g/cm^3), calculated (measured)	Z	R	Significant bond lengths and nonbonded distances (Å), and bond and interplanar angles (deg)	
Cycloveratril(CVT)-4,5,11,12,18,19-hexamethoxy[1.1.1]orthocyclophane : benzene : H_2O (1 : 0.5 : 1) complex (165)[158]									
Monoclinic	$C2/c$	Mo$K\alpha$	$a = 33.908(9)$ $b = 9.629(3)$ $c = 22.748(5)$ $\beta = 134.02(1)$	—	1.26 (1.24)	8	0.057	$a = 1.53$ $e = 1.40$ $f = 1.40$ $g = 1.38$	$i = 1.40$ $v = 1.38$ $\gamma = 112.5$ $\lambda = 118.9$

Comments: The host molecules have an umbrella (cone) shape, with the methyl groups of —O—Me approximately coplanar with the aromatic rings. The cyclononatrienoid ring has a crown conformation, and the cage cavity has an approximate hourglass shape (formed by the syn-oriented benzenoid rings and their appended O—CH_3 groups), which the guest molecules occupy. The H_2O molecules probably form bifurcated H bonds with the O atoms of —OMe. (Of two monoclinic α and β phases known for cycloveratril the α phase is adopted for aromatic or bulky guests, whereas the β phase is adopted for threadlike guests.)

Crystal system	Space group	Radiation	Cell dimensions (Å)	Cell volume (Å^3)	Density (g/cm^3), calculated (measured)	Z	R	Significant bond lengths and nonbonded distances (Å), and bond and interplanar angles (deg)	
4,5,11,12,18,19-Hexahydroxy[1.1.1]orthocyclophane : 2-propanol (1 : 2) complex (166)[159]									
Triclinic	$P\bar{1}$	Mo$K\alpha$	$a = 9.605(2)$ $b = 9.819(2)$ $c = 13.737(3)$ $\alpha = 95.36(2)$ $\beta = 94.30(2)$ $\gamma = 99.54(2)$	1,267	1.28	2	0.075	$a = 1.523(5)$ $e = 1.398(5)$ $f = 1.403(5)$ $g = 1.383(5)$	$i = 1.383(5)$ $v = 1.394$ $\gamma = 113.2(3)$ $\lambda = 118.8(3)$

Comments: The host adopts a crown conformation of the cyclononatrienoid ring with the three aromatic rings pointing toward the bottom of the crown, producing a conical shape for the molecule. All the —OH groups are not in the aromatic plane. One ring pair is coplanar with the aromatic ring, but the H atoms point away (anti) from one another. Another ring pair is coplanar, but the H atoms point in the same direction (syn), giving rise to an H bond. In the third ring one —OH group is coplanar, and the other is twisted 77° from the aromatic plane. These groups destroy the threefold symmetry and are important in the construction of the cavity channel. The two 2-propanol guests are disordered and are included into the channels (series of cavities) by H bonding to the hosts, where they sit in the concave portion of the orthocyclophane structure and at the junction of the OH groups of four hosts.

VI. MULTILAYERED PHANES

[2.2](2,5)Furano(2,5)[2.2]paracyclophane (167)[160,161]

								Furano ring	Paracyclo ring
Monoclinic	$C2/c$	CuKα	$a = 28.066(2)$	10,769.9	1.21	24	0.055	$a = 1.496$	$a = 1.506$
			$b = 16.018(1)$		(1.20)			$b = 1.570$	$b = 1.578$
			$c = 26.753(2)$					$e = 1.375$	$e = 1.390$
			$\beta = 116.43(1)$					$f = 1.349$	$f = 1.382$
								$g = 1.430$	
								$\beta = 3.8$	
								$\gamma = 114.0$	$\gamma = 113.0$
								$\lambda = 108.6$	$\lambda = 116.9$
								$\theta = 18.6$	

Central ring

Furanoid side		Paracyclo side
$a = 1.505$		$a = 1.507$
	$e = 1.389$	
	$f = 1.405$	
$\gamma = 108.3$		$\gamma = 112.2$
	$\lambda = 116.4$	

Comments: The furanoid ring is not quite planar but envelope-shaped and is inclined 18.6° (θ) to the central aromatic ring. The central ring is distorted into a twist-boat shape as in other triple-layered phanes. The nonbonded distances between the O atom and the C-13 and C-16 atoms are 2.860 and 2.820 Å, whereas those to the C-14 and C-15 atoms are 3.054 and 2.927 Å, respectively. The outer benzenoid ring is boat-shaped and is twisted slightly with respect to the central aromatic ring about an axis normal to the planes of these rings (see Chapter 10, Fig. 13).

168

169

170

Crystal system	Space group	Radiation	Cell dimensions (Å)	Cell volume (Å³)	Density (g/cm³), calculated (measured)	Z	R	Significant bond lengths and nonbonded distances (Å), and bond and interplanar angles (deg)

4,7-Dimethyl[2.2](2,5)thiopheno(2,5)[2.2]paracyclophane (168)[160,161]

Crystal system	Space group	Radiation	Cell dimensions (Å)	Cell volume (Å³)	Density (g/cm³), calculated (measured)	Z	R
Monoclinic	$P2_1/c$	CuKα	$a = 15.126(1)$ $b = 10.342(1)$ $c = 13.321(1)$ $\beta = 101.87(1)$	2,039.3	1.213	4	0.066

Thiopheno ring	Paracyclo ring
$a = 1.490(6)$	$a = 1.517(5)$
$b = 1.578(6)$	$b = 1.586(5)$
$e = 1.720(4)$	$e_{6,7} = 1.390(5)$
	$e_{5,6} = 1.381(5)$
$f = 1.356(6)$	$f = 1.384(5)$
$g = 1.406$	$u = 1.523(5)$
$\beta = 8.0$	
$\gamma = 114.5(4)$	$\gamma = 112.0(3)$
$\lambda = 108.6(3)$	$\lambda = 117.7(4)$
$\theta = 6.8$	

Central ring

Thiopheno side	Paracyclo side
$a = 1.505(5)$	$a = 1.507(4)$

Comments: The thiophenoid ring is only slightly distorted from planarity ($\beta = 8.0°$; envelope shape) and is stacked nearly parallel to ($\theta = 6.8°$) the central aromatic ring. The thiophenoid ring is also laterally displaced with respect to the central aromatic ring in such a way that the thiophenoid H atom at C-20 comes into close contact with the bridge methylene H atoms at C-10. The central aromatic ring is distorted into a twist-boat shape as in other triple-layered phanes. The nonbonded distance between the S ...

atoms is 3.122 Å, whereas those to the C-14 and C-15 atoms are 3.197 and 3.049 Å, respectively. The outer benzenoid ring is boat-shaped and is twisted slightly with respect to the central aromatic ring about an axis normal to the planes of these rings (see Chapter 10, Fig. 13).

γ = 111.9(3)	γ = 111.7(3)
λ = 116.0(3)	

[2.2](2,6)Pyridino(2,5)[2.2]paracyclo(2,6)pyridinophane-1,9,17,25-tetraene (169)[162]

								Outer rings	Inner ring
Monoclinic	$P2_1/c$	CuKα	20°C	1,731.3	1.28 (1.29)	4	0.049		
			a = 9.233(9)					a = 1.475(4)	a = 1.479(3)
			b = 8.801(9)					b = 1.337(4)	
			c = 22.29(2)					e = 1.340(3)	e,g = 1.382(3)
			β = 107.08(5)					f = 1.391(4)	f = 1.409(3)
								g = 1.373(5)	β^* = 0.08–0.10
								γ = 127.8	γ = 119.1
								λ = 120.5	λ = 115.3
								θ = 75	

Comments: There is a noncrystallographic but reasonably precise twofold axis through the C-12 and C-15 atoms. The pyridino rings are essentially planar but are inclined 75° (θ) to the central aromatic ring. The N atoms lie 2.52 Å from the mean plane of the central ring and 5.04 Å from one another. The N · · · N vector is inclined 1.8° to the normal of the benzenoid ring and passes about 0.10 Å from the center of this ring. The central benzenoid ring is distorted into a twist-boat conformation with torsion angles of 19, 19, and 38° for the C-11—C-12, C-12—C-13, and C-13—C-14 bonds. Atoms C-1, C-10, C-17, and C-26 are displaced 0.08–0.10 Å from the plane of the three neighboring aromatic atoms.

[2.2]Paracyclo(4,6)[2.2]metaparacyclophane (uu isomer) (170)[163]

								Outer rings (para type)	Inner ring (meta type)
—	—	—	—	—	—	—	—		
								a = 1.512	a = 1.521
								b = 1.576	
								e = 1.396	e,g = 1.393
								f = 1.390	f = 1.402
								γ = 108.7	β,δ = 4.3‡
								λ = 117.5	
								θ = 10.4	

Comments: The two outer benzenoid rings are boat-shaped, whereas the inner ring is distorted into a slight chair conformation. The C-12 and C-15 atoms are bent out of the C-11,C-13,C-14,C-16 plane by 4.3° in opposite directions from their nearest paracyclo ring. The two paracyclo rings are oriented anti to one another through the central benzenoid ring (i.e., one above and one below the central ring) (see Chapter 10, Fig. 12).

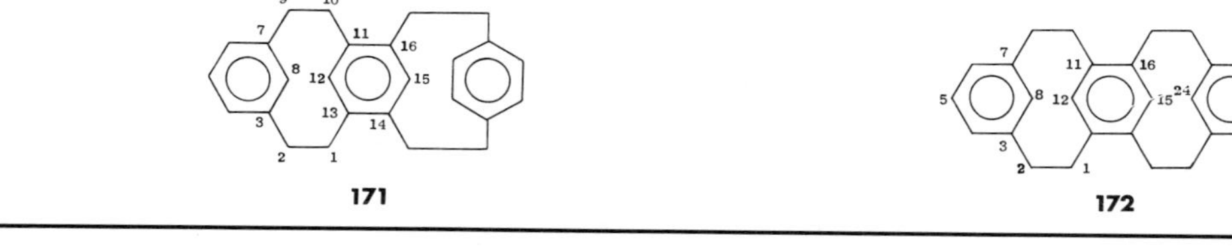

Crystal system	Space group	Radiation	Cell dimensions (Å)	Cell volume (Å^3)	Density (g/cm^3), calculated (measured)	Z	R	Significant bond lengths and nonbonded distances (Å), and bond and interplanar angles (deg)	
[2.2]Metacyclo(4,6)metaparacyclophane (ud isomer) (171)[163]									
—	—	—	—	—	—	—	—	Outer rings	
								Para ring	Meta ring
								$a = 1.512$	$a = 1.508$
								$b = 1.567$	$b = 1.567$
								$e_{inner} = 1.389$	$e = 1.395$
								$e_{outer} = 1.403$	
								$f_{inner} = 1.380$	$f = 1.394$
								$f_{outer} = 1.389$	$g = 1.396$
									$\beta = 9.5$
									$\delta = 4.6$
								$\gamma = 109.1$	$\gamma = 110.7$
								$\lambda = 117.0$	$\lambda = 118.0$
								$\theta = 27.6$	$\theta = 11.9$

Inner ring	
Para ring side	Meta ring side
$a = 1.533$	$a = 1.516$
$b = 1.567$	$b = 1.567$
$e = 1.397$	$e = 1.402$
$f = 1.406$	
$\beta = 12.5$	$\beta = 14.2$
$\gamma = 114.7$	$\gamma = 112.9$
$\lambda = 117.7$	$\lambda = 117.5$

Comments: The outer rings and the inner ring are boat-shaped. The outer rings are syn to one another through the central benzenoid ring (i.e., the outer metacyclo and paracyclo rings are on the same face of the central ring). The C-12 and C-15 atoms of the central ring are displaced out of the C-11,C-13,C-14,C-16 plane [14.2 and 12.5° (β), respectively] away from the outer rings (see Chapter 10, Fig. 11).

[2.2]Metacyclo(4,6)[2.2]metacyclophane (ud isomer) (172)[59]

								Outer rings	Inner ring
Monoclinic	$P2_1/c$	CuKα	−160°C	1,909.9(2)	1.777 (1.776)	4	0.048		
			$a = 11.659(1)$					$a = 1.506$	$a = 1.515$
			$b = 11.715(1)$					$b = 1.565$	
			$c = 14.678(1)$					$e = 1.392$	$e,g = 1.400$
			$\beta = 107.69(1)$					$f = 1.392$	$f = 1.401$
								$g = 1.387$	
								$\beta = 10.0$	$\beta,\delta = 1.35$
								$\gamma = 110.8$	$\gamma = 112.7$
								$\delta = 4.9$	
								$\lambda = 118.0$	$\lambda = 117.7$
								$q = 2.667$	
								$\theta = 1.3; 10.4$	

Comments: The outer and inner aromatic rings are boat-shaped. Both outer aromatic rings are on the same face of the inner aromatic ring (syn) (the molecule has approximate *mm*2 symmetry). The inner ring is more severely deformed than the outer rings. The C-12—H and C-15—H vectors make angles of 10.0 and 9.4°, respectively, with the plane of the nearest three aromatic ring carbons and are directed toward the outer rings. Correspondingly, the C-5—H, C-8—H, C-24—H, and C-21—H vectors are directed toward the inner ring and make angles with the nearest three aromatic ring C atoms of 4.5, 9.5, 8.2, and 3.7°, respectively. The reason for the difference in the angle of inclination (θ) of the two outer rings relative to the inner ring is the nonbonded repulsion between the C-8—H · · · H—C-24 atoms (nonbonded distance, 2.22 Å) (see Chapter 10, Fig. 9).

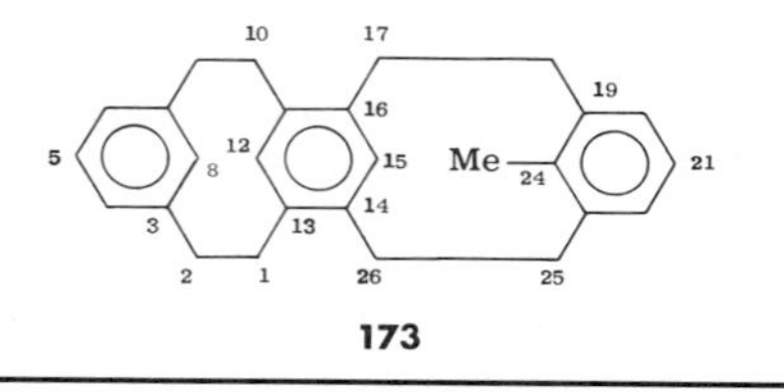

173

Crystal system	Space group	Radiation	Cell dimensions (Å)	Cell volume (Å³)	Density (g/cm³), calculated (measured)	Z	R	Significant bond lengths and nonbonded distances (Å), and bond and interplanar angles (deg)
24-Methyl[2.2]metacyclo(4,6)[2.2]metacyclophane (ud isomer) (173)[164]								
Monoclinic	$P2_1/c$	Ni-Filtered CuKα	$a = 10.090(2)$ $b = 14.501(2)$ $c = 13.677(3)$ $\beta = 101.09(2)$	1,063.8(7)	1.192 (1.200)	4	0.055	See below

Outer rings

Unsubstituted	Substituted
$a = 1.510(3)$	$a = 1.510(3)$
$b = 1.559(5)$	$b = 1.559(5)$
$e = 1.390$	$e = 1.402$
$f = 1.388(2)$	$f = 1.388(2)$
$g = 1.374(6)$	$g = 1.374(6)$
	$u = 1.529$
$\beta = 10.6$	$\beta = 11.4$
$\gamma = 111.1$	$\gamma = 110.0$
$\delta = 5.7$	$\delta = 4.5$
$\lambda = 117.7$	$\lambda = 118.9$
$\theta = 16.3$	$\theta = 22.5$
$q = 2.681$	$q = 2.715$

Inner ring

$a = 1.518(3)$
$b = 1.559(5)$

Comments: All the aromatic rings are distorted into boat shapes, with the central ring most distorted and the unbsubstituted ring least distorted. The outer rings are positioned on the same face of the central ring, giving rise to an Me · · · H—C-8 nonbonded distance of only 1.92 Å. The dihedral angle (θ) that the substituted ring plane makes with the central ring plane is more than 25% greater than that made by the unsubstituted ring due to the bulky methyl substituent (disordered). The C-24—Me vector, however, is bent 5° out of the nearest three-ring-C-atom plane toward the central ring. Similar out-of-plane deformations are observed for the C-5—H, C-8—H, and C-21—H vectors and are 5.9, 8.6, and 11.9°, respectively. The C-12—H and C-15—H vectors are also bent out of their nearest three-ring-C-atom planes by 7.8 and 8.4°, respectively, and are directed toward the outer aromatic ring (see Chapter 10, Fig. 10a).

$\beta_{subs.side} = 14.7$
$\beta_{unsubs.side} = 14.8$
$\gamma = 114.0(2)$
$\lambda = 117.4(1)$

24-Methyl[2.2]metacyclo(4,6)[2.2]metacyclophane (uu isomer) (173)[165]

Monoclinic	$P2_1/c$	Ni-Filtered CuKα	$a = 7.651(1)$ $b = 19.856(1)$ $c = 14.556(1)$ $\beta = 117.84(1)$	—	1.197 (1.20)	4	0.059

Outer rings

Unsubstituted	Substituted
$a = 1.510(2)$	$a = 1.510(2)$
$b = 1.575(2)$	$b = 1.575(2)$
$e = 1.394(5)$	$e = 1.409$
$f = 1.392(1)$	$f = 1.392(1)$
$g = 1.385(4)$	$g = 1.385(4)$
	$u = 1.507$
$\beta = 9.9$	$\beta = 12.4$
$\gamma = 110.4(2)$	$\gamma = 110.4(2)$
$\delta = 4.8$	$\delta = 5.4$
$\lambda = 118.5(4)$	$\lambda = 118.5(4)$
$\theta = 3.1$	$\theta = 12.1$
$q = 2.592$	$q = 2.656$

Inner ring

$a = 1.516(1)$
$b = 1.575(2)$
$e,g = 1.396(2)$
$f = 1.405(1)$
$\beta_{subs.side} = 6.0$
$\beta_{unsubs.side} = 5.0$
$\gamma = 110.5(4)$
$\lambda = 117.9(1)$

Comments: The outer aromatic rings are positioned on opposite faces of the central ring and are distorted into boat shapes, with the substituted ring (C-24—Me) being more deformed than the other. The latter ring is inclined 12.1° (θ) with respect to the central ring, whereas the former ring is inclined 3.1° (θ) to the central ring. The C-8—H and C-5—H vectors are inclined toward the central ring from the plane of the nearest three-ring atoms by 7.0 and 3.3°, respectively, whereas the C-24—Me and the C-21—H vectors are inclined toward the central ring from the plane of their nearest three-ring atoms by 2.7 and 4.4°, respectively. The central benzenoid ring is deformed into a chair conformation with C-12 and C-15 directed away from their proximate outer rings. The C-12—H and C-15—H vectors, however, are displaced toward the proximate outer rings from the plane of their nearest three-ring atoms by 3.2 and 1.8°, respectively. The difference in the two sides of the central ring is due to the methyl substituent in one of the outer rings (see Chapter 10, Fig. 10b).

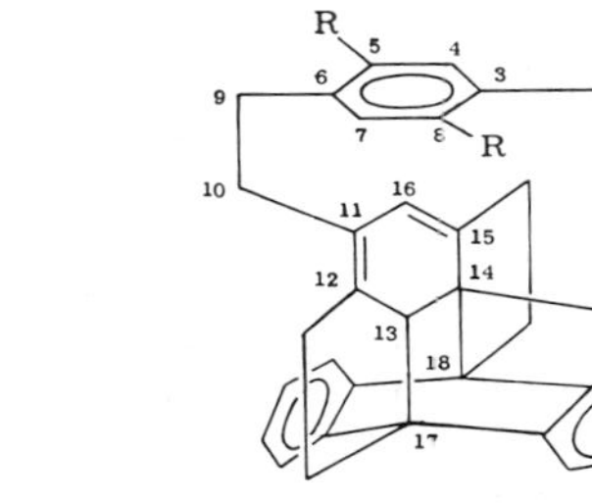

174

175

Crystal system	Space group	Radiation	Cell dimensions (Å)	Cell volume (Å³)	Density (g/cm³), calculated (measured)	Z	R	Significant bond lengths and nonbonded distances (Å), and bond and interplanar angles (deg)
			12-Bromo[2.2]paracyclo(2,4)[2.2]metaparacyclophane (174)[166]					
—	—	—	—	—	—	—	—	Outer rings Ring A: $a_{22,25} = 1.522$, $a_{19,18} = 1.509$, $b_{25,26} = 1.602$, $b_{18,17} = 1.568$, $e = 1.407$, $f = 1.375$, $\gamma_{25} = 109.4$, $\gamma_{18} = 108.3$, $\lambda = 117.2$ Ring B: $a_{2,3} = 1.526$, $a_{6,9} = 1.513$, $b_{1,2} = 1.590$, $b_{9,10} = 1.583$, $e = 1.393$, $f = 1.394$, $\gamma_2 = 108.0$, $\gamma_9 = 109.2$, $\lambda = 118.3$ Inner ring $a_{14,26}$ $a_{14,26} = 1.546$ $a_{15,1} = 1.517$

								$a_{16,17} = 1.527$ $a_{11,10} = 1.536$ b (see above) e, f, g (avg) $= 1.401$ $y = 1.914$ $\gamma_{26} = 110.4$ $\gamma_1 = 111.6$ $\gamma_{17} = 111.8$ $\gamma_{10} = 112.0$ $\lambda_{14} = 119.4$ $\lambda_{16} = 117.7$	$\lambda_{15} = 120.2$ $\lambda_{11} = 117.5$

Comments: The outer aromatic rings are on opposite faces of the inner benzenoid ring (see Chapter 10, Fig. 8, and for comparison Fig. 2).

$C_{34}H_{30}$ (paracyclophane derivative: R = H) (175)[167,168]

Monoclinic (benzene)	$P2_1/c$	MoKα	$a = 15.896(3)$ $b = 8.937(2)$ $c = 18.575(3)$ $\beta = 120.31(1)$	2,278(6)	1.279 (1.275)	4	0.081	$a = 1.513; 1.515$ $b = 1.562; 1.581$ $c = 1.571; 1.519$ $e = 1.384$ $f = 1.390$ $p_{6,11} = 2.718$ $p_{3,14} = 3.115$ $q_{7,12} = 3.100$; $q_{8,13} = 3.266$ $q_{4,15} = 3.274$; $q_{5,16} = 3.056$	$\alpha = 13.2$ $\beta = 8.3; 16.2$ $\gamma = 112.9; 108.9$ $\lambda = 117.4$

Comments: The compound is the intramolecular 4 + 2 adduct between the strained central benzenoid ring and the bridged portion of the anthracenoid ring in triple-layered [2.2](9,10)anthraceno(2,5)[2.2]paracyclophane. The C-13—C-17 and C-14—C-18 bonds are considerably long (1.567 and 1.600 Å, respectively). The cyclohexadienoid moiety is almost perfectly planar. The paracyclobenzenoid ring is boat-shaped. Crystals of the dimethyl derivative (R = CH_3) of the title compound have similar cell constants and belong to the same space group.

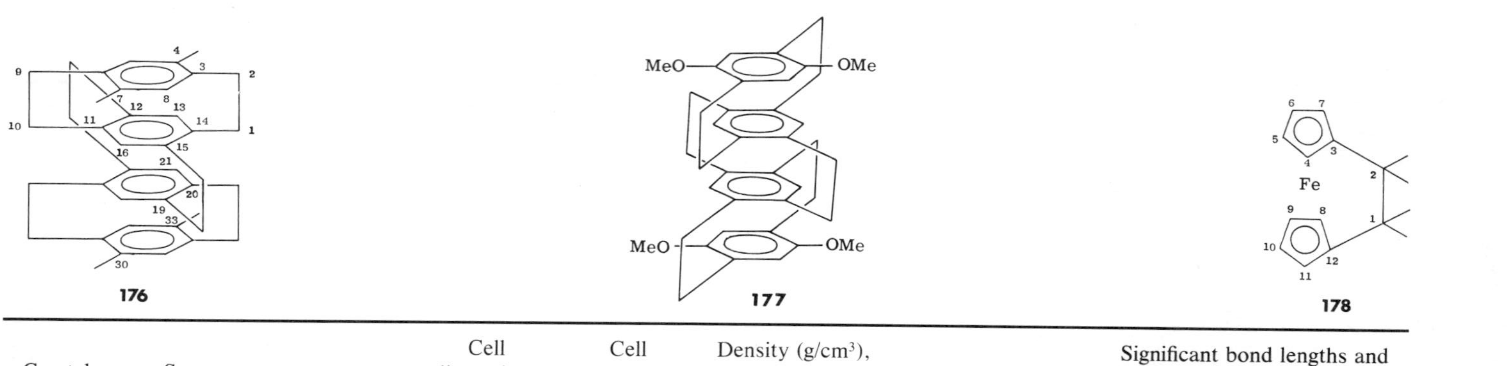

Crystal system	Space group	Radiation	Cell dimensions (Å)	Cell volume (Å³)	Density (g/cm³), calculated (measured)	Z	R	Significant bond lengths and nonbonded distances (Å), and bond and interplanar angles (deg)	
4,7,30,33-Tetramethyl quadruple-layered [2.2]paracyclophane (176)[169,170]									
Monoclinic (toluene)	$C2/c$	Zr-Filtered MoKα	a = 24.399 b = 8.924 c = 13.672 β = 95.53	2,693.1	1.176 (1.18)	4	0.084	Outer rings (para type)	Inner rings (meta type)
								$a = 1.52$	$a = 1.53$
								$b = 1.61$	$b = 1.57$
								$e = 1.41$	$e,g = 1.39$
								$f = 1.39$	$f = 1.40$
								$u = 1.54$	$\alpha^*_{outer} = 0.758$
								$\alpha = 12.4$	$\alpha^*_{inner} = 0.737$
								$\beta = 11.4$	
								$\gamma = 112.1$	$\gamma_{outer} = 111.6$
								$\lambda = 119.0$	$\gamma_{inner} = 113.4$
								$p = 2.79$	$\lambda = 116.6$
								$q_{4,13} = 3.10$	$q_{13,21} = 3.31$
								$q_{8,15} = 3.22$	$p_{15,19} = 2.74$
									$p_{14,20} = 3.31$

Comments: The outer benzenoid rings are boat-shaped. The H atoms of these rings are displaced out of the plane of the nearest three C atoms toward the inner aromatic rings, whereas the methyl groups are displaced out of their corresponding plane (nearest three C atoms) away from the inner aromatic rings. The outer rings are twisted 8° with respect to the inner rings around a normal to their mean planes. The inner rings are distorted into twist-boat shapes (see Chapter 10, Table I). The dihedral angle formed between the planes of the inner ring (C-12,C-13,C-14 and C-11,C-16,C-15) is 13.4°. The displacements of the C atoms of the inner ring from their least-squares plane are considerably greater than those for the outer ring because of the two upper and two lower bridges. The H atoms of the C atoms of the inner rings are located on the mean plane of these rings and are not displaced from these planes. The inner rings exhibit a parallel displacement with respect to one another (0.15 Å) in order to reduce eclipsing of the ring atoms and π–π repulsion (see Chapter 10, Fig. 1).

Tetramethoxy quadruple-layered [2.2]paracyclophane (177)[171]

Monoclinic	$P2_1/c$	—	$a = 8.31$ $b = 21.35$ $c = 9.32$ $\beta = 106$	—	—	2	—	—

VII. FERROCENOPHANES

1,1,2,2-Tetramethyl[2]ferrocenophane (178)[172]

Monoclinic	$P2_1/c$	CuKα	$a = 7.756$ $b = 10.97$ $c = 15.41$ $\beta = 92.63$	—	1.36 (1.35)	4	0.079	$a = 1.55$ $b = 1.58$ $e = 1.45$ $f = 1.43$ $g = 1.41$ $n = 1.63$ $o_3 = 1.97$ $o_4 = 2.04$ $o_5 = 2.08$ $u = 1.56$ $p = 2.70$ $q_{7,8} = 3.14$ $q_{6,9} = 3.63$	$\beta = 11$ $\gamma = 110$ $\lambda = 107$ $\pi = 23.2$ $\rho = 9$

Comments: The aromatic rings are planar, and the C atoms in each ring do not eclipse one another ($\rho = 9°$). The ring planes are not parallel to one another because of the two-carbon bridge and are tilted 23.2° (π) with respect to one another, causing the C-3 and C-12 atoms to be closer together than the C-6 and C-9 atoms. The dihedral angle in the bridge (around C-1—C-2) is ~26°, is a compromise between the ideal gauche conformation (60°) and the eclipsed conformation (0°), which would occur if the atoms in the two rings were eclipsed.

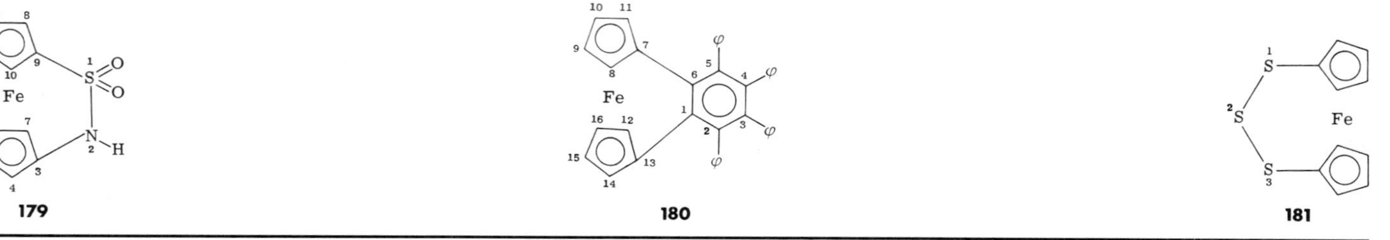

Crystal system	Space group	Radiation	Cell dimensions (Å)	Cell volume (Å³)	Density (g/cm³), calculated (measured)	Z	R	Significant bond lengths and nonbonded distances (Å), and bond and interplanar angles (deg)
2-Aza-1-thia[2]ferrocenophane 1,1-dioxide (179)[173]								
Monoclinic	$P2_1/c$	MoKα	22°C $a = 10.566(7)$ $b = 11.802(7)$ $c = 7.671(6)$ $\beta = 93.35(5)$	955.0	—	4	0.051	S side / N side $a = 1.740(9)$ / $a = 1.419(11)$ $b = 1.671(7)$ $e = 1.435(13)$ / $e = 1.424(13)$ $f = 1.392(17)$ / $f = 1.394(14)$ $g = 1.365(15)$ / $g = 1.407(13)$ S—O = 1.430(8) $n = 1.64$ $o_9 = 1.962(8)$ / $o_3 = 1.983(8)$ $o_{10} = 2.006(9)$ / $o_4 = 2.007(9)$ $o_{11} = 2.081(10)$ / $o_5 = 2.083(8)$ $p = 2.78(1)$ $q_{4,10} = 3.313(1)$ $q_{5,11} = 3.65(1)$ $\gamma = 104.5(4)$ / $\gamma = 112.5(6)$ $\lambda = 105.8(9)$ / $\lambda = 104.9(9)$ $\pi = 23$

Comments: The aromatic rings are planar, and the atoms of each ring are intermediate between being staggered and eclipsed. The angle of twist about the N—S bond is 29° (angle between planes defined by atoms N,S,C-9 and S,N,C-3), so the rings are not directly above one another.

2,3,4,5-Tetraphenyl-1,6-benzo[2]ferrocenophane (180)[174]

Monoclinic (hexane/ CH_2Cl_2)	$P2_1/a$	CuKα	a = 24.342(7) b = 10.959(5) c = 11.397(5) β = 112.69(3)	2,805.0	1.337 [1.38(10)]	4	0.054	a = 1.511(7) b = 1.388(9) e = 1.439(8) f = 1.433(8) g = 1.406(10) n = 1.637 o_7 = 1.976(6) o_8 = 2.034(7) o_9 = 2.080(7) p = 2.729(9) $q_{8,14}$ = 3.142(10) $q_{11,12}$ = 3.066(10) $q_{9,15}$ = 3,653(10) $q_{10,16}$ = 3.627(10)	β = 14.9 γ = 116.4(5) λ = 108.4(5) π = 23.7 ρ = ~0

Comments: All the aromatic rings are planar. The atoms of the two cyclopentadienyl rings eclipse one another almost perfectly (rotation around the C-1—C-6 bond is 3°), and the ring planes are tilted 23.7° (π) with respect to one another. These ring planes are also almost perpendicular to the hexasubstituted ring and form angles of 90.6 and 94.9° to the plane of this benzenoid ring. The other four benzenoid rings form interplanar angles of about 62.7° with respect to the hexasubstituted ring.

1,2,3-Trithia[3]ferrocenophane (181)[175]

Monoclinic	$P2_1/c$	CuKα	a = 9.628(3) b = 9.347(4) c = 11.408(4) β = 96.70(3)	1,019.5(3)	1.825 [1.83(2)]	4	0.096	a = 1.757(10) b = 2.049(4) e = 1.41(1) n = 1.66 o = 2.044(10)	γ = 102.8(4) λ = 108.9(9) π = 2.85 ρ = 0.08

Comments: The molecule adopts an eclipsed conformation of the cyclopentadienyl groups (ρ = 0.08°), which seems to be controlled by the trisulfide bridge. The S atoms directly attached to the rings are coplanar with the rings. The angle formed by the intersection of planes defined by S-1,Fe,S-3 and S-1, S-2, S-3 is 110.9°.

182

183

184

Crystal system	Space group	Radiation	Cell dimensions (Å)	Cell volume (Å^3)	Density (g/cm^3), calculated (measured)	Z	R	Significant bond lengths and nonbonded distances (Å), and bond and interplanar angles (deg)	
			1,3-Dithia-2-selena[3]ferrocenophane (182)[176]						
Monoclinic (hexane)	$P2_1/c$	MoKα	$a = 9.682(3)$ $b = 9.364(3)$ $c = 11.609(4)$ $\beta = 95.61(2)$	1,047.5	2.074	4	0.038	$a = 1.766$ $b = 2.195$ $e = 1.429$ $f = 1.418$ $g = 1.428$ $n = 1.653$ $o = 2.056$	$\alpha^* = 0.04$ $\gamma = 101.8(2)$ $\lambda = 108.4(4)$ $\pi = 2.4$ $\rho = 1.5$

Comments: The geometry is similar to that of 1,2,3-trithia[3]ferrocenophane. The two sets of ring C atoms nearly eclipse one another ($\rho = 1.5°$). The S atoms are displaced slightly ($\alpha^* = 0.04$ Å) from the mean ring plane in the direction opposite to the ring to which it is not bound. The rings are slightly tilted toward one another ($\pi = 2.4°$) so that C-5 and C-12 come into close contact. The angle formed by planes Fe,S-1,S-3 and S-1,Se-2,S-3 is 112.2°.

[3]Ferrocenophan-1-one (183)[177]

Monoclinic (heptane)	$P2_1/a$	—	$a = 22.981(2)$	—	1.61	4	0.067	C=O side	CH_2 side
			$b = 7.381(1)$		(1.60)				
			$c = 5.833(1)$					$a = 1.493$	$a = 1.532$
			$\beta = 93.38(2)$					$b = 1.508$	$b = 1.529$
								$e = 1.424$	$e = 1.433$
								$f = 1.420$	$f = 1.414$
								$g = 1.428$	$g = 1.432$
								$v = 1.211$	
								$n = 1.640$	
								$o_{10} = 2.007$	$o_4 = 2.030$
								$o_{11} = 2.027$	$o_5 = 2.037$
								$o_{12} = 2.070$	$o_6 = 2.046$
								$p = 3.103$	
								$q_{5,11} = 3.227$	
								$q_{6,12} = 3.421$	
								$\beta^* = 0.301$	$\beta^* = 0.123$
								$\gamma = 117.7$	$\gamma = 112.4$
								$\lambda = 108.1$	$\lambda = 106.6$
								$\pi = 8.8$	
								$\rho = 11.8$	

Comments: The aromatic rings are planar, and the C atoms in each ring do not fully eclipse one another (ρ = 11.8°). Because of the three-carbon bridge, the ring planes are not parallel (π = 8.8°). The bridge C atoms bound directly to the rings are displaced from the aromatic planes to which they are attached toward the opposite aromatic ring.

[3]Ferrocenophane^{+} : TCNE^{-} (184)[178]

Monoclinic	$P2_1/c$	MoKα	$a = 6.833(3)$	—	—	4	0.099	$a = 1.53(2)$	$o_4 = 2.06(2)$
			$b = 13.385(9)$					$b = 1.51(2)$	$o_{5,8} = 2.07(1)$
			$c = 12.214(6)$					$e = 1.43(2)$	$o_{6,7} = 2.11(2)$
			$\beta = 90.56(4)$					$f = 1.41(2)$	$\gamma = 117(1)$
								$g = 1.43(2)$	$\lambda = 108(1)$

Crystal system	Space group	Radiation	Cell dimensions (Å)	Cell volume (Å³)	Density (g/cm³), calculated (measured)	Z	R	Significant bond lengths and nonbonded distances (Å), and bond and interplanar angles (deg)
			5-(1-Phenyl-1-hydroxypropyl)[3](1,1′)ferrocenophane (185)[179]					
Monoclinic	$P2_1/n$	CuKα	$a = 11.42(1)$ $b = 13.25(1)$ $c = 11.48(1)$ $\beta = 97.75(20)$	—	1.42	4	0.071	See below

Substituted ring	Unsubstituted ring
$a = 1.539(25)$	$a = 1.560(32)$
$b = 1.569(22)$	$b = 1.552(26)$
$e_{4,5} = 1.456(18)$	$e_{10,11} = 1.399(30)$
$e_{4,8} = 1.451(21)$	$e_{9,10} = 1.476(29)$
$f_{5,6} = 1.431(21)$	$f_{11,12} = 1.539(30)$
$f_{7,8} = 1.437(18)$	$f_{9,13} = 1.445(31)$
$g = 1.463(17)$	$g = 1.439(30)$
$u = 1.521(17)$	
$n = 1.643$	
$o_4 = 2.017(12)$	$o_{10} = 2.021(17)$
$o_5 = 2.028(10)$	$o_{11} = 2.083(16)$
$o_6 = 2.073(10)$	$o_{12} = 2.093(17)$
$o_7 = 2.069(11)$	$o_{13} = 2.070(16)$
$o_8 = 2.059(12)$	$o_9 = 2.053(16)$

$\beta^* = 0.04$	$\beta^* = 0.15$
$\gamma = 111.6(1.4)$	$\gamma = 113.4(1.7)$
$\lambda = 107.6(1.0)$	$\lambda = 111.1(1.5)$

$\pi = 10$

$\rho \cong 0$

Comments: The cyclopentadienyl rings are planar, and the atoms in each ring eclipse those in the other ring. The ring planes are tilted 10° with respect to one another, and C-1 and C-3 are displaced from their adjacent aromatic plane (β^*).

5-Methyl-3-phenyl[3](1,1′)ferrocenophan-1-one (186)[180]

Orthorhombic	*Aba*2	CuKα	$a = 14.45(2)$ $b = 25.08(3)$ $c = 8.13(1)$	—	1.45	8	0.061

C=O side	Ph side
$a = 1.493(24)$	$a = 1.541(24)$
$b = 1.512(34)$	$b = 1.557(27)$
$e_{12,13} = 1.390(24)$	$e_{4,5} = 1.454(24)$
$e_{9,13} = 1.454(28)$	$e_{4,8} = 1.391(28)$
$f_{11,12} = 1.442(26)$	$f_{5,6} = 1.475(25)$
$f_{9,10} = 1.470(28)$	$f_{7,8} = 1.480(21)$
$g = 1.437(26)$	$g = 1.389(26)$
$v = 1.235(25)$	$u_3 = 1.594(26)$
	$u_5 = 1.516(26)$
$o_{13} = 1.999(19)$	$o_4 = 2.033(18)$
$o_{12} = 2.061(20)$	$o_5 = 2.052(16)$
$o_{11} = 2.102(21)$	$o_6 = 2.091(20)$
$o_{10} = 2.108(20)$	$o_7 = 2.064(21)$
$o_9 = 2.057(23)$	$o_8 = 2.020(21)$
$p = 3.079(28)$	
$q_{5,12} = 3.230(28)$;	$q_{8,9} = 3.256(34)$
$q_{6,11} = 3.503(30)$;	$q_{7,10} = 3.486(31)$
$\beta^* = 0.282$	$\beta^* = 0.082$
$\gamma = 117.9(1.7)$	$\gamma = 115.0(1.5)$
$\lambda = 109.4(1.7)$	$\lambda = 108.4(1.6)$

$\pi = 10.6$

$\rho = \sim 0$

Comments: The cyclopentadienyl rings are nearly planar and are tilted (π) 10.6° with respect to one another. The five-membered-ring C atoms are very nearly eclipsed, and the bridging atoms attached to the five-membered ring are displaced from the mean ring planes in a direction opposite from the Fe atom (β^*). The carbonyl group plane is inclined 42° to the plane of its appended five-membered ring. The C atom of the methyl substituent is coplanar with the ring to which it is appended.

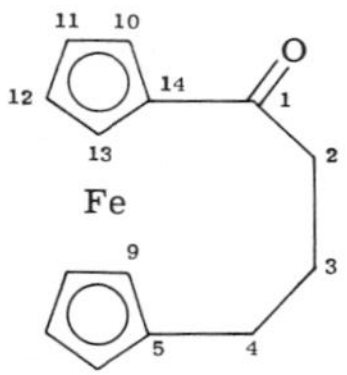

187 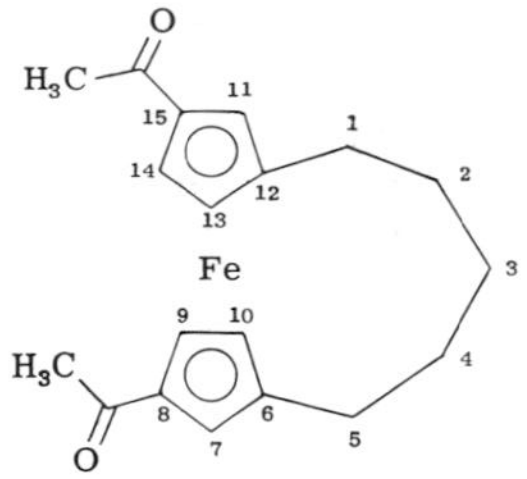 188 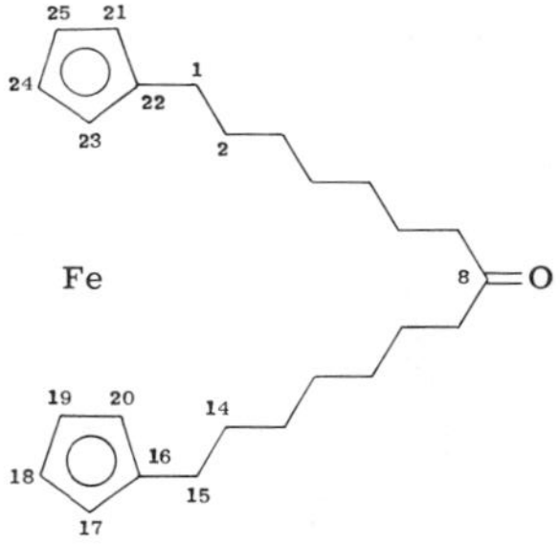 189

Crystal system	Space group	Radiation	Cell dimensions (Å)	Cell volume (Å^3)	Density (g/cm^3), calculated (measured)	Z	R	Significant bond lengths and nonbonded distances (Å), and bond and interplanar angles (deg)
[4]Ferrocenophan-1-one (187)[181]								
Orthorhombic	$P2_12_12_1$	CuKα	a = 5.785(3) b = 11.090(3) c = 17.115(8)	—	1.537	4	0.04	Carbonyl side / Saturated side a = 1.457(10) / a = 1.510(12) b = 1.507(10) / b = 1.432(14) c = 1.484(13) e = 1.434(10) / e = 1.393(10) f = 1.409(10) / f = 1.386(12) g = 1.411(10) / g = 1.413(12) v = 1.221(9) α^* = 0.33 / α^* = 0.02 γ = 120.2(6) / γ = 121.2(8) λ = 107.9(6) / λ = 107.6(7) π = 4.4(1) ρ = 36

Comments: The structure is poorly resolved due to high thermal motion. The cyclopentadienyl groups are nearly parallel and perfectly staggered, with the 4° angle of inclination (π) approximately oriented at right angles to the general plane of the aliphatic bridge. The carbonyl plane (C-2,C-14,O) is not coplanar with its neighboring cyclopentadienyl ring but is twisted 18.3(1)° with respect to it. The strain in the molecule is attributed to the twisting about this junction to obtain conjugation and maintain the staggered orientation of the two cyclopentadienyl rings.

3,4′-Diacetyl[5]ferrocenophane (188)[182]

Monoclinic	$P2_1/a$	MoKα	$a = 8.739(2)$ $b = 19.080(4)$ $c = 9.987(3)$ $\beta = 109.5$	1.570.0(6)	1.43	4	0.043	$a = 1.50(1)$ $b = 1.52(1)$ $e = 1.52(1)$ $e_{6,7} = 1.43(1)$ $e_{6,10} = 1.42(1)$ $f_{7,8} = 1.42(1)$ $f_{9,10} = 1.41(1)$ $g_{8,9} = 1.44(1)$ $u = 1.48(1)$ $n = 1.68$

$o_6 = 2.094(9)$ $o_7 = 2.045(9)$ $o_8 = 2.047(8)$ $o_9 = 2.043(9)$ $o_{10} = 2.064(8)$ $\beta^* = 0.095; 0.175$ $\gamma = 115.1(9)$ $\lambda = 105.3(7)$ $\pi = 3.9$ $\rho = 17.4$

Comments: Atoms C-1 and C-5 are displaced out of the cyclopentadienyl ring plane away from the cyclophane cavity. The aromatic rings are planar.

[15]Ferrocenophan-8-one (189)[183]

Monoclinic	$P2_1/c$	MoKα	$a = 25.528(4)$ $b = 7.6340(8)$ $c = 11.258(2)$ $\beta = 91.6$	2,193(1)	1.23 (1.24)	4	0.054	$a = 1.51$ $b = 1.50$ $c = 1.52$ $d = 1.52$ $e = 1.42$ $f = 1.42$ $g = 1.40$ $v = 1.23$ $n = 1.652$

$o_{16} = 2.068$ $o_{17} = 2.042$ $o_{18} = 2.035$ $\gamma = 116.4(6)$ $\lambda = 106.7(6)$ $\pi = 4.7$ $\rho = 4.6$

$\beta^* = 0.141; 0.099\ (5.0°; 3.8°)$
$p_{16,22} = 3.40$
$p_{1,15} = 3.74$

Comments: The C-1 and C-15 atoms are displaced out of the cyclopentadienyl ring plane away from the cyclophane cavity.

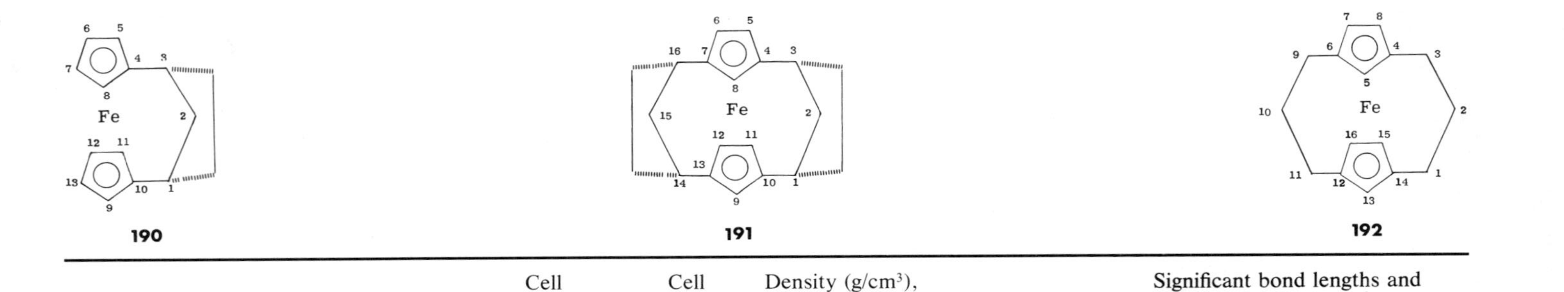

Crystal system	Space group	Radiation	Cell dimensions (Å)	Cell volume ($Å^3$)	Density (g/cm^3), calculated (measured)	Z	R	Significant bond lengths and nonbonded distances (Å), and bond and interplanar angles (deg)

1,3-Ethano[3](1,1′)ferrocenophane (190)[184]

Crystal system	Space group	Radiation	Cell dimensions (Å)	Cell volume ($Å^3$)	Density	Z	R	3-Bridge side	4-Bridge side
Orthorhombic	*Pbca*	MoKα	a = 16.250(7) b = 14.266(5) c = 10.337(2)	2,250.4	1.49 (1.46)	8	0.057		
								a = 1.493(30)	
								b = 1.547(27); 1.651(28)	b = 1.544(30)
									c = 1.605(30)
								e = 1.437(30)	e = 1.429(31)
								f = 1.414(32)	f = 1.419(33)
								g = 1.417(30)	
								n = 1.65	
								o_4 = 2.016(20)	
								o_8 = 2.039(16)	o_5 = 2.058(25)
								o_7 = 2.061(22)	o_6 = 2.057(21)
								p = 3.037(28)	
								$q_{8,9}$ = 3.196(24)	$q_{4,9}$ = 3.255(30)
								$q_{2,7}$ = 3.452(30)	$q_{3,8}$ = 3.482(29)
								β^* = 0.13(2)	
								λ = 106.8(185)	
								π = 11.1	
								ρ = ~0	

Comments: The aromatic rings are planar, and their atoms totally eclipse one another. The tilting angle between these two rings is 11.1°, and it is shown that this angle (π) is related to the number of atoms in the bridge (a bridge of three atoms has a tilting angle of about 9 to 11°). The bridging cyclopentanoid ring is distorted, but its mean plane is almost perpendicular (92.5°) to the aromatic ring planes. The nonbonded distance between the C-1 and C-3 atoms is 2.504(26) Å.

1,3;14,16-Bisethano[3](1,1′)[3](3,3′)ferrocenophane (191)[185]

Monoclinc	$C2/c$	MoKα	a = 20.119(5)	—	1.49	4	0.059	3-Bridge side / 4-Bridge side
			b = 5.974(1)		[1.46(3)]			a = 1.532(3)
			c = 14.456(9)					b = 1.551(3) / b = 1.547(4)
			β = 125.47(2)					c = 1.555(3)
								e = 1.441(3)
								f = 1.427(3)
								g = 1.437(3)
								n = 1.608
								o_8 = 2.008(2)
								o_4 = 2.008(1)
								o_5 = 2.051(2)
								p = 3.080(4)
								$q_{8,9}$ = 2.968(4)
								$q_{5,11}$ = 3.478(4)
								α^* = 0.235
								β = 4.60
								λ = 106.35(14)
								π = 16.95
								ρ = 0

Comments: The molecule has C_{2v} symmetry, with a twofold axis passing through the Fe atom and perpendicular to the C-5,C-6,C-11,C-12 plane. The two sets of aromatic ring atoms perfectly eclipse one another. The bridging atoms (C-4 and C-7) of the aromatic rings are displaced 0.021 Å out of their aromatic ring plane toward the Fe atom, and therefore the aromatic rings are not perfectly planar. These rings have an envelope shape, as is observed by the 0.067 Å displacement of the C-8 atom away from the Fe atom out of the C-4, C-5,C-6,C-7 plane. The bridging cyclopentanoid rings are identical and have a C_s envelope conformation. The angle between the planes of the four-carbon bridge and the three-carbon bridge (about C-1 and C-3) is 42.06°. The cyclopentanoid rings have a syn relationship. The nonbonded distance between the C-1 and C-3 atoms is 2.459 Å.

[3](1,1′)[3](3,3′)Ferrocenophane (192)[186]

Monoclinic	$C2/c$	MoKα	a = 23.91(2)	—	—	16	0.013	Average of equivalent values
			b = 9.07(1)					o = 2.01
			c = 23.22(2)					$p_{4,14} = p_{6,12}$ = 3.09
			β = 94.68(25)					$p_{7,16} = p_{8,15}$ = 3.36
								q = 3.01
								α^* = 0.21
								π = 9
								ρ = 0

Comments: There is positional disorder of the central methylene units of the bridges, and there are two crystallographically independent molecules in the crystal. The two sets of ring atoms eclipse one another.

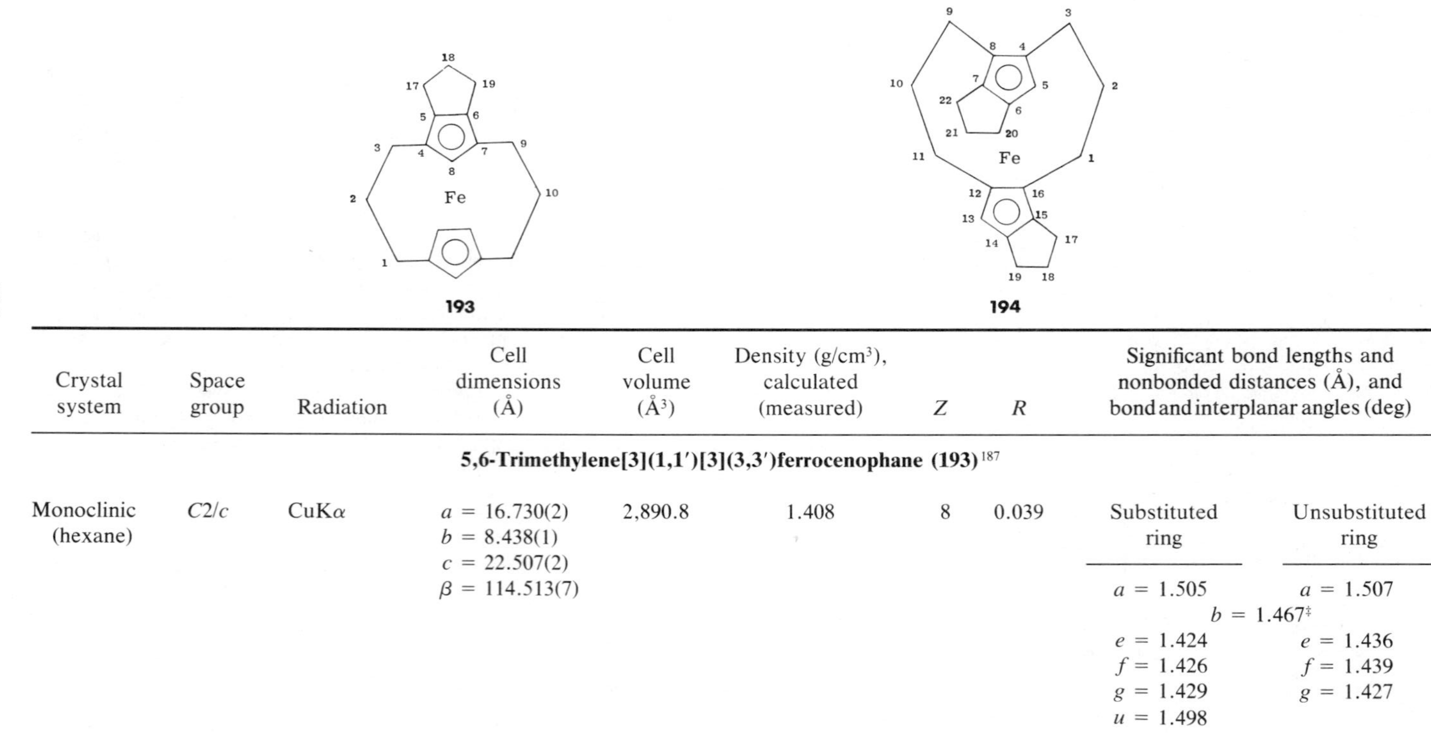

Crystal system	Space group	Radiation	Cell dimensions (Å)	Cell volume (Å³)	Density (g/cm³), calculated (measured)	Z	R	Significant bond lengths and nonbonded distances (Å), and bond and interplanar angles (deg)	
					5,6-Trimethylene[3](1,1′)[3](3,3′)ferrocenophane (193)[187]				
Monoclinic (hexane)	$C2/c$	CuKα	a = 16.730(2) b = 8.438(1) c = 22.507(2) β = 114.513(7)	2,890.8	1.408	8	0.039	Substituted ring	Unsubstituted ring
								a = 1.505	a = 1.507
								b = 1.467‡	
								e = 1.424	e = 1.436
								f = 1.426	f = 1.439
								g = 1.429	g = 1.427
								u = 1.498	

Comments: The central atoms (C-2 and C-10) of the methylene bridges bridging the cyclopentadienyl units are disordered. The cyclopentadienyl rings are slightly nonplanar. The central C atom of the homoannular ring C-18 is bent approximately 0.38 Å from the plane of its neighboring cyclopentadienyl ring toward the Fe atom.

3,4;4′,5′-Bis(trimethylene)[3](1,1′)[3](2,2′)ferrocenophane (194)[188]

Monoclinic (benzene)	$P2_1/c$	Nb-Filtered MoKα	$a = 8.649(4)$ $b = 21.623(2)$ $c = 9.120(1)$ $\beta = 107.21(3)$	—	1.412	4	0.035	Bridge opposite fused ring	Bridge adjacent to fused ring
								$a = 1.504(3)$	$a = 1.505(4)$
								$b = 1.536(4)$	$b = 1.524(4)$
								$e_{4,5} = 1.442(3)$	
								$e_{4,8} = 1.449(3)$	
								$f_{5,6} = 1.419(3)$	
								$f_{7,8} = 1.428(3)$	
								$g = 1.420(3)$	
								$u = 1.503(3)$	
								$n = 1.616$	
								$o_4 = 1.997(2)$	
								$o_5 = 2.032(2)$	
								$o_6 = 2.046(2)$	
								$o_7 = 2.029(2)$	
								$o_8 = 2.014(2)$	
								$\gamma = 114.7(2)$	$\gamma = 114.9(2)$
								$\lambda = 108.1(2)$	$\lambda = 106.6(2)$
								$\pi = 11.1$	
								$\rho = 0$	

Comments: The aromatic rings are planar, their atoms eclipse one another, and their mean planes are tilted by 11.1° (π) with respect to one another. The two homoannular rings are in a pseudo-ortho relationship and their central C atoms (C-18 and C-21) are bent 0.4 Å out of the appended aromatic ring plane toward the Fe atom. The central C atoms of the bridges (C-2 and C-10) are syn-oriented with respect to one another toward the homoannular rings.

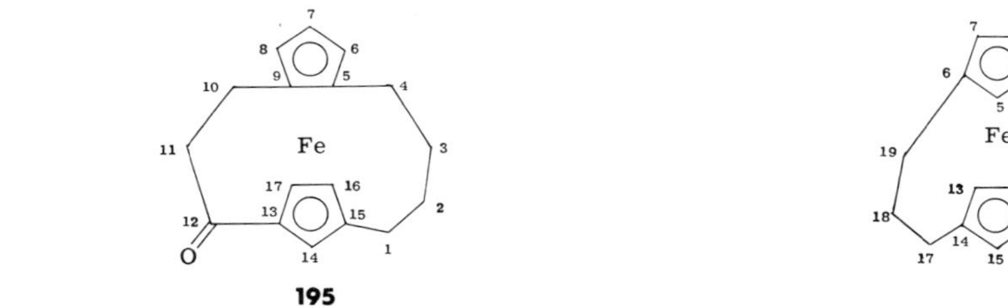

195

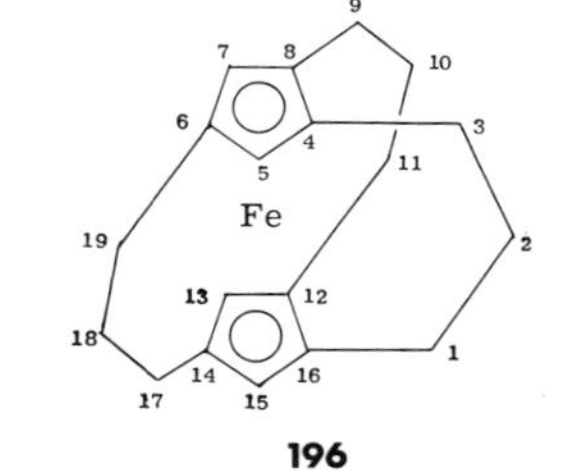

196

Crystal system	Space group	Radiation	Cell dimensions (Å)	Cell volume (Å³)	Density (g/cm^3), calculated (measured)	Z	R	Significant bond lengths and nonbonded distances (Å), and bond and interplanar angles (deg)
[4](1,1′)[3](3,2′)Ferrocenophan-12-one (195) [189]								
Monoclinic	$P2_1/n$	—	a = 7.455(5) b = 19.976(9) c = 9.154(6) β = 107.87(4)	1.51	—	4	0.033	1,2 ring Keto bridge side / Tetramethylene side a = 1.512(5) / a = 1.504(4) b = 1.548(5) / b = 1.531(5) c = 1.536(6) (tetramethylene side) e = 1.440(4) f = 1.424(4) / f = 1.418(4) g = 1.411(5) / g = 1.410(5) $o_{9,5}$ = 2.007(3) $o_{8,6}$ = 2.043(3) o_7 = 2.066(3) α^* = 0.085 / α^* = 0.044 γ = 116.8(3) / γ = 116.8(3) λ = 106.7(3) / λ = 107.5(3)

1,3 ring

Keto bridge side	Tetramethylene side
$a = 1.473(4)$	$a = 1.502(5)$
$b = 1.516(5)$	$b = 1.529(6)$
$v = 1.215$	$c = 1.536(6)$
$e = 1.431(4)$	$e = 1.419(4)$
$f = 1.433(4)$	$f = 1.431(5)$
$g = 1.412(4)$	
$o_{13} = 1.995(3)$; $o_{15} = 2.040(3)$	
$o_{14} = 2.010(3)$	
$o_{17} = 2.044(3)$; $o_{16} = 2.061(3)$	
$\alpha^* = 0.495$	$\alpha^* = 0.101$
$\gamma = 117.1(3)$	$\gamma = 115.9(3)$
$\lambda = 106.8(3)$	$\lambda = 106.7(3)$
$\pi = 12.2$	
$\rho = 27$	

Comments: The three-carbon bridge causes distortion in the molecule. The rings are inclined 12.2° (π) with respect to one another, causing the C · · · Fe distances to the rings to be different. The ring atoms are also not perfectly staggered with respect to one another, and the atoms α to the rings are displaced from the mean plane of the rings toward the Fe atom. The bridging aliphatic chains are oriented syn with respect to one another.

[3](1,1′)[3](2,2′)[3](4,4′)Ferrocenophane (196)[190]

								1,1′;2,2′ bridge	4,4′ bridge
Monoclinic (hexane)	$C2/c$	Zr-Filtered MoKα	$a = 30.899(6)$	5,818.9	1.399 (1.37)	16	0.063	$a = 1.512(10)$	$a = 1.523(10)$
			$b = 9.416(1)$					$b = 1.555(12)$	$b = 1.511(26)^{\ddagger}$
			$c = 25.145(1)$						$e = 1.437(10)$
			$\beta = 127.31(1)$						$f = 1.438(9)$
									$g = 1.443(9)$
								$n = 1.573$	
								$o = 1.985(6)$	$o = 1.978(6)$
								$o_{5,7} = 2.008(6)$	
								$p_{5,15} = 3.194$	
								$p = 3.095$	$p = 3.150$
								$\gamma = 113.2(7)$	$\gamma = 114.6(12)$
								$\lambda = 107.9(6)$	$\lambda = 108.2(7)$
								$\pi = 2.4$	

Comments: There are two crystallographically independent molecules with similar geometries. The aromatic rings are nearly parallel ($\pi = 2.4°$), are eclipsed, and are slightly distorted from being planar. The central C atoms of the 1,1′ and 2,2′ bridges are oriented in opposite directions from one another, whereas the central C atom of the 4,4′ bridge is disordered in two equivalent positions. The distance between the two rings (3.15 Å) is significantly shorter than that found in ferrocene.

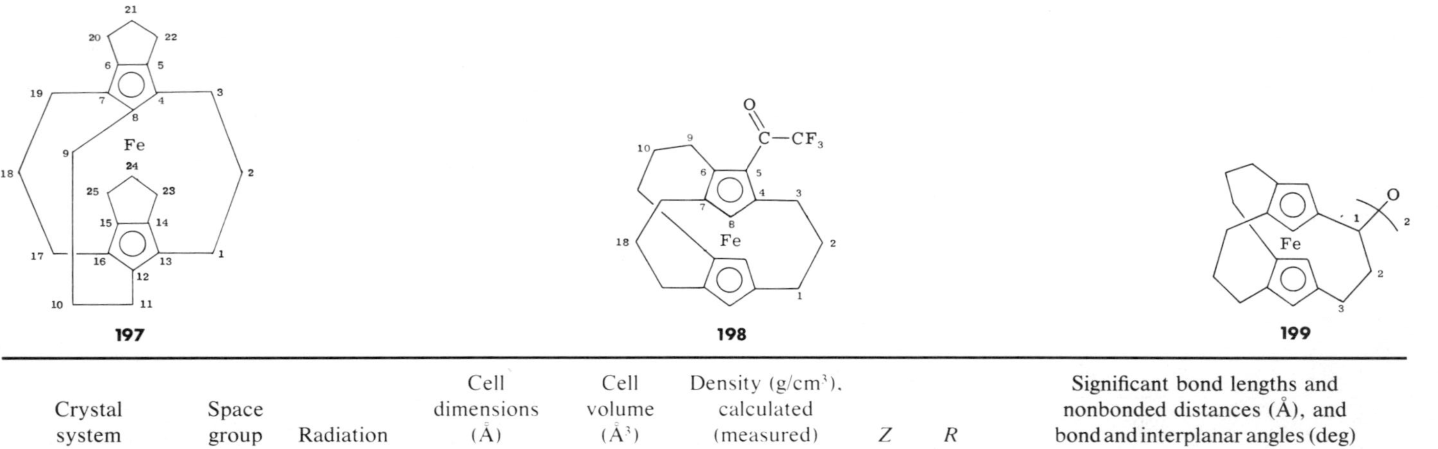

Crystal system	Space group	Radiation	Cell dimensions (Å)	Cell volume (Å^3)	Density (g/cm^3), calculated (measured)	Z	R	Significant bond lengths and nonbonded distances (Å), and bond and interplanar angles (deg)	
4,5;4′,5′-Bis(trimethylene)[3](1,1′)[3](2,2′)[3](3,3′)ferrocenophane (197)[191]									
Monoclinic	$P2_1/n$	CuKα	a = 9.486(2)	1.842	—	4	0.08	Central bridge	1,3 bridges
			b = 12.134(2)					a = 1.499(5)	a = 1.511(5)
			c = 16.024(2)					b = 1.499(7)	b = 1.527(6)
			β = 93.12(1)					e = 1.453(5)	
								f = 1.427(5)	
								g = 1.423(4)	
								u = 1.495(5)	
								n = 1.60	
								o_6 = 2.039(4)	
								o_7 = 1.997(4)	
								o_8 = 1.988(3)	
								γ = 117.5(4)	γ = 114.4(3)
								λ = 108.2(3)	λ = 106.8(3)
								π = 17.5	
								ρ = 0	

Comments: The two cyclopentadienyl ring atoms are eclipsed, and the angle of tilt (π) is the largest found in any three-atom-bridged ferrocenophane (17.5°). The aromatic rings are distorted slightly from planarity, with the C-8 and C-12 atoms being displaced out of their mean aromatic planes in a direction away from the Fe atom. The appended five-membered rings are almost coplanar with the aromatic rings to which they are attached. The central C atoms (C-2 and C-18) of the 1,3-trimethylene bridges are oriented syn with respect to one another and point toward the direction of the two homoannulated rings.

Triclinic (vapor diffusion of H_2O into EtOH)	$P\bar{1}$	CuKα	$a = 8.228(2)$ $b = 13.810(3)$ $c = 15.391(3)$ $\alpha = 99.43(1)$ $\beta = 99.10(2)$ $\gamma = 90.29(2)$	1,702.7	1.574	4	0.086	Substituted ring $a = 1.52$ $b = 1.56$ e, f, g (avg) = 1.46 $u_5 = 1.47$ C=O = 1.21 $n = 1.564(3)$	Unsubstituted ring $a = 1.51$ $b = 1.55$ e, f, g (avg) = 1.45 $n = 1.590(2)$

Comments: The structural features are similar to those of [3](1,1′)[3](2,2′)[3](4,4′)ferrocenophane. The atoms of the cyclopentadienyl rings eclipse one another. The C-10 and C-18 atoms are directed away from one another, whereas C-2 is directed toward C-18.

1,1′-Bis-[3](1,1′)[3](3,3′)[3](4,4′)ferrocenophano ether (199)[187]

Monoclinic (vapor diffusion of pentane into xylene; crystallized in either D or L crystals)	$P2_1$	CuKα	$a = 15.402(1)$ $b = 7.895(1)$ $c = 12.398(1)$ $\beta = 108.56(2)$	1,429(1)	1.452	2	0.063	Substituted bridge	Unsubstituted bridge
								Ether side $a = 1.52$ $b = 1.57$	Ether side $a = 1.51$ $b = 1.57$
								$e = 1.44$ $f = 1.46$ $g = 1.44$	
								Nonether side $a = 1.53$ $b = 1.54$	Nonether side $a = 1.50$ $b = 1.55$
								$e = 1.43$ $f = 1.47$ $g = 1.46$	

Comments: The two ferrocenophano units are roughly related by C_2 symmetry.

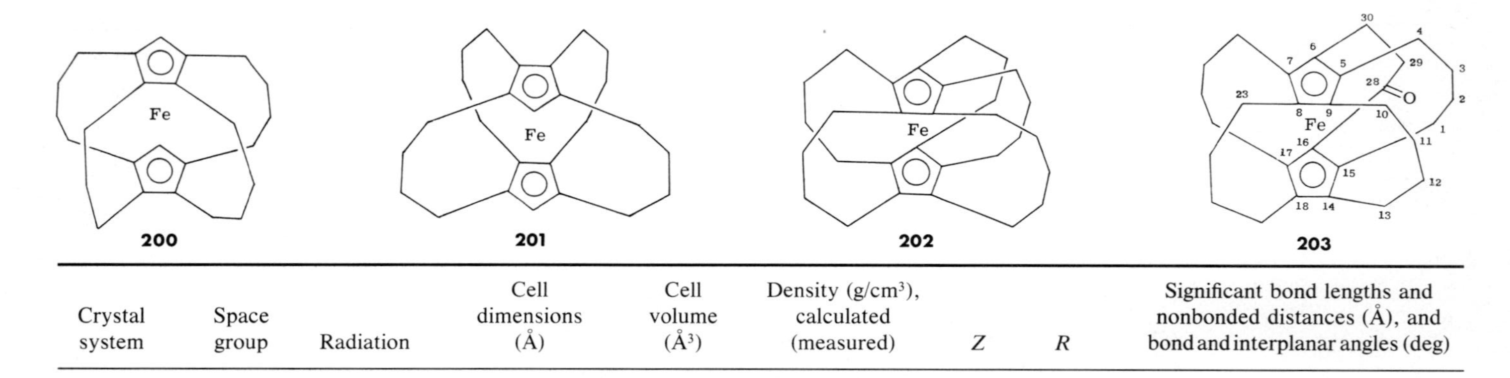

Crystal system	Space group	Radiation	Cell dimensions (Å)	Cell volume (Å³)	Density (g/cm³), calculated (measured)	Z	R	Significant bond lengths and nonbonded distances (Å), and bond and interplanar angles (deg)
			[4](1,1′)[4](3,3′)[4](5,5′)[3](4,4′)Ferrocenophane (200) [192]					
Monoclinic (hexane)	$P2_1/n$	MoKα	$a = 17.833(10)$ $b = 8.521(5)$ $c = 12.305(6)$ $\beta = 97.11(3)$	—	—	4	0.078	Trimethylene bridge: $a = 1.561$, $b = 1.625$, $n = 1.618$, $o = 2.023$, $\gamma = 116.6$, $\lambda = 108.8$ Tetramethylene bridge: $a = 1.498$, $b = 1.449$, $c = 1.433$, $\gamma = 120.4$, $\lambda = 106.9$ e, f, g (avg) $= 1.426$ $\pi = 5.7$

Comments: The planes of the aromatic rings are inclined 5.7° (π) with respect to one another, and the corresponding atoms of the two rings are eclipsed. The bridges have different conformations. All the bridges are oriented toward the nonbridged C atoms of the ring. The C atoms in the tetramethylene bridges bound directly to the five-membered rings are displaced above the ring planes, whereas the corresponding atoms in the trimethylene bridge are displaced below the ring planes.

[4_4]Ferrocenophane (201) [193]

(No data provided)

[4][4][4][4][3](1,2,3,4,5)Ferrocenophane (202) [194]

Tetragonal	$I\bar{4}2d$	MoKα	Ambient temperature a = 15.127(2) c = 18.905(2)	—	—	8	0.060	—

Comments: There is a disorder in the trimethylene chain. The five bridging chains have similar conformations, and there seems to be no distinction between the tetra- and trimethylene units. The central C—C bond in the tetramethylene chains are unacceptably short (~1.4 Å). The crystal structure suggests a rotational disorder about the pseudo-fivefold axis, which consists of four-fifths of a tetramethylene chain and one-fifth of a trimethylene chain.

[4][4][4][4][3](1,2,3,4,5)Ferrocenophan-28-one (203) [194]

Tetragonal	$I\bar{4}2d$	MoKα	a = 15.106(3) c = 19.069(3)	—	—	8	0.079	—

Comments: There is a disorder in the trimethylene chain and no sign of the C═O oxygen atom in the electron density map. The five bridging chains have similar conformations, and there seems to be no distinction between the tetra- and trimethylene units. The central C—C bond in the tetramethylene chains are unacceptably short (~1.4 Å). The crystal structure suggests a rotational disorder about the pseudo-fivefold axis, which consists of four-fifths of a tetramethylene chain and one-fifth of a trimethylene chain.

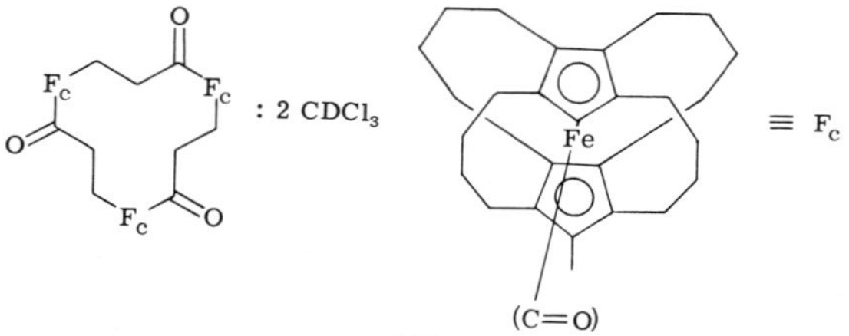

F_c : 2 $CDCl_3$ ≡ F_c (C=O)

204

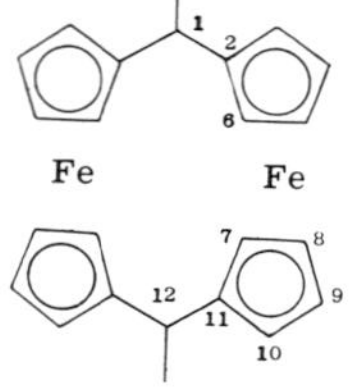

205

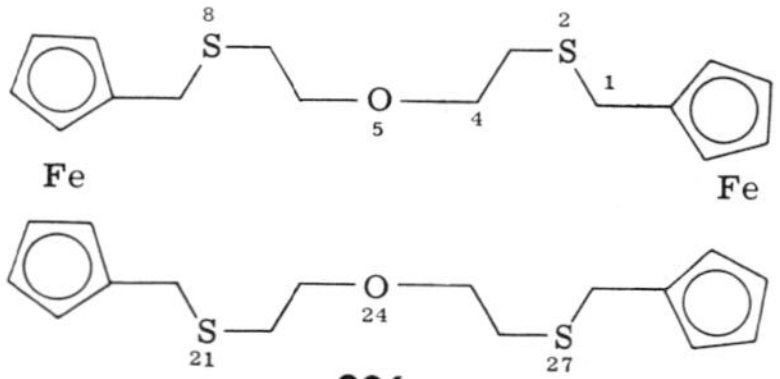

206

Crystal system	Space group	Radiation	Cell dimensions (Å)	Cell volume (Å³)	Density (g/cm^3), calculated (measured)	Z	R	Significant bond lengths and nonbonded distances (Å), and bond and interplanar angles (deg)
6,9,12-Trioxo[3.3.3](1,1′)[4](2,2′)[4](3,3′)[4](4,4′)[4](5,5′)ferrocenophane–2 $CDCl_3$ (204)[195]								
Orthorhombic ($CDCl_3$)	$P2_1nb$	CoKα	23°C a = 23.30(2) b = 21.69(2) c = 14.94(1)	7,550(18)	1.416 (1.399)	4	0.083	Ferrocenophano unit Four-C bridge: a = 1.53(2), b = 1.54(2), c = 1.53(2), α^* = 0.15, γ = 116.4, λ = 107.5 Three-C bridge: a = 1.52(2), b = 1.55(2), c = 1.51(2), d = 1.51(2), e = 1.44(2), f = 1.45(2), g = 1.44(2), v = 1.21(1), α^* = 0.15, γ = 110, λ = 109.5 π = 2.83

Comments: The three [4_4]ferrocenophano units are bridged by 3 three-carbon bridges, forming a cage. The macrocyclic structure has pseudo-D_3 symmetry, with a screw conformation of the tetramethylene bridges. No anisotropic thermal vibrations were observed for these C atoms even though they are commonly observed in ferrocenophanes. The tetramethylene bridges have novel dextro- and levo-screw conformations. The crystal contains two $CDCl_3$ molecules per trimeric unit, which fill the macrocyclic cavity by directing the C—D bonds into it toward the O atoms of the trimethylene chain [O · · · D, 3.13(3) Å]. The $CDCl_3$ seems to play an important role for crystallization since the trimer does not crystallize from other common solvents. The cyclopentadienyl units are planar, but the α-bridge carbons are displaced 0.15 Å (α^*) from the aromatic planes away from the Fe atom in the tetramethylene bridge and toward the Fe atom in the trimethylene bridge. The aromatic rings in each ferrocenophane unit are almost parallel to one another ($\pi = 2.83°$).

1,12-Dimethyl[1.1]ferrocenophane (205)[196,197]

Monoclinic	$P2_1/c$	CoKα	$a = 18.14(3)$	—	1.57	4	0.083	$a = 1.510(14)$	$\gamma = 116.7(9)$
			$b = 6.10(2)$		(1.51)			$e = 1.407(25)$	$\lambda = 107.0(7)$
			$c = 18.67(3)$					$m = 4.6$	$\pi = 2.7$
			$\beta = 119.67(10)$					$n = 1.65$	$\rho = 22.4$
								$o = 2.05$	$\sigma = 31$

Comments: The molecule has a twofold axis of rotation and exists in a syn conformation (the one-carbon bridges overlap one another). The methyl groups are exo to the cavity. The cyclopentadienyl rings in each ferrocenoid unit are rotated 22.4° with respect to one another (about the normal to the ring planes) in order to adopt a staggered orientation of the atoms of these rings. These rings are also tilted 2.7° with respect to one another so that the C-3 · · · C-10 and C-6 · · · C-7 nonbonded distances through the cavity are nonequal (3.29 and 3.39 Å, respectively). The Fe—ring distance is 1.65 Å (average distance to the C atoms of the ring, 2.05 Å), and the nonbonded Fe · · · Fe distance is 4.6 Å. The planes of the cyclopentadienyl groups linked through the methine bridge are twisted 31° with respect to one another. The endo C-1 and C-12 H atoms were calculated to be 2.54 Å from one another.

5,24-Dioxa-2,8,21,27-tetrathia[9.9]ferrocenophane (206)[198]

Monoclinic	$P2_1/c$	MoKα	$a = 17.985(4)$	1,522	1.52	2	0.068	$a = 1.485(12)$	$g = 1.417(13)$
(ethylene			$b = 9.035(4)$		(1.50)			$b = 1.827(9)$	$n = 1.659$
chloride)			$c = 9.475(3)$					$c = 1.805(11)$	$o = 2.042(10)$
			$\beta = 98.60(2)$					$d = 1.494(15)$	$\alpha^* = 0.085$
								C-4—O-5 = 1.426(12)	$\gamma = 111.2(9)$
								$e = 1.425(13)$	$\lambda = 106(1)$
								$f = 1.431(18)$	$\rho = 26$

Comments: The cyclopentadienyl rings are planar but twisted with respect to one another ($\rho = 26°$) at an angle rarely found in ferrocenophanes.

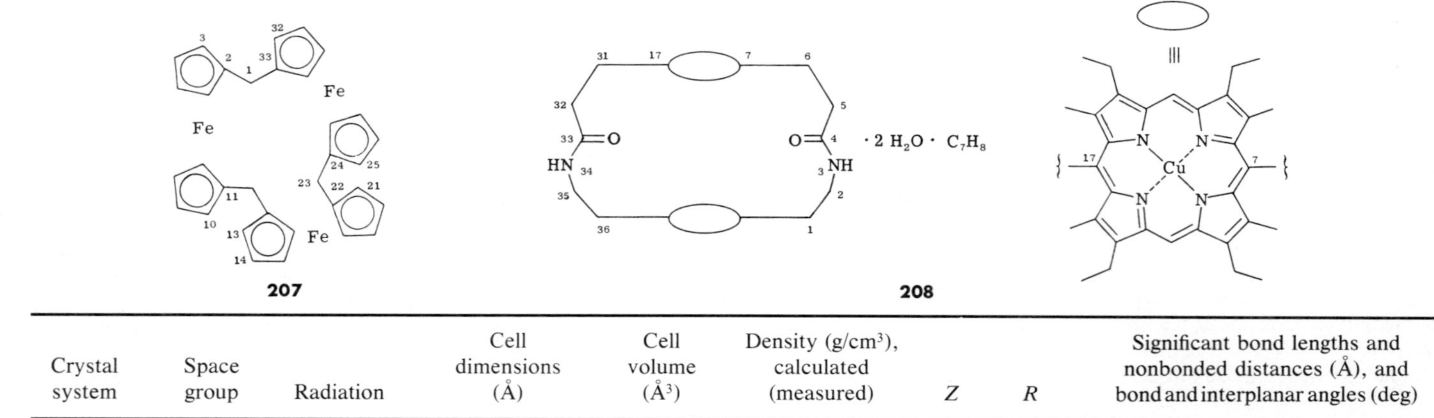

Crystal system	Space group	Radiation	Cell dimensions (Å)	Cell volume (Å³)	Density (g/cm³), calculated (measured)	Z	R	Significant bond lengths and nonbonded distances (Å), and bond and interplanar angles (deg)	
[1³]Ferrocenophane (207)[199]									
Monoclinic (CS_2)	$P2_1c$	Ni-Filtered CuKα	a = 6.008(3) b = 18.906(9) c = 22.53(1) β = 108.52(10	2,427	1.62 [1.63(3)]	4	0.030	a = 1.52 e = 1.42 m = 6.09(9) n = 1.65(1) o = 2.05	γ = 113.8 λ = 107.3 π = 2.6–4.6 ρ = 19; 11.5; 24.5 σ = 53; 61.5; 24.5

Comments: Three fereocenoid units are bridged by three methylene units to form a macrocyclic system. The Fe atoms are found at the apexes of a nearly perfect equilateral triangle (plane p), with the bridging methylene units and their appended H atoms pointing toward the inside of the cavity. Plane p forms angles of 24.6, 17.6, and 28.2° with the principal axis of each ferrocenoid unit. Nonbonded interactions between the H atoms on C-1, C-12, and C-23, which are directed into the center of the molecule, range from 2.13 to 2.81 Å, and those between the H atoms at C-10 and C-14 (C-21 and C-25; C-32 and C-3) range from 2.39 to 2.85 Å. It is these nonbonded interactions that are considered to be dominant in determining the conformation in these lower [1^n]ferrocenophanes.

3,34-Diaza[6.6](7,17)porphyrinophane-4,33-dione (2 Cu(II) · 2 H_2O · C_7H_8) (208)[200]

Monoclinic	$P2_1/c$	MoKα	−140°C	3,409	1.315 [1.29(1)]	2	0.063	—
			a = 11.878(6) b = 13.304(7) c = 23.725(13) β = 114.60(2)					

Comments: The precision is compromised by disorder and/or high thermal motion, but the molecule exhibits the porphyrin units linked at the 7,17 positions by the amide bridge. A perfect overlapping "face-to-face" structure is not observed, but rather the porphyrin units are laterally displaced with respect to one another in order to achieve a close approach of the units and to remove the large intraannular cavity that would be obtained from an overlay placing the central Cu atoms along the normal to both rings. The lateral displacement is almost parallel to the C-7,C-17 vector at the bridging atoms of the porphyrin units and displaces the Cu atoms by 4.95 Å. The Cu · · · Cu distance is 6.332 Å and the porphyrin planes are separated by an average distance of 3.87 Å. The porphyrin planes appear as curved surfaces with a slight saddle shape. Because of the lateral displacement a good portion of the bridging chain lies over the porphyrin rings.

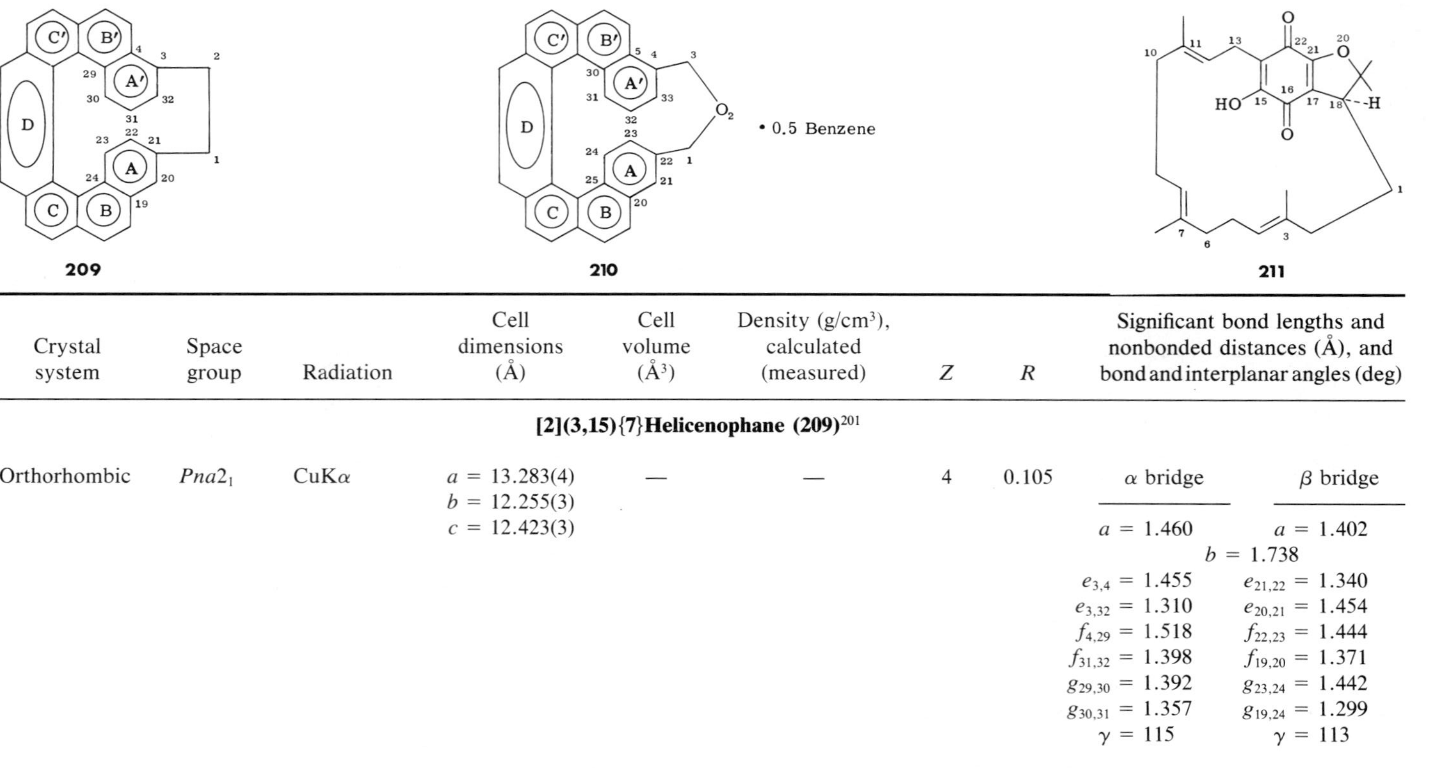

Crystal system	Space group	Radiation	Cell dimensions (Å)	Cell volume (Å³)	Density (g/cm³), calculated (measured)	Z	R	Significant bond lengths and nonbonded distances (Å), and bond and interplanar angles (deg)	
[2](3,15){7}Helicenophane (209)[201]									
Orthorhombic	$Pna2_1$	CuKα	a = 13.283(4) b = 12.255(3) c = 12.423(3)	—	—	4	0.105	α bridge	β bridge
								a = 1.460	a = 1.402
								b = 1.738	
								$e_{3,4}$ = 1.455	$e_{21,22}$ = 1.340
								$e_{3,32}$ = 1.310	$e_{20,21}$ = 1.454
								$f_{4,29}$ = 1.518	$f_{22,23}$ = 1.444
								$f_{31,32}$ = 1.398	$f_{19,20}$ = 1.371
								$g_{29,30}$ = 1.392	$g_{23,24}$ = 1.442
								$g_{30,31}$ = 1.357	$g_{19,24}$ = 1.299
								γ = 115	γ = 113

Comments: The helical conformation can easily be seen in the dihedral angles between the fused aromatic ring plane. The four dihedral angles associated with the B and B′ rings average 15.4°, and the other two associated with the D and C(C′) rings average 10.9°. The two monofused rings at the termini of the helix (A, A′) overlap one another almost perfectly, and the rings have a twisted conformation. The C-1—C-2 bond distance of 1.74 Å is probably one of the largest C—C bond lengths in a cyclophane bridge and is probably due to the spring action of the helicene.

2-Oxa[3](3,15){7}helicenophane · ½ benzene (210)

Monoclinic	$P2_1/c$	MoKα	a = 8.799(2)	2,308	—	4	0.105	α bridge	β bridge
			b = 12.461(3)						
			c = 21.388(5)					a = 1.522	a = 1.510
			β = 99.38(3)					b = 1.421	b = 1.468
								$e_{4,5}$ = 1.411	$e_{21,22}$ = 1.349
								$e_{4,33}$ = 1.400	$e_{22,23}$ = 1.437
								$f_{5,30}$ = 1.422	$f_{20,21}$ = 1.433
								$f_{32,33}$ = 1.397	$f_{23,24}$ = 1.410
								$g_{30,31}$ = 1.397	$g_{20,25}$ = 1.389
								$g_{31,32}$ = 1.399	$g_{24,25}$ = 1.437
								γ = 116	γ = 112

Comments: The helical conformation can easily be seen in the dihedral angles between the fused aromatic ring planes. The four dihedral angles associated with B and B′ rings average 12.2°, and the other two associated with the D and C(C′) rings average 11.6°. The two monofused rings at the termini of the helix (A, A′) overlap one another almost perfectly, and the nonbonded distances between atoms C-31 · · · C-25, C-5 · · · C-23, and C-30 · · · C-24 range between 2.95 and 3.53 Å. These rings have a twisted conformation. The C—O—C angle is 117°. This is 6–7° larger than normal C—O—C angles and is probably due to the spring action of the helicene.

15-Hydroxy-3,7,11,19,19-pentamethyl[13](3,6)2,3-dihydrobenzofuranophane (211)[202]
[from *S. tridentinus* (Boletales)]

Orthorhombic	$P2_12_12_1$	Ni-Filtered CuKα	a = 29.133(5)	4,643	1.731(1)	8	0.034	Average for both structures‡	
			b = 14.628(5)		[1.16(1)]				
			c = 10.896(4)					$a_{13,14}$ = 1.515	$e_{14,22}$ = 1.467
								$b_{12,13}$ = 1.506	$e_{14,15}$ = 1.344
								$c_{11,12}$ = 1.322	$f_{21,22}$ = 1.491
								$d_{10,11}$ = 1.511	$f_{15,16}$ = 1.506
								γ = 111.2	$g_{21,17}$ = 1.338
								λ = 118.3	$g_{16,17}$ = 1.439

Comments: This naturally occurring cyclophane has two independent molecules in the crystal with rather similar structures. The major difference lies in the torsion angles around C-6—C-7 and C-10—C-11, which have values that vary from −33.7 to +11.3° and +13.7 to −103.1°, respectively. The *p*-benzoquinoid portion is planar, and the dihydrofuranoid ring is almost planar, with the C-19 atom raised 0.2 Å from the mean quinoid plane. This latter plane is almost perpendicular to the macrocyclic ring (mean plane). The macrocyclic ring approximates an elipse with a major axis of ~9 Å (C-1 · · · C-10, 8.73 Å) and a minor axis of about 3.5 Å (C-3 · · · C-17, 3.46 Å). All the double bonds in the bridge have the trans configuration.

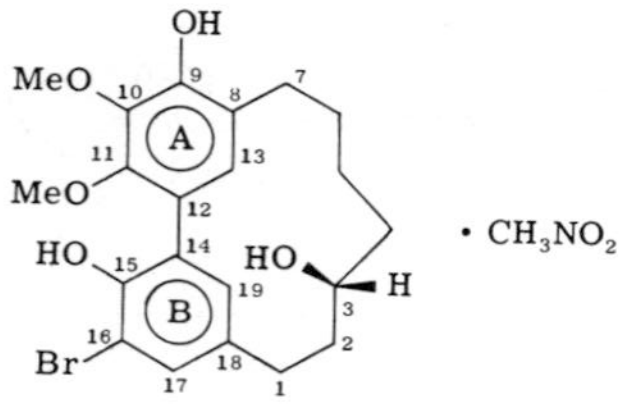

212 · CH_3NO_2

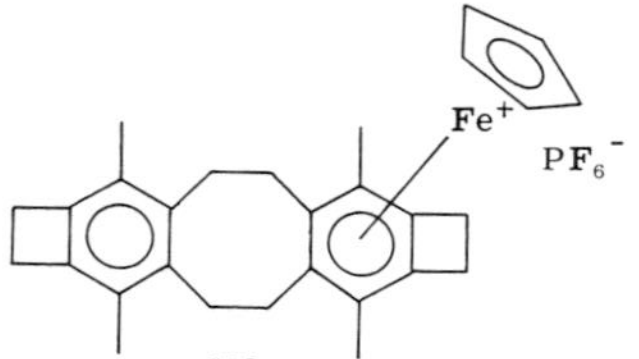

213

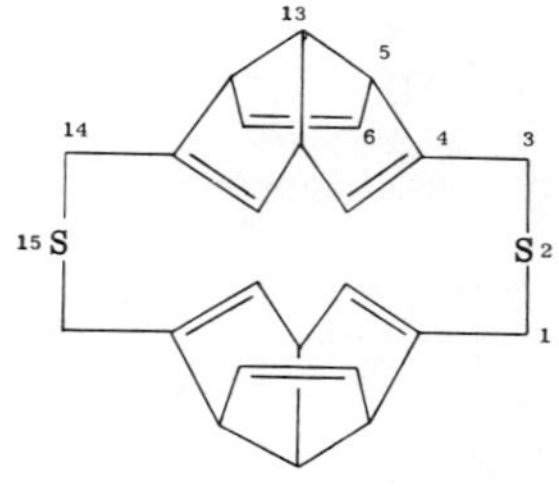

214

Crystal system	Space group	Radiation	Cell dimensions (Å)	Cell volume (Å³)	Density (g/cm³), calculated (measured)	Z	R	Significant bond lengths and nonbonded distances (Å), and bond and interplanar angles (deg)	
16-Bromo-3,9,15-trihydroxy-10,11-dimethoxy[7](3,3′)biphenylophane · nitromethane (16-bromomyricanol) (212)[203]									
Monoclinic	$P2_1$	MoKα	$a = 6.02(1)$	1,159	1.43	2	0.057	Br-Substituted ring	Di-OMe-substituted ring
			$b = 17.34(2)$		(1.38)			$a = 1.50$	$a = 1.48$
			$c = 11.22(2)$					$b = 1.52$	$b = 1.52$
			$\beta = 98.4(2)$					$c = 1.50$	$c = 1.52$
								$d = 1.57$	$d = 1.56$
								$e_{18,19} = 1.39$	$e_{8,13} = 1.39$
								$e_{17,18} = 1.39$	$e_{8,9} = 1.43$
								$f_{14,19} = 1.40$	$f_{12,13} = 1.37$
								$f_{16,17} = 1.36$	$f_{9,10} = 1.37$
								$g_{14,15} = 1.41$	$g_{11,12} = 1.39$
								$g_{15,16} = 1.39$	$g_{10,11} = 1.38$
									$v_{C\text{-}3—O} = 1.44$
								$\gamma = 115$	$\gamma = 115$
								$\lambda = 116$	$\lambda = 115$

Comments: The 16-bromo derivative of the naturally occurring material myricanol crystallizes with one CH_3NO_2 molecule as a 1 : 1 solvate. The most striking feature of this naturally occurring cyclophane is the distortion of the biphenyloid moiety because of meta–meta bridging. Thus, whereas both aromatic rings are nearly planar, C-12 and C-9 are displaced above the mean plane of ring B by 0.43 and 1.27 Å, respectively, and C-17 is displaced above the mean plane of ring A by 0.25 Å (C-14 lies in the plane of ring A). The dihedral angle between the aromatic planes is 33° (i.e., the rings are twisted about the C-12—C-14 bond). There are two intramolecular H bonds (C-15—OH · · · O—C-11, 2.66 Å; C-9—OH · · · O—C-10, 2.75 Å) and one intermolecular H bond (C-3—OH · · · O—C-9, 2.79 Å).

Cyclopentadienyl(4,7,12,15-tetramethyl-5,6;13,14-dicyclobuta[2.2]orthocyclophano)iron(II) hexafluorophosphate (213)[204]

(No data provided)

***anti*-2,15-Dithia[3.3](2,6)triquinacenophane (214)**[205]

Triclinic (CHCl₃) (When crystallized rapidly, the compound gave crystals with $Z = 2$; these were not subjected to X-ray analysis)	$P\bar{1}$	—	$a = 7.679(2)$ $b = 7.766(3)$ $c = 9.730(3)$ $\alpha = 111.91(3)$ $\beta = 93.58(3)$ $\gamma = 111.62(3)$	—	1.28	1	0.04	—

Comments: The molecule is centrosymmetric, and the triquinaceno units are oriented anti to one another. The S atoms are oriented anti as well.

Acknowledgments

I would like to thank Dr. Steven Madison and Ms. Ruth Chen for their valuable assistance, my research group (Dr. S. M., Dr. S. R., Dr. L. G., D. G., R. S., S. N., A. H., M. DeB., Y. H., and A. M.) for their patience, and my family (Lillian, Ilana, Galie, and Shlomit) for their love and understanding while I wrote this chapter. Special thanks to Professor Bruce Foxman for helpful comments and to Mrs. Evangeline Goodwin for her professional typing of the manuscript. The Camille and Henry Dreyfus Foundation is gratefully acknowledged for its support through a Dreyfus Teacher–Scholar Award (1979–1984).

תם ונשלם שבח לאל בורא עולם

REFERENCES

1. C. J. Brown and A. C. Farthing, *Nature (London)* **164,** 915–916 (1949).
2. C. J. Brown, *J. Chem. Soc.* pp. 3265–3270 (1953).
3. K. Lonsdale, H. J. Milledge, and K. V. Krishna Rao, *Proc. R. Soc. London, Ser. A* **255,** 82 (1960).
4. K. Lonsdale and H. J. Milledge, *Nature (London)* **184,** 1545–1549 (1959).
5. H. Hope, J. Bernstein, and K. N. Trueblood, *Acta Crystallogr., Sect. B* **B28,** 1733 (1972).
6. Y. Kai, H. Goto, N. Yasuoka, and N. Kasai, *Acta Crystallogr., Sect. A* **A34,** 5292 (1978).
7. M. Nagel and R. Allmann, *Chem. Ber.* (in press).
8. P. Goldstein, *Am. Crystallogr. Assoc., Abstr. Pap., Summer Meet.* p. 209 (1974).
9. K. Parker, R. C. Helgeson, E. Maverick, and K. N. Trueblood, *Am. Crystallogr. Assoc., Abstr. Pap., Winter Meet.* p. 49 (1973).
10. V. Taglieber and H. A. Staab, *Chem. Ber.* **110,** 3366 (1977).
11. H. Irngartinger and B. Merkert, *Eur. Crystallogr. Meet.* p. 442 (1977).
12. H. A. Staab, M. Jörns, and C. Krieger, *Tetrahedron Lett.* **20,** 2513 (1979).
13. H. Irngartinger, R.-D. Acker, W. Rebafka, and H. A. Staab, *Angew. Chem., Int. Ed. Engl.* **13,** 674 (1974).
14. H. A. Staab and V. Schwendemann, *Angew. Chem., Int. Ed. Engl.* **17,** 756 (1978).
15. T. Mizuma, H. Matsuora, Y. Kai, N. Yasuoka, and N. Kasai, *Bull. Chem. Soc. Jpn.* **55,** 979 (1982).
16. H. A. Staab, J. Ippen, C. Tao-pen, C. Krieger, and B. Starker, *Angew. Chem., Int. Ed. Engl.* **19,** 66 (1980).
17. J. Ippen, C. Tao-pen, B. Starker, D. Schweitzer, and H. A. Staab, *Angew. Chem., Int. Ed. Engl.* **19,** 67 (1980).
18. A. W. Hanson, *Cryst. Struct. Commun.* **10,** 559–563 (1981).
19. Y. Kai, N. Yasuoka, and N. Kasai, *Acta Crystallogr., Sect. B* **B34,** 2840 (1978).
20. M. G. Newton, T. J. Walter, and N. L. Allinger, *J. Am. Chem. Soc.* **95,** 5652 (1973).
21. N. L. Allinger, T. J. Walter, and M. G. Newton, *J. Am. Chem. Soc.* **96,** 4588 (1974).
22. K. Harata, T. Aono, K. Sakabe, N. Sakabe, and J. Tanaka, *Acta Crystallogr., Sect. A* **A28,** S14 (1972).
23. T. Aono, K. Sakabe, N. Sakabe, C. Katayama, and J. Tanaka, *Acta Crystallogr., Sect. B* **B31,** 2389 (1975).

24. R. M. Weiss, *Diss. Abstr. Int. B* **37** (10), 5117 (1977).
25. H. Ueda, C. Katayama, and J. Tanaka, *Bull. Chem. Soc. Jpn.* **54,** 891–896 (1981).
26. T. Kaneda and S. Misumi, *Bull. Chem. Soc. Jpn.* **50,** 3310 (1977).
26a. E. A. Truesdale, R. S. Hutton, and J. Bernstein, *Abstr. Pap., 179th Meet., Am. Chem. Soc.,* March 1980.
26b. J. Bernstein and A. Maltz, personal communication.
27. P. K. Gantzel and K. N. Trueblood, *Acta Crystallogr.* **18,** 958 (1965).
28. K. N. Trueblood and J. Bernstein, *Am. Crystallogr. Assoc., Abstr. Pap., Winter Meet.* p. 42 (1970).
29. J. Bernstein and K. N. Trueblood, *Acta Crystallogr., Sect. B* **B27,** 2078 (1971).
30. R. Benn, N. E. Blank, M. W. Haenel, J. Klein, A. R. Koray, K. Weidenhammer, and M. L. Ziegler, *Angew. Chem., Int. Ed. Engl.* **19,** 44 (1980).
31. N. E. Blank, M. W. Haenel, A. R. Koray, K. Weidenhammer, and M. L. Ziegler, *Acta Crystallogr., Sect. B* **B36,** 2054–2059 (1980).
32. A. R. Koray, M. L. Ziegler, E. Norman, and M. W. Haenel, *Tetrahedron Lett.* **20,** 2465–2466 (1979).
33. H. A. Staab, A. Döhling, and C. Krieger, *Liebigs Ann. Chem.* pp. 1052–1064 (1981).
34. P. N. Swepston, S. T. Lin, A. Hawkins, S. Humphrey, S. Siegel, and A. W. Cordes, *J. Org. Chem.* **46,** 3754–3756 (1981).
35. B. K. Balbach, A. R. Koray, A. Okur, P. Wuelknitz, and M. L. Manfred, *J. Organomet. Chem.* **212,** 77–94 (1981).
36. J. S. Ricci, Jr. and I. Bernal, *J. Chem. Soc., Chem. Commun.* p. 1453 (1969).
37. C. L. Coulter and K. N. Trueblood, *Acta Crystallogr.* **16,** 667 (1963).
38. A. W. Hanson, *Acta Crystallogr., Sect. B* **B27,** 197 (1971).
39. L. H. Weaver and B. W. Matthews, *J. Am. Chem. Soc.* **96,** 1581 (1974).
40. A. W. Hanson and M. Rohrl, *Acta Crystallogr., Sect. B* **B28,** 2287 (1972).
40a. V. Boekelheide and R. A. Hollins, *J. Am. Chem. Soc.* **95,** 3201–3208 (1973).
41. A. W. Hanson, *Cryst. Struct. Commun.* pp. 1243–1247 (1980).
42. J. Jureiw, T. Skorachodowa, J. Merkuschew, W. Winter, and H. Meier, *Angew. Chem., Int. Ed. Engl.* **20,** 269 (1981).
43. A. W. Hanson, K. Huml, and E. W. Macauley, *Am. Cryst. Assoc., Abstr. Pap., Summer Meet.* p. 69 (1970).
44. A. W. Hanson and E. W. Macauley, *Acta Crystallogr., Sect. B* **B28,** 1255 (1972).
45. A. W. Hanson and T. S. Cameron, *J. Chem. Res., Synop.* pp. 336–337 (1980).
45a. P. F. L. Schirch, *Diss. Abstr. Int. B* **42,** 1897–1898 (1981).
45b. P. F. L. Schirch and V. Boekelheide, *J. Am. Chem. Soc.* **103,** 6873–6878 (1981).
45c. Y. Sekine and V. Boekelheide, *J. Am. Chem. Soc.* **103,** 1777–1785 (1981).
46. R. Weiss and K. N. Trueblood, *Am. Crystallogr. Assoc., Abstr. Pap., Summer Meet.* p. 251 (1974).
47. H. Irngartinger, J. Hekeler, and B. M. Lang, *Chem. Ber.* (in press).
48. A. W. Hanson, *Acta Crystallogr., Sect. B* **B33,** 2003 (1977).
49. A. W. Hanson, *Cryst. Struct. Commun.* **10,** 313–317 (1981).
50. A. W. Hanson, *Cryst. Struct. Commun.* **10,** 751–756 (1981).
51. A. W. Hanson, *Cryst. Struct. Commun.* **10,** 195–200 (1981).
52. K. Mirsky, K. Trueblood, E. F. Maverick, and L. Grossenbacher, *Am. Crystallogr. Assoc., Ser. 2* **7,** 24 (1980).
53. E. Ljungström, O. Lindqvist, and O. Wennerström, *Acta Crystallogr., Sect. B* **B34,** 1889 (1978).
54. G. Ghemard, C. Souleau, C. Seigheraert, and E. Cuingnel, *J. Appl. Crystallogr.* **14,** 465 (1981).

55. K. Odashima, A. Itai, Y. Iitaka, and K. Koga, *J. Am. Chem. Soc.* **102,** 2504–2505 (1980).
56. Y. Ito, S. Miyata, M. Nakatsuka, T. Saegusa, M. Takamoto, and Y. Wada, *J. Chem. Soc., Chem. Commun.* pp. 375–376 (1982).
57. G. Sawitzki and H. Rau, *Liebigs Ann. Chem.* pp. 993–998 (1981).
58. C. J. Brown, *J. Chem. Soc.* p. 3278 (1953).
59. Y. Kai, N. Yasuoka, and N. Kasai, *Acta Crystallogr., Sect. B.* **B33,** 754 (1977).
60. K. Mislow, M. Brzechffa, H. W. Gschwend, and R. T. Puckett, *J. Am. Chem. Soc.* **95,** 621 (1973).
61. A. W. Hanson, *Acta Crystallogr.* **15,** 956 (1962).
62. M. Mathew, *Acta Crystallogr., Sect. B* **B24,** 530 (1968).
63. H. A. Staab, C. P. Herz, A. Döhling, and C. Krieger, *Chem. Ber.* **113,** 241 (1980).
64. I. Goldberg, *Acta Crystallogr., Sect. B* **B32,** 41–46 (1976).
65. I. Goldberg, *Acta Crystallogr., Sect. B* **B31,** 2592–2600 (1975).
66. I. N. Reinhoudt and H. J. den Hertog, Jr., *Tetrahedron Lett.* **22,** 2513–2516 (1981).
67. M. Mathew and A. W. Hanson, *Acta Crystallogr., Sect. B* **B24,** 1680 (1968).
68. A. W. Hanson and K. Huml, *Acta Crystallogr., Sect. B* **B25,** 2310 (1969).
69. A. W. Hanson and K. Huml, *Acta Crystallogr., Sect. B* **B27,** 459 (1971).
70. S. Kiryu and W. Nowacki, *Z. Kristallogr.* **142,** 99 (1975).
71. S. Kiryu and W. Nowacki, *Z. Kristallogr.* **142,** 108 (1975).
72. W. Anker, G. W. Bushnell, and R. H. Mitchell, *Can. J. Chem.* **57,** 3080–3087 (1979).
73. G. W. Bushnell and R. H. Mitchell, *Can. J. Chem.* **60,** 362–367 (1982).
74. T.-L. Chan, C.-K. Chan, K.-W. Ho, J. S. Tse, and T. C. W. Mak, *J. Cryst. Mol. Struct.* **7,** 199 (1977).
75. B. R. Davis and I. Bernal, *J. Chem. Soc. B* p. 2307 (1971).
76. N. Bresciani-Pahor, M. Calligaris, and L. Randaccio, *Acta Crystallogr., Sect. B* **B36,** 632–638 (1980).
77. F. Bottino, S. Foti, S. Pappalardo, and N. Bresciani-Pahor, *Tetrahedron Lett.* **20,** 1171 (1979).
78. F. Bottino and S. Pappalardo, *Tetrahedron* **36,** 3095 (1980).
79. N. B. Pahor, M. Calligaris, L. Randaccio, F. Bottino, and S. Pappalardo, *Gazz. Chim. Ital.* **110,** 227–231 (1980).
80. D. Taylor, *Aust. J. Chem.* **31,** 1235 (1978).
81. A. W. Hanson and M. Rohrl, *Acta Crystallogr., Sect. B* **B28,** 2032 (1972).
82. A. W. Hanson, *Acta Crystallogr., Sect. B* **B31,** 2352 (1975).
83. B. Nilsson, *Acta Chem. Scand.* **22,** 732–747 (1968).
84. K. J. Palmer, R. Y. Wong, L. Jurd, and K. Stevens, *Acta Crystallogr., Sect. B* **B32,** 847–852 (1976).
85. G. D. Andreeti, R. Ungaro, and A. Pochini, *J. Chem. Soc., Chem. Commun.* p. 1005 (1979).
86. G. D. Andreeti, R. Ungaro, and A. Pochini, *J. Chem. Soc., Chem. Commun.* p. 533 (1981).
87. I. Ueda and W. Nowacki, *Z. Kristallogr.* **139,** 70 (1974).
88. I. Ueda, F. E. Scarbrough, and W. Nowacki, *Z. Kristallogr.* **140,** 169 (1974).
89. A. Itai, Y. Tanaka, and Y. Iitaka, *Am. Crystallogr. Assoc., Proc. and Abstr.* p. PA32 (1979).
90. H. Irngartinger, *Acta Crystallogr., Sect. B* **B32,** 696–702 (1976).
91. K. N. Trueblood, C. B. Knobler, E. Maverick, B. C. Helgeson, S. B. Brown, and D. J. Cram, *J. Am. Chem. Soc.* **103,** 5594–5596 (1981).
92. I. Goldberg, *Acta Crystallogr., Sect. B.* **B34,** 3387 (1978).

93. I. Goldberg, *J. Am. Chem. Soc.* **102,** 4106 (1980).
94. D. S. Lingenfelter, R. C. Helgeson, and D. J. Cram, *J. Org. Chem.* **46,** 393–406 (1981).
95. Y. Fujise, T. Shiokawa, Y. Mazaki, Y. Fukazawa, M. Fujii, and S. Ito, *Tetrahedron Lett.* **23,** 1601 (1982).
96. K. Ibata, H. Shimanoushi, and Y. Sasada, *Acta Crystallogr., Sect. B* **B31,** 482–489 (1975).
97. A. Kawamata, Y. Fukazawa, Y. Fujise, and S. Ito, *Tetrahedron Lett.* **23,** 1083 (1982).
98. S. Ito, *Pure Appl. Chem.* **54,** 957–974 (1982).
99. N. Kato, Y. Fukazawa, and S. Ito, *Tetrahedron Lett.* **20,** 1113 (1979).
100. M. W. Heanel, *Chem. Ber.* **111,** 1789 (1978).
101. H. Irngartinger and A. Goldmann, *Z. Kristallogr.* **149,** 97 (1979).
102. A. Fratini, private communication.
103. J. H. Golden, *J. Chem. Soc.* p. 3741 (1961).
104. A. Wada and J. Tanaka, *Acta Crystallogr., Sect. B* **B33,** 355–360 (1977).
105. T. Toyoda and S. Misumi, *Tetrahedron Lett.* **19,** 1479 (1978).
106. A. Iwama, T. Toyoda, M. Yoshida, T. Otsubo, Y. Sakata, and S. Misumi, *Bull. Chem. Soc. Jpn.* **51,** 2988 (1978).
107. A. Dunand, J. Ferguson, M. Puza, and G. B. Robertson, *J. Am. Chem. Soc.* **102,** 3524–3530 (1980).
108. N. Kato, H. Matsunaga, S. Oeda, Y. Fukazuawa, and S. Ito, *Tetrahedron Lett.* **20,** 2419 (1979).
109. Y. Fukazawa, M. Aoyagi, and S. Ito, *Tetrahedron Lett.* **20,** 1055 (1979).
110. Y. Fukazawa, M. Aoyagi, and S. Ito, *Tetrahedron Lett.* **19,** 1067 (1978).
111. Y. Fukazawa, M. Sohikawa, and S. Ito, *Tetrahedron Lett.* **23,** 2129 (1982).
112. H. Irngartinger, R. G. H. Kinstetter, C. Krieger, H. Rodewald, and H. A. Staab, *Tetrahedron Lett.* **18,** 1425 (1977).
113. H. A. Staab and R. G. H. Kinstetter, *Liebigs Ann. Chem.* pp. 886–898 (1979).
114. Y. Kai, F. Hama, N. Yasuoka, and N. Kasai, *Acta Crystallogr., Sect. B* **B34,** 1263 (1978).
115. F. Frolow and P. Keehn, *Tetrahedron* (in press).
116. N. B. Pahor, M. Calligaris, and L. Randaccio, *J. Chem. Soc., Perkin Trans.* **2,** 42 (1978).
117. B. Kamenar and C. K. Prout, *J. Chem. Soc.* p. 4838 (1965).
118. A. Halverson, S. Rosenfeld, B. M. Foxman, and P. M. Keehn, *Tetrahedron* (in press).
119. B. Taskhodzaev, L. G. Voroncova, and F. O. Alasev, *Zh. Strukt. Khim.* **18,** 316–318 (1977).
120. Z. V. Todres, N. G. Furmanova, S. C. Avagyan, Yu. T. Struchkov, and D. N. Kursanov, *Phosphorus Sulfur* **5,** 309–313 (1979).
121. M. Corson, B. M. Foxman, and P. M. Keehn, *Tetrahedron* **34,** 1641 (1978).
122. J. L. Atwood, W. E. Hunter, C. Wong, and W. W. Paudler, *J. Heterocycl. Chem.* **12,** 433–438 (1975).
123. N. B. Pahor, M. Calligaris, and L. Randaccio, *J. Chem. Soc., Perkin Trans.* **2,** 38 (1978).
124. D. Taylor, *Aust. J. Chem.* **31,** 1953–1957 (1978).
125. F. Hama, Y. Sakata, and S. Misumi, *Tetrahedron Lett.* **22,** 1123–1126 (1981).
126. K. Doyama, F. Hama, Y. Sakata, and S. Misumi, *Tetrahedron Lett.* **22,** 4101–4104 (1981).
127. I. Goldberg and H. Rezmovitz, *Acta Crystallogr., Sect. B* **B34,** 2894 (1978).
128. G. Sawitzki and H. G. von Schnering, *Chem. Ber.* **112,** 3104–3109 (1979).
129. Z. P. Haque, M. McPautlin, and P. A. Tasker, *Inorg. Chem.* **18,** 2920 (1979).

130. Z. P. Haque, D. C. Liles, M. McPartlin, and P. A. Tasker, *Inorg. Chim. Acta* **23,** L21 (1977).
131. G. R. Newkome, A. Nayak, F. Fronczek, T. Kawato, H. C. R. Taylor, L. Meade, and W. Mattice, *J. Am. Chem. Soc.* **101,** 4472–4447 (1979).
132. G. R. Newkome, D. K. Kohli, and F. R. Fronczek, *J. Chem. Soc., Chem. Commun.* pp. 9–11 (1980).
133. G. R. Newkome, D. K. Kohli, F. R. Fronczek, B. J. Hales, E. E. Case, and G. Chiari, *J. Am. Chem. Soc.* **102,** 7608–7610 (1980).
134. G. R. Newkome, T. Kawato, F. R. Fronczek, and W. H. Benton, *J. Org. Chem.* **45,** 5423–5425 (1980).
135. F. R. Fronczek, S. F. Watkins, and G. R. Newkome, *J. Chem. Soc., Perkin Trans.* **2,** 877–882 (1981).
136. F. Fronczek, A. Nayak, and G. R. Newkome, *Acta Crystallogr., Sect. B* **B35,** 775 (1979).
137. F. R. Fronczek, V. K. Majestic, G. R. Newkome, W. C. Hunter, and J. L. Atwood, *J. Chem. Soc., Perkin Trans.* **2,** 331–335 (1981).
138. G. R. Newkome, S. J. Garbis, V. H. Majestic, and F. R. Fronczek, *J. Org. Chem.* **46,** 833–839 (1981).
139. G. R. Newkome, V. Majestic, F. Fronczek, and J. L. Atwood, *J. Am. Chem. Soc.* **101,** 1047–1048 (1979).
140. I. R. Hanson and M. R. Truter, *J. Chem. Soc., Perkin Trans.* **2,** 1–11 (1981).
141. I. R. Hanson, O. G. Parsons, and M. R. Truter, *J. Chem. Soc., Chem. Commun.* p. 486 (1979).
142. J. A. Herbert and M. R. Truter, *J. Chem. Soc., Perkin Trans.* **2,** 1253–1258 (1980).
143. J. I. Owen, *J. Chem. Soc., Perkin Trans.* **2,** 12–18 (1981).
144. I. R. Hanson, I. G. Parsons, and M. R. Truter, *Acta Crystallogr., Sect. B* **B38,** 448–451 (1982).
145. P. A. Tasker, J. Trotter, and L. F. Lindoy, *J. Chem. Res., Synop.* pp. 328–329 (1981).
146. D. J. Williams, and D. Lawton, *Tetrahedron Lett.* **16,** 111 (1975).
147. D. Lawton and H. M. Powell, *J. Chem. Soc.* p. 2339 (1958).
148. J. Allamand and R. Gerdil, *Cryst. Struct. Commun.* **10,** 33 (1981).
149. R. Gerdil and J. Allemand, *Tetrahedron Lett.* **20,** 3499 (1979).
150. R. A. Yellin, B. S. Green, N. Knossow, and G. Tsoucaris, *Tetrahedron Lett.* **21,** 387 (1980).
151. S. Brunie, A. Navaza, G. Tsoucaris, J. P. Declercq, and G. Germain, *Acta Crystallogr., Sect. B* **B33,** 2645 (1977).
152. R. A. Yellin, S. Brunie, B. S. Green, M. Knossow, and G. Tsoucaris, *J. Am. Chem. Soc.* **101,** 7529 (1979).
153. F. E. Elhadi, D. W. Ollis, J. F. Stoddart, D. J. Williams, and K. A. Woode, *Tetrahedron Lett.* **21,** 4215–4218 (1980).
154. A. M. Van Herk, K. Goubitz, A. R. Overbeck, and C. H. Stam, *Acta Crystallogr., Sect. B* **B38**(2), 490–494 (1982).
155. S. J. Edge, D. W. Ollis, J. S. Stephanatou, J. F. Stoddart, D. J. Williams, and K. A. Woode, *Tetrahedron Lett.* **22,** 2229–2232 (1981).
156. D. W. Ollis, J. S. Stephanatou, J. F. Stoddart, G. G. Unal, and D. J. Williams, *Tetrahedron Lett.* **22,** 2225–2228 (1981).
157. D. W. Ollis, J. S. Stephanatou, J. F. Stoddart, A. Quick, D. Rogers, and D. J. Williams, *Angew. Chem., Int. Ed. Engl.* **15,** 757 (1976).
158. S. Cerrini, E. Giglio, F. Mazza, and N. V. Pavel, *Acta Crystallogr., Sect. B* **B35,** 2605 (1979).

159. J. A. Hyatt, E. N. Duesler, D. Y. Curtin, and I. C. Paul, *J. Org. Chem.* **45,** 5074 (1980).
160. T. Otsubo, S. Mizogami, N. Osaka, Y. Sakata, and S. Misumi, *Bull. Chem. Soc. Jpn.* **50,** 1841 (1977).
161. Y. Kai, J. Watanabe, N. Yasuoka, and N. Kasai, *Acta Crystallogr., Sect. B* **B36,** 2276–2281 (1980).
162. A. W. Hanson, *Acta Crystallogr., Sect. B* **B33,** 2657 (1977).
163. N. Kannen, T. Otsubo, Y. Sakata, and S. Misumi, *Bull. Chem. Soc. Jpn.* **49,** 3203 (1976).
164. Y. Kai, F. Hama, N. Yasuoka, and N. Kasai, *Acta Crystallogr., Sect. B* **B34,** 3422 (1978).
165. F. Hama, Y. Kai, N. Yasuoka, and N. Kasai, *Acta Crystallogr., Sect. B* **B33,** 3905 (1977).
166. H. Horita, Y. Koizomi, T. Otsubo, Y. Sakata, and S. Misumi, *Bull Chem. Soc. Jpn.* **51,** 2668 (1978).
167. T. Toyoda, A. Iwama, Y. Sakata, and S. Misumi, *Tetrahedron Lett.* **16,** 3203–3206 (1975).
168. T. Toyoda, A. Iwama, T. Otsubo, and S. Misumi, *Bull. Chem. Soc. Jpn.* **49,** 3300 (1976).
169. H. Mizuno, K. Nishiguchi, T. Otsubo, S. Misumi, and N. Morimoto, *Tetrahedron Lett.* **13,** 4981 (1972).
170. H. Mizuno, K. Nishiguchi, T. Toyoda, T. Otsubo, S. Misumi, and N. Morimoto, *Acta Crystallogr., Sect. B* **B33,** 329 (1977).
171. H. A. Staab and U. Zapf, *Angew. Chem., Int. Ed. Engl.* **17,** 757 (1978).
172. M. B. Laing and K. N. Trueblood, *Acta Crystallogr.* **19,** 373 (1965).
173. R. A. Abramovitch, J. L. Atwood, M. L. Good, and B. A. Lampert, *Inorg. Chem.* **14,** 3085 (1975).
174. K. Yasufuku, K. Aoki, and H. Yamataki, *Inorg. Chem.* **16,** 624 (1977).
175. B. R. Davis and I. Bernal, *J. Cryst. Mol. Struct.* **2,** 107 (1972).
176. A. G. Osborne, R. E. Hollands, J. A. K. Howard, and R. F. Bryan, *J. Organomet. Chem.* **205,** 395–406 (1981).
177. N. D. Jones, R. E. Marsh, and J. H. Richards, *Acta Crystallogr.* **19,** 330 (1965).
178. D. A. Lemenovskii, R. A. Stukan, B. N. Tarasevich, Yu. L. Slovokhotov, M. Yu Antipin, A. E. Kalinin, and Yu. T. Struchkov, *Koord. Khim.* **7,** 240–249 (1981).
179. C. Lecomte, Y. Dusausoy, J. Protas, C. Moise, and J. Tirouflet, *Acta Crystallogr., Sect. B* **B29,** 488 (1973).
180. C. Lecomte, Y. Dusausoy, J. Protas, and C. Moise, *Acta Crystallogr., Sect. B* **B29,** 1127 (1973).
181. T. S. Cameron, and R. E. Cordes, *Acta Crystallogr., Sect. B* **B35,** 748–750 (1979).
182. T. N. Sal'nikova, V. G. Andrianov, M. Yu Antipin, and Yu. T. Struchkov, *Koord. Khim.* **3,** 939 (1977).
183. T. N. Sal'nikova, V. G. Andrianov, and Yu. T. Struchkov, *Koord. Khim.* **3,** 768–782 (1977).
184. P. Batail, D. Grandjean, D. Astruc, and R. Dabard, *J. Organomet. Chem.* **102,** 79 (1975).
185. P. Batail, D. Grandjean, D. Astruc, and R. Dabard, *J. Organomet. Chem.* **110,** 91 (1976).
186. I. C. Paul, *J. Chem. Soc., Chem. Commun.* p. 377 (1966).
187. M. Hillman, B. Gordon, N. Dudek, R. Fajer, E. Fugita, and J. Gaffney, *J. Organomet. Chem.* **194,** 229–256 (1980).
188. M. Hillman and E. Fujita, *J. Organomet. Chem.* **155,** 99 (1978).

189. Yu. T. Struchkov, G. G. Aleksandrov, A. Z. Kreindlin, and M. I. Rylinskaya, *J. Organomet. Chem.* **210,** 237–245 (1981).
190. M. Hillman and E. Fujita, *J. Organomet. Chem.* **155,** 87 (1978).
191. L. D. Spalding, M. Hillman, and G. J. B. Williams, *J. Organomet. Chem.* **155,** 109 (1978).
192. M. Hisatome, Y. Kawaziri, K. Yamakawa, and Y. Iitaka, *Tetrahedron Lett.* **20,** 1777 (1979).
193. Y. Kawajiri, M. Hisatome, and K. Yamakawa, *12th Symp. Struct. Org. Chem., 1979* pp. 281–284 (1979).
194. M. Hisatome, Y. Kawajiri, and K. Yamakawa, *Tetrahedron Lett.* **23,** 1713–1716 (1982).
195. M. Hisatome, Y. Kawajiri, K. Yamakawa, K. Mamiya, Y. Harada, and Y. Iitaka, *Inorg. Chem.* **21,** 1345–1348 (1982).
196. J. S. McKechnie, B. H. Bersted, I. C. Paul, and W. E. Watts, *J. Organomet. Chem.* **8,** 29 (1967).
197. J. S. McKechnie, C. A. Maier, B. Bersted, and I. C. Paul, *J. Chem. Soc., Perkin Trans.* **2,** 138 (1973).
198. J. Dizymala, Z. Kaluski, A. Ratajczak, and B. Czech, *Bull. Acad. Pol. Sci., Ser. Sci. Chim.* **28,** 621 (1980).
199. S. J. Lippard and G. Martin, *J. Am. Chem. Soc.* **92,** 7291 (1970).
200. J. P. Collman, A. O. Chong, G. B. Jameson, R. T. Oakley, E. Rose, E. R. Schmittou, and J. A. Ibers, *J. Am. Chem. Soc.* **103,** 516–533 (1981).
201. M. Joly, N. Defay, R. H. Martin, J. P. Declerq, G. Germain, B. Soubrier-Payen, and M. Van Meerssche, *Helv. Chim. Acta* **60,** 537 (1977).
202. H. Besl, H.-J. Hecht, P. Luger, V. Pasupathy, and W. Steglich, *Chem. Ber.* **108,** 3675–3691 (1975).
203. M. J. Begley, R. U. M. Campbell, L. Crombie, B. Tuck, and D. A. Whiting, *J. Chem. Soc. C* pp. 3634–3642 (1971).
204. B. E. Eaton, E. D. Laganis, and V. Boekelheide, *Proc. Natl. Acad. Sci. U.S.A.* **78,** 6564–6566 (1981).
205. W. D. Roberts and G. Shoham, *Tetrahedron Lett.* **22,** 4895 (1981).

CHAPTER 4

Nuclear Magnetic Resonance Properties and Conformational Behavior of Cyclophanes

REGINALD H. MITCHELL

Department of Chemistry
University of Victoria
Victoria, British Columbia, Canada

ISBN 0-12-403001-7

I. INTRODUCTION

Since the appearance of Smith's book,[1] "Bridged Aromatic Compounds," in 1964, cyclophane chemistry has undergone a virtual explosion. This has happened in part perhaps because of the widespread introduction of nmr spectroscopy as a common organic characterization technique. The chapter on nmr in Smith's book[1] (p. 407) was only a minor one, and yet today the word *cyclophane* is almost synonymous with an unusual ^{1}H-nmr spectrum. Although the assignment of structure of cyclophanes was relatively straightforward throughout the 1970s, current easy access to high-field nmr instrumentation suggests that a rapid expansion of our knowledge of cyclophane properties will again occur. This is because very detailed stereochemical assignments, which were often impossible with the older low-field instruments, can now be made.

This chapter concentrates on the ways in which nmr techniques have been used to assign stereochemistry in the fixed phanes and to elucidate what conformational effects are observed in the mobile phanes. The latter area is far from precise and is likely to show considerable development in the next decade.

Because of the sheer volume of material published it has not been possible to cover all examples of cyclophane nmr in this chapter. For example, multibridged, multilayered and multistepped cyclophanes have been reviewed by Vögtle[2] and are discussed in Chapters 10 and 11 of this book. These cyclophanes are referred to here only when their nmr spectra suggest special stereochemical properties that are different from those of other types of phanes. Nevertheless, a representative coverage of the literature to December 1981 has been attempted. The nomenclature of phanes used here is that of Vögtle.[3]

II. SIMPLE FIXED PHANES

A. Metacyclophanes

1. Chemical Shift Reference Points

a. Saturated Bridges

The upfield shift of the internal protons H_i in [2.2]metacyclophane **(1)** to δ 4.25[4] (4.17)[5] from their normal position of $\delta \approx 7.0$ in *m*-xylene is caused by the shielding of the opposite benzene ring and is not only well investigated,[6–6b] but well substantiated in terms of ring current theory.[7] Moreover, this explanation is supported by the structure of [2.2]metacyclophane in the crystal state, which is now known by X-ray determination to be stepped.[8] The nonequivalent AA′BB′ bridge protons also have a fixed, staggered arrangement between −80 and +190°C, with H_{ax} at δ 2.04 and H_{eq} at δ 3.05.[6a,9] The analogous 8,16-dimethyl derivative **2** behaves similarly,[5,10] with the internal methyl protons Me_i appearing at δ 0.56, considerably shielded from those of 1,2,3-trimethylbenzene (δ 2.15). Like **1,** it also has the stepped anti structure, as determined by X-ray studies.[11] There appears to be no significant barrier to rotation of the —CH_3 groups of **2** in solution; however, their introduction does deshield the axial bridge protons by about 0.5 ppm.[9]

The difference in shielding from their respective models of H_i in **1** ($\Delta\delta$ = 2.75 ppm) and of CH_3 in **2** ($\Delta\delta$ = 1.6 ppm) is interesting in that it probably reflects in part the somewhat different conformational geometries of **1** and **2,** in which the C-8–C-16 distance is somewhat greater (0.13 Å) for **2** than **1,** together with the fact that the methyl protons will be farther out of plane, along the *Z* axis, and hence less shielded.[7]

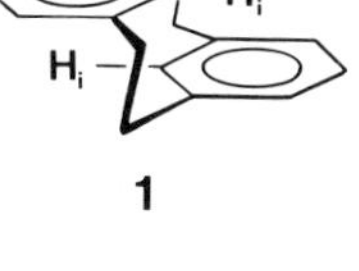

H_{ax} H_{eq} H_i H_i **1**

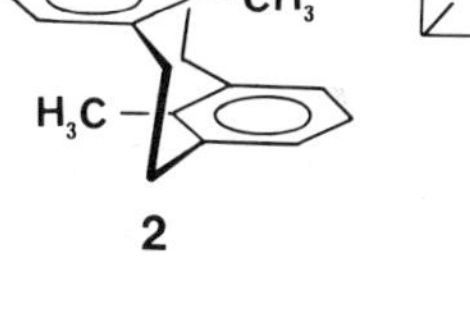

H_{ax} H_{eq} CH_3 H_3C **2** Z

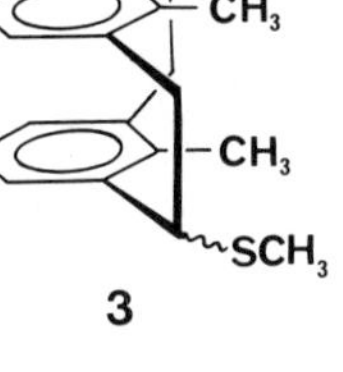

SCH_3 CH_3 CH_3 SCH_3 **3**

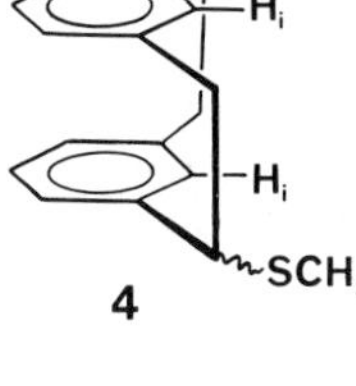

SCH_3 H_i H_i SCH_3 **4**

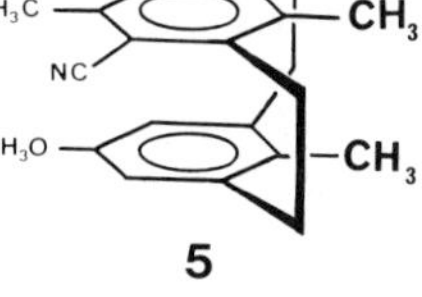

NC H_3C NC CH_3 CH_3O CH_3 **5**

The *syn*-cyclophanes isomeric to **1** and **2** are not yet known. Mitchell and Boekelheide,[12] however, have prepared their bridge thiomethyl derivatives (**3** and **4**). The H_i of **4** appear at $\delta \approx 7.3$ and CH_3 of **3** appear at $\delta \approx 2$ and thus do not show the shielding of the corresponding anti derivatives, which behave similarly to **1** and **2** and show H_i at δ 4.4–5.3 and CH_3 at δ 0.5–1.0. More recently, Boekelheide reported[13] the *syn*-[2.2]metacyclophane **5,** which does not have substituents on the bridges and in which CH_3 appear at δ 2.10 and 2.22.

b. Unsaturated Bridges (Metacyclophanenes)

The introduction of a double bond into one or both of the bridges of **1** or **2** still leaves the molecule conformationally rigid, but it has a rather dramatic effect on the chemical shift of the H_i. The chemical shifts are given in Table I.[12,14]

CH_3

H

6

This effect, which is very much greater for H_i than Me_i, has been explained[14] as an anisotropy effect of the double bond. However, if one considers the geometry of these systems, in which the protons (H_i) of **1** are closer to the edge of the opposite ring than are the methyl groups (Me_i) of **2** (i.e., in a region of space where the magnetic field is more rapidly changing[7]), then it is quite possible that the small change in geometry[15] in going from **1** to the diene may pull H_i out of the shielding into the deshielding region of the opposite ring. Consequently, care must be taken

TABLE I

Chemical Shifts of the Internal Protons (H_i) and Internal Methyl Protons (Me_i) for Compounds 1, 2, and 6 and Their Monoenes and Dienes

Compound	δH_i (ppm)	δMe_i (ppm)
1	4.17	—
1-1-Ene	5.62	—
1-1,9-Diene	7.90	—
2	—	0.56
2-1-Ene	—	0.79
2-1,9-Diene	—	1.52
6	3.72	0.48
6-1-Ene	4.90	0.78

in assigning stereochemistry by the ^{1}H-nmr analysis of compounds containing bridge double bonds.

c. Ring Substituents

Generally, the addition of substituents to the cyclophane rings has only a small effect on the chemical shift of the internal substituents. For example, in the tetramethyl derivative **7** (Y = H_i) H_i are at δ 4.20 (in **1,** δ 4.25), and in **8** (Y = Me_i) Me_i are at δ 0.48 (cf. **2,** δ 0.56).[9] In the dimethoxy derivative **9,** Me_i are at δ 0.75, surprisingly 0.19 ppm *deshielded* from

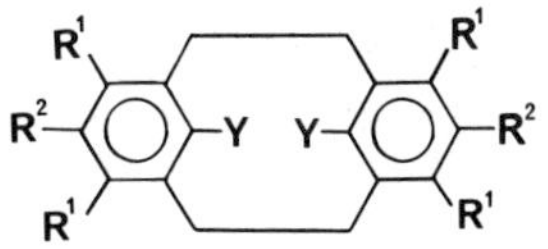

7 Y = H, R^1 = CH_3, R^2 = H
8 Y = CH_3, R^1 = CH_3, R^2 = H
9 Y = CH_3, R^1 = H, R^2 = OCH_3

those of **2.**[16] Small effects have been observed for a number of 5,13[9] and 4,14[17] derivatives. Effects occur on the methylene bridge protons as well.[17] Substituents on the bridges are also subject to shifts. For example, in **10** the equatorial methyl appears at δ 1.48, whereas the axial methyl is at δ 0.85. The internal proton H-16 is deshielded to δ 4.50 relative to H-8, δ 4.25.[18]

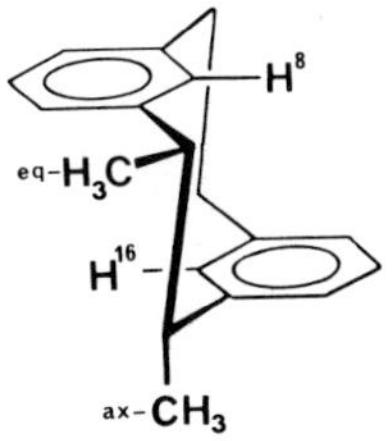

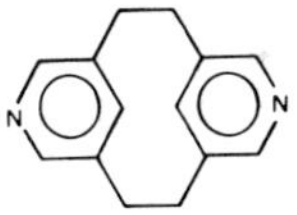

10 **11**

An internal substituent such as a heteroatom, for example, in the [2.2]pyridinophane **11,** similarly deshields H_i to δ 4.51, as might be expected for an electron-withdrawing group.[19]

2. Stereochemical Assignments in Fixed Metacyclophanes

a. anti-*Cyclophanes*

On the basis of the data given above, many [2.2]metacyclophanes have been assigned anti stereochemistry. Examples of these are given in Table

TABLE II
[1]H-nmr Data for Selected Fixed *anti*-[2.2]Metacyclophanes

Structure	Compound number	R	δH_i (ppm)	References
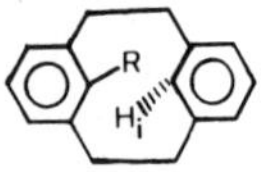	**12**	F	4.47	20
	13	Cl	3.94	20
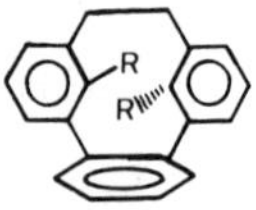	**14**	H	5.44	21
	15	CH_3	0.67	22
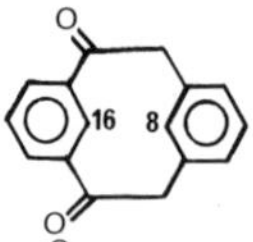	**16**	H-16	4.22	23
		H-8	5.91	23
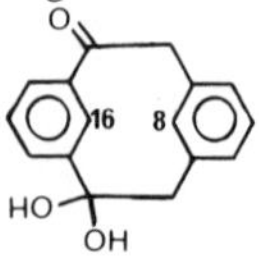	**17**	H-16	4.71	23
		H-8	5.02	23
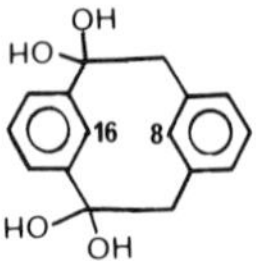	**18**	H-16	5.32	23
		H-8	4.25	23
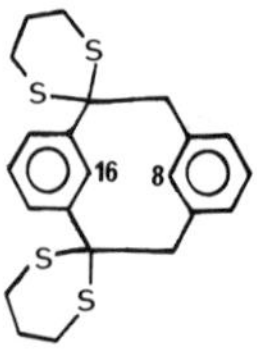	**19**	H-16	6.5	23
		H-8	3.8	23
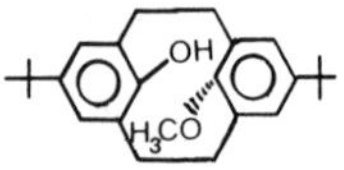	**20**	CH_3O—	2.95	24
		ArH	7.1	24
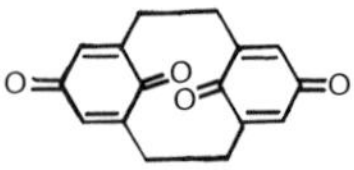	**21**	ArH	6.44	24
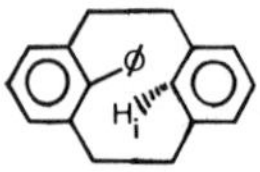	**22**	H_i	3.57	25
		ϕ	See Section III,C,1	

TABLE II (*Continued*)

Structure	Compound number	R	δH_i (ppm)	References
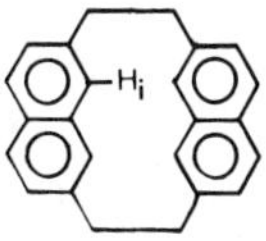	**23**	H_i	5.02	26, 27
	24	H_i	3.10	28
		$H_{i'}$	5.14	28
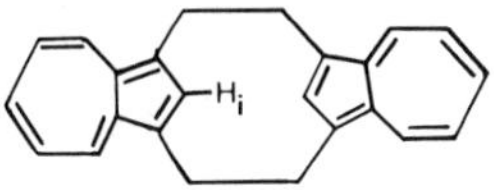	**25**	H_i	4.39	28
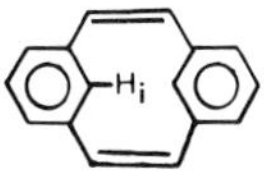	**26**	H_i	7.90	12
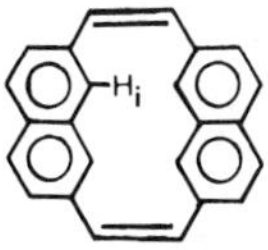	**27**	H_i	7.08	27

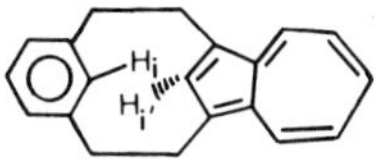

II (structures **12–27**) with the pertinent [1]H-nmr data. Some of these deserve comment. In [2.2]metacyclophanes that bear a substituent at the 8 position or functionality at the 1 and/or 10 positions, four possible geometries can be envisioned: **A, B, C,** or **D.** The preference for a particular geometry would depend on the size and electronic nature of X and on the hybridization at the 1,10 positions. With a large X substituent such as $—CH_3$ **(6),** —Cl **(13),** or —ϕ **(22),** the H_i are pushed well into the cavity of

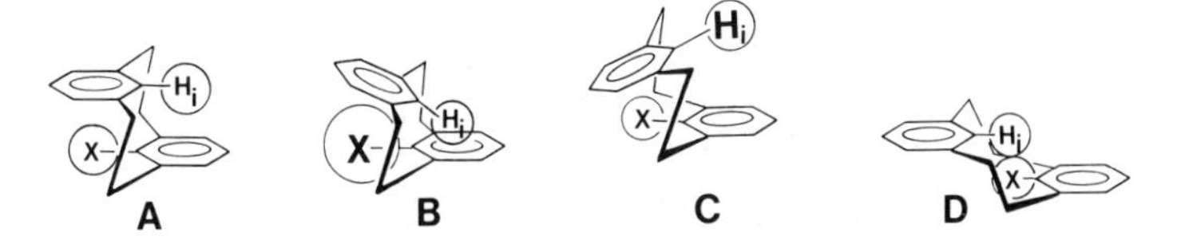

the opposite π cloud (geometry **B**) and hence appear more shielded (at δ 3.72, 3.94, and 3.52, respectively) than does H_i for **1** (δ 4.25, geometry **A**). With sp^2 hybridization at positions 1,10 as in **16,** geometry **B, C,** or **D** might be preferred. We have already mentioned that [2.2]metacyclophan-1-ene and [2.2]metacyclophane-1,9-diene both have H_i considerably less shielded than for **1,** which could be caused by a flattening of the molecule as in geometry **D** or by a bond anisotropy effect. In the case of **16,** H-8 is substantially deshielded to δ 5.91, whereas H-16 remains relatively normal (cf. **1**) at δ 4.22, which suggests that geometry **B** is the more likely. (It seems unlikely that a bond anisotropy effect would, with sp^2 hybridization at 1,10, leave H-16 unchanged but deshield H-8.) As the hybridization of position 10 and then 1 is converted from sp^2 to sp^3 in compounds **17** and **18,** respectively, H-16 becomes more deshielded and H-8 more shielded. This suggests that **18** has geometry **C.** The effect is even more pronounced for the bulky dithiane groups in **19.** For compounds of type **20** and **21,** assignment of stereochemistry is more difficult. However, a shielding of the —OMe in **20** can be observed, which is consistent with the anti structure, and because **20** can be converted[24] to **21,** it also was assigned the anti stereochemistry. Compound **22** is one of the first compounds to be reported[25] in which the internal substituent (ϕ) shows restricted rotation. This is discussed in Section III,C,1.

Of the [2.2]phanes with rings larger than benzene, the 2,7-naphthalenophane **23** has been isolated in both[26] anti and syn conformers (see below); *anti*-**23** has H_i at δ 5.02,[26,27] somewhat less shielded than for **1,** as might be expected for an α-naphthalene proton. The azulenophanes[28] **24** and **25** are interesting in that, because **25** must have geometry **A,** H_i are almost at identical chemical shifts to **1,** but **24** must distort to geometry **B,** in which H_i are pushed well into the π cloud of the opposite azulene ring (because of the geometry at the bridges), leaving $H_{i'}$ much less directed into its opposite cavity and hence at δ 5.14. When X-ray structures of these compounds become available, it will be interesting to see if this argument is supported by ring current theory.

Very few cyclophanenes and cyclophanedienes have been reported (see also Table I). However, in the naphthalenophane **27,** H_i are strongly deshielded from those in **23** ($\Delta\delta$ = 2.06 ppm), analogous to those of **26** from **1** ($\Delta\delta$ = 3.65 ppm).

b. syn-*Cyclophanes*

Very few additional *syn*-[2.2]metacyclophanes are known. The *syn*-naphthalenophane **28,** in which H_i are at δ 6.95, has the methylene bridge protons as a singlet at δ 3.14. Unless an accidental degeneracy of chemical shift occurs between H_{ax} and H_{eq} (which seems unlikely), this would not have been expected (the bridges almost always show AA′BB′ spectra).

H_i H_{ax} H_{eq}

28

A conformational mobility in which H_{ax} and H_{eq} are exchanged also seems unlikely. Possibly the structure is twisted sufficiently to make $\delta H_{ax} \approx \delta H_{eq}$. An X-ray structure determination seems warranted.

Compound **5** was mentioned on p. 242, and clearly compounds **29A–29C** are similarly syn isomers, as shown by the chemical shifts of

X, Me, Me, X, MeO, Me

5 X = CN
29A X = CHO
29B X = CH_2OH
29C X = CH_2OMe

Me_i ($\delta \approx 2$), because the corresponding anti conformers are all known[13] and have Me_i at $\delta \approx 0.6$. At 196°C *syn*-**5** converts thermally to *anti*-**5.**

Halo-substituted metacyclophanes are of special interest, because in the thiacyclophane series the presence of a halogen atom leads to a marked increase in yield of the syn conformer (see p. 253). In those cases in which two halogens are present as internal substituents, assignment of

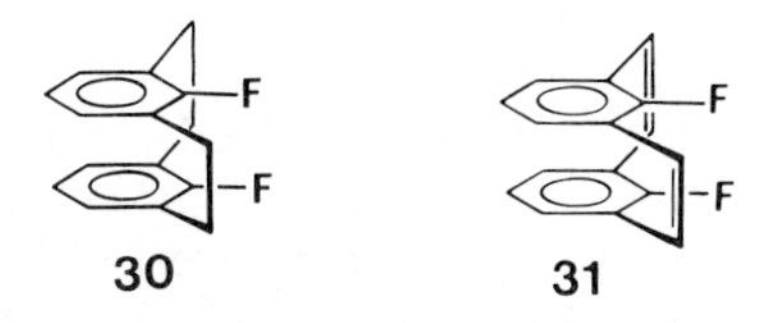

30 **31**

stereochemistry is not trivial. Boekelheide[29] has assigned syn stereochemistry to **30** and **31** on the basis of dipole moment studies. However, for X-ray data that indicate **31** is anti in the solid state see ref. 184.

c. Heterophanes

A stereochemical assignment problem similar to that of **30** exists for heterophanes of type **32.** In fact when X = N, the molecule is mobile[30] and hence is discussed in Section III. The monopyridinometacyclophane **33,** however, is fixed and has the anti configuration with H_i at δ 4.40.[31] The

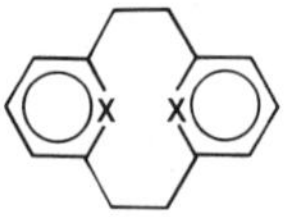

32 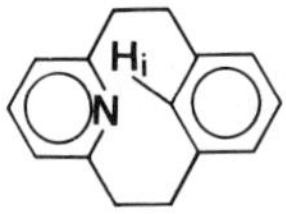 33 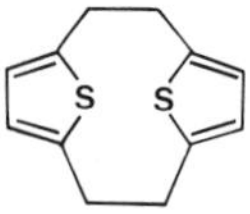 34 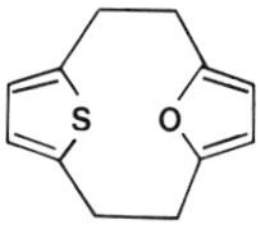 35

analogous five-membered-ring heterocycles **34**[32] and **35**[31] are also reported to be fixed in the anti configuration on the basis of their AA′BB′ bridges and Raman spectra. A lengthy tabulation of heterophane spectral data has appeared.[33]

d. Staircase [2.2]*Metacyclophanes*

These compounds are dealt with in detail in Chapter 10; however one or two examples are included here because the assignment of stereochemistry is interesting. Two compounds exist[34,34a] for the double metacyclophanes **36, 37** ($R^1 = R^2 = H$), **38, 39** ($R^1 = R^2 = CH_3$), and **40, 41** ($R^1 = H$; $R^2 = CH_3$), which have been assigned the up–up (uu) and up–down

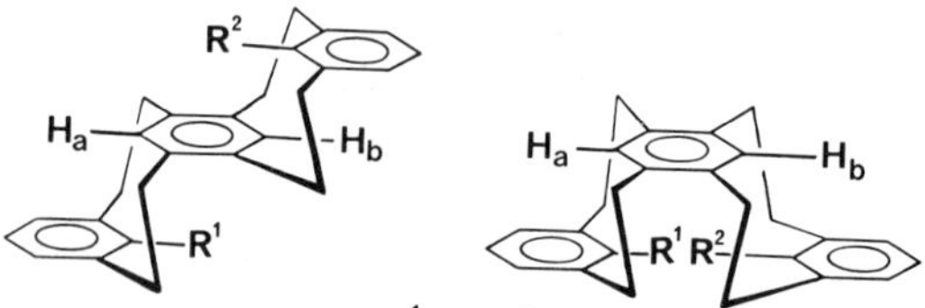

	R^1	R^2	
36	H_c	H_c	37
38	CH_3	CH_3	39
40	H_c	CH_3	41

(ud) stereochemistry in each case. That all compounds are anti is obvious from the shielding of all internal substituents. The ud isomer **41** was assigned its structure on the basis of an NOE enhancement of H_c when the CH_3 was irradiated, whereas **40** showed no such effect. In the case of **40** H_a is found at δ 3.85, H_b at δ 3.59, H_c at δ 4.21, and CH_3 at δ 0.58, whereas for *ud*-**41** H_a is at δ 3.67, H_b at δ 3.57, H_c at δ 5.64, and CH_3 at δ 1.07. Clearly, H_c in *ud*-**41** suffers steric deshielding by the adjacent —CH_3, which is likewise deshielded. This can then be used to assign stereochemistry of *uu*-**38** (CH_3 at δ 0.58) and *ud*-**39** (CH_3 at δ 1.07) as well as *uu*-**36** (H_c at δ 4.41) and *ud*-**37** (H_c at δ 5.03). It is interesting that the uu isomers readily (~100°C) convert to the thermodynamically more stable ud isomers on heating. This is in contrast to the barrier for [2.2]metacyclophane, which indicates no substantial conversion at 200°C.[6a]

e. Higher Phanes

Few fixed [*m.n*] type of phanes seem well documented. Staab[35] reported syn and anti conformers of the [3.3]metacyclophane **42**, which do

42

	δ Me$_i$	δ Me$_o$
syn	3.46	3.67
anti	3.14	3.78

not interconvert on heating to 190°C. His assignment was based on the fact that *anti*-**42** showed $\Delta\delta$ for the methoxyl protons of 0.64 ppm, whereas $\Delta\delta$ for *syn*-**42,** was 0.19 ppm. It is further supported by similar differences in $\Delta\delta$ for *syn*-**43** (0.05 ppm) and *anti*-**43** (0.42 ppm) and the fact that *syn*-**43** can be converted to *syn*-**44** and hence to the cage compound **45.**[36]

43

	δ Me$_i$	δ Me$_o$
syn	3.58	3.63
anti	3.40	3.82

44 **45**

The ferrocenophane **46** has been assigned[37] the syn stereochemistry shown, partly because H_a appears at δ 6.57. Caution may be required in such circumstances, however, because of the deshielding observed in the cyclophanedienes **26** and **27.**

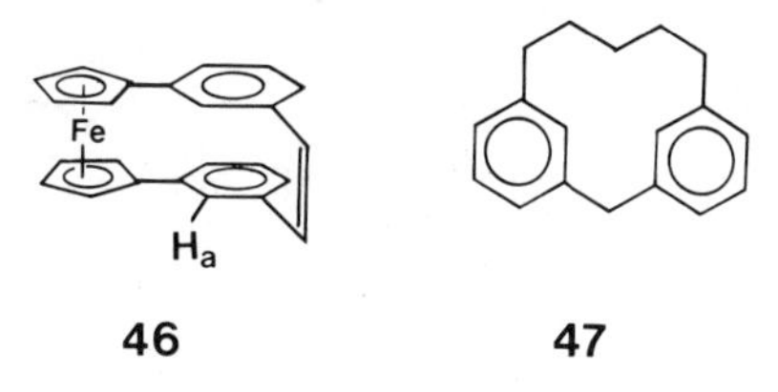

The [5.1]metacyclophane **47** would appear to have a fixed conformation from its reported[38] ^{1}H-nmr spectrum, in which a high-field 2H quartet is observed at δ 1.12–0.80.

B. Thiametacyclophanes

1. Chemical Shift Reference Points

a. Thia[2.2]phanes

Very few thia[2.2]metacyclophanes are known. Vögtle[39] prepared 1,10-dithia[2.2]metacyclophane **(48)** and 1,12-dithia[2.2]naphthalenophane **(49),** and both exist in fixed anti conformations (to at least 190°C), as

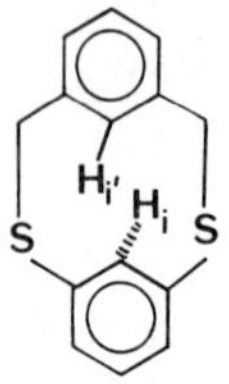

48 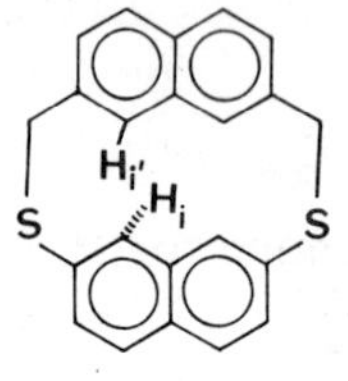 **49**

clearly indicated by their shielded H_i at δ 5.50 (H_i) and δ 4.48 (H_i') for **48** and δ 5.52 (H_i) and δ 5.00 ($H_{i'}$) for **49.** The assignments for the substituted phanes **50–53** are therefore also anti.[4]

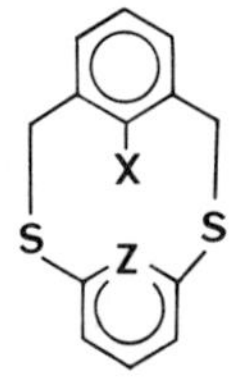

	Z	X	δX	δZ
50	CH	CH_3	0.59	4.95
51	CH	OCH_3	–	5.20
52	CH	CN	–	5.07
53	N	CH_3	1.59	–

With a dithia bridge, as in the tetrathiacyclophanes **54** and **55,** the assignment is more difficult. In **54** H_i appear at δ 7.85 and H_o at δ 6.43, and in **55** H_i appear at δ 7.88 and H_o at δ 7.01. The authors[40] assign both the

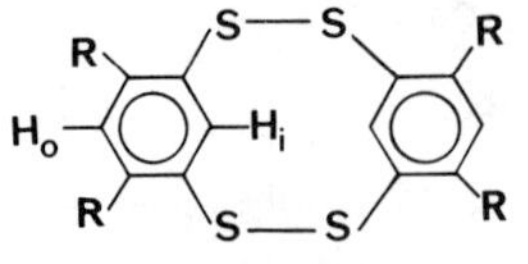

54 R = OCH_3

55 R = CH_3

anti conformation. If **54** and **55** adopt geometry **D,** this would not be unreasonable, especially in light of the longer C—S (1.82 Å) and S—S (2.03 Å) bond lengths in comparison with C—C (1.54 Å), which could substantially change the geometry; compare the [2.2]dienes in Section II,A,1,b. This then eliminates the need to involve a deshielding effect of the S atoms or a syn assignment, both of which would be unusual. An X-ray structure determination, however, would certainly be useful. These authors have more recently reported[41] the larger octathia[2.2.2.2]metacyclophanes **56–58** in which three geometries, **E** (basket), **F** (saddle), and **G**

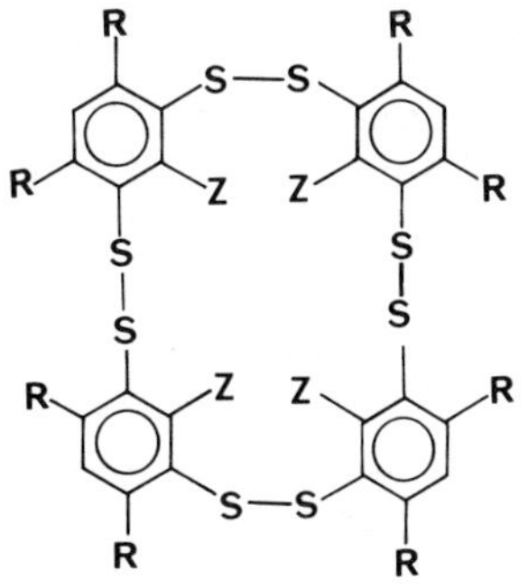

	Z	R
56	H	CH_3
57	H	Cl
58	CH_3	CH_3

(crown), are considered possible. Because H_o would experience significant ring current shielding only in the saddle form **F,** and by comparison with models only **56** (δH_o 6.66) shows a 0.24-ppm shielding, it is assigned the saddle stereochemistry. In both **57** and **58** H_o are normal at δ 7.14 and 6.94, respectively, <0.01 ppm different from model compounds. Because **58** has H_o very similar to those of **55** (δ 6.93) it is assigned the crown structure **G.** However, further structural verification of these interesting compounds would be desirable.

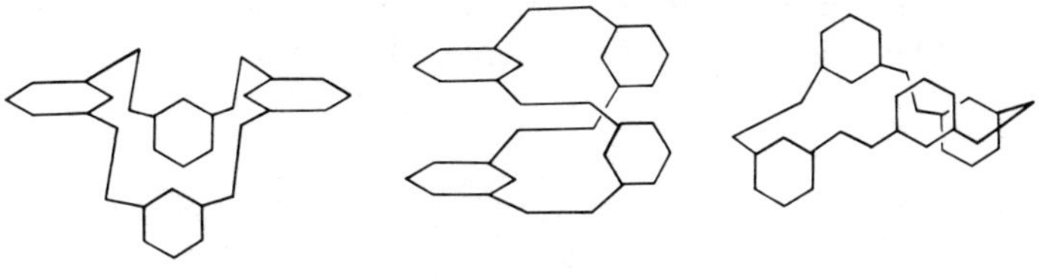

E **F** **G**

b. Thia[3.3]phanes

Whereas few thia[2.2]phanes are known, the literature on thia[3.3]phanes is extensive. We demonstrated[12] in 1970 that 9,18-dimethyl-2,11-dithia[3.3]metacyclophane exists in syn **(59)** and anti **(60)**

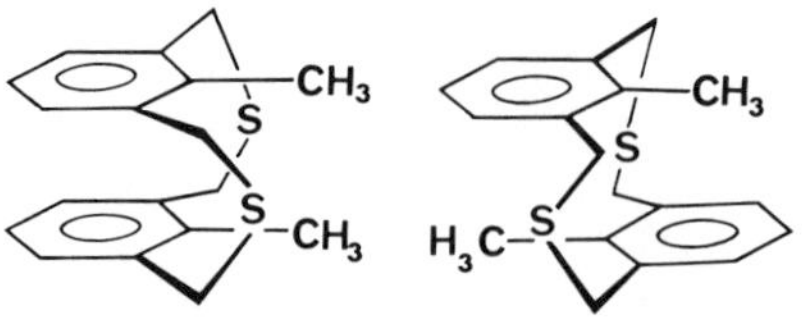

59 60

conformers, which do not interconvert below 200°C. As in the case of the [2.2]phane **2,** the internal methyl protons of *anti*-**60** are shielded at δ 1.30 by 1.24 ppm from those of *syn*-**59** (δ 2.54), which are relatively normal for a toluene [those of 2,6-bis(mercaptomethyltoluene) appear at δ 2.32].

Even one internal methyl substituent is sufficient to allow the isolation of a discrete anti or syn isomer; for example, in *anti*-**61**[42] H_i appear at δ 5.50 and Me_i at δ 2.18. The reduced shielding of Me_i relative to **60** suggests that **61** adopts geometry **B** (or **D**).

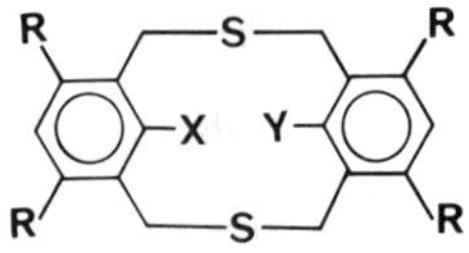

	X	Y	R
61	Me_i	H_i	H
62	Me_i	Me_i	Me

Whereas the methylene bridge signals for *anti*-**60** and *syn*-**62** appear as singlets at δ 3.68 and 3.72, respectively (possibly because of accidental chemical shift degeneracy), those of *anti*-**61** appear as a multiplet at δ 4.1–3.4, and those of *syn*-**59** as an AB quartet at δ 4.00 and 3.80. In fact, in nearly all substituted examples, e.g., *anti*-**62,** in which Me_i appear[43] at δ 1.14, the bridges appear as AB patterns that are not temperature dependent over the range −100 to +150°C. This suggests that the bridges are either immobile or undergo a very easy wobble of the type **H, I,** or **J.** A

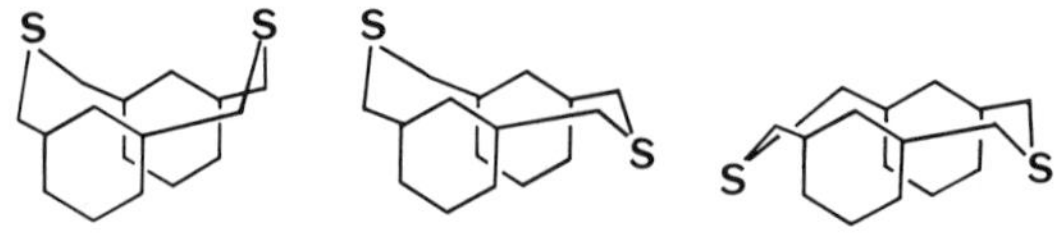

H I J

slight broadening of the bridge signals of *syn*-**62** at low temperatures,[43] together with X-ray data from a related system,[44] suggests the latter. In the case of *anti*-**62** Sato,[43] using NOE data, was able to assign the chemical shifts of the axial bridge protons as δ 3.64 and those of the equatorial ones as δ 3.50.

2. Stereochemical Assignments in Fixed Thiametacyclophanes

a. [2.2]Phanes

Because so few of these are known, they are discussed in Section II,B,1,a.

b. [3.3]Phanes

By the use of the data presented in the preceding section the assignment of fixed phanes is relatively straightforward, especially when both isomers exist. Table III (structures **63–76**) lists some of the examples thus assigned. When electron-withdrawing groups such as halo, nitro, and cyano are present (e.g., as in **64, 70,** and **71**), the yields of the syn isomers increase substantially. Very bulky groups, as in **65,** decrease the yields of syn isomers. When a sufficiently large group is present (e.g., ϕ, as in **73**), although the phane ring stereochemistry is fixed, the spectra obtained are temperature dependent and demonstrate a restricted rotation of the substituent. This is discussed in Section III. For compounds such as **75** and **76,** the assignment of *syn*- or *anti*-phane stereochemistry is relatively straightforward. The assignment of the relative regiochemistry of the naphthalene rings as transoid to **75** and cisoid to **76** was considerably more difficult.[53] It was made by careful analysis of the aryl and bridge protons with the help of molecular models, such that interaction between equatorial bridge hydrogens and *peri*-naphthalene hydrogens is minimal. For example, the preferred conformer of **75** requires that H-1_{eq} be out of the naphthalene plane to avoid interaction with H-20 and thus that H-1_{ax} be twisted in the plane, toward the shielding region of the opposite ring. Because H-1 (and H-12) are α-naphthyl substituents, they normally appear deshielded with respect to H-3 and H-14. Thus, H-1_{eq},H-12_{eq} are at δ 4.28 and H-1_{ax},H-12_{ax} are at δ 4.08; similarly, H-3_{ax},H-14_{ax} are at δ 3.37 and H-3_{eq},H-14_{eq} are at δ 3.67. A similar argument for **76,** which minimizes the H-9_{eq},H-12_{eq} interaction, requires that H-16_{eq},H-14_{eq} do interact. Thus, this isomer has to skew more to avoid this, and thus H-12_{eq},H-14_{eq} are at δ 4.19, H-12_{ax},H-14_{ax} are at δ 3.91, H-1_{eq},H-3_{eq} are at δ 3.90, and H-1_{ax},H-3_{ax} are at δ 3.82. On average this requires that H-3_{ax},H-14_{ax}

TABLE III

Assigned Stereochemistries of Some Thia[3.3]metacyclophanes

Structure	Compound number		Stereochemistry	δX_i (ppm)	References
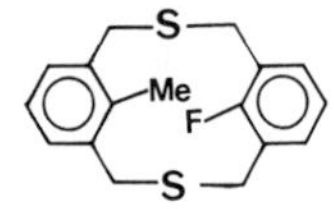	**63**		Syn	2.41	29
			Anti	1.49	29
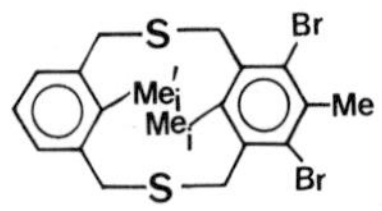	**64**		Syn	2.60, 2.67	45
			Anti	1.10, 1.63	45
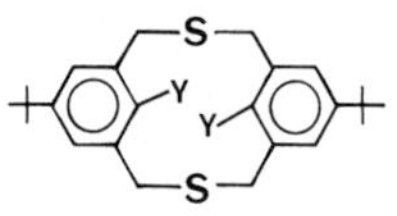	**65A**	Y = CH_3	Anti	1.25	46
	65B	Y = CH_3CH_2	Anti	0.63, 1.58	46
	65C	Y = $CH_3CH_2CH_2$	Anti	0.68, 0.8–1.07, 1.68	46
	65D	Y = OCH_3	Anti	3.20	46
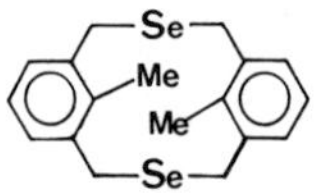	**66**		Anti	1.77	47
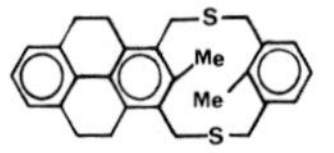	**67**		Syn	2.48, 2.44	48
			Anti	1.38, 1.18	48
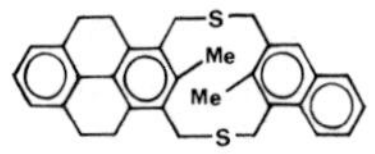	**68**		Syn	2.63, 2.49	48, 49
			Anti	1.5, 0.72	48, 49
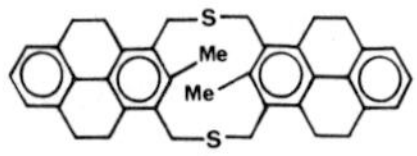	**69**		Syn	2.14	48
			Anti	1.32	48
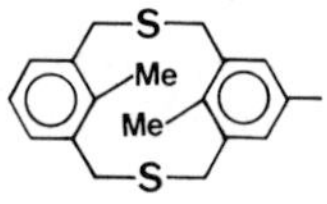	**70**		Syn	3.64, 2.52	50
			Anti	1.41, 1.29	50
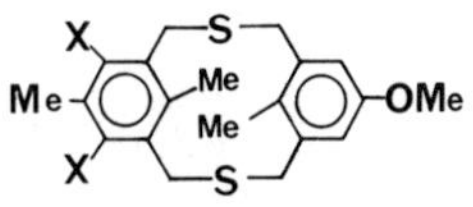	**71**	X = Br	Syn	2.5	50
			Anti	1.42, 1.32	50
	72	X = CN	Syn	2.6	50
			Anti	1.44, 1.32	50

TABLE III (*Continued*)

Structure	Compound number	Stereo-chemistry	δX_i (ppm)	References
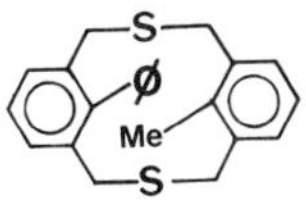	**73**	Syn	Me_i 2.37 ϕ, temperature variable (see p. 273)	51
		Anti	Me_i 1.46 ϕ (see p. 273)	51
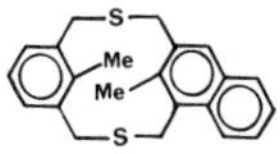	**74**	Syn	2.45, 2.62	52
		Anti	0.92, 1.42	52
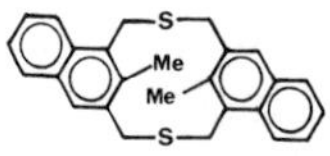	**75**	Syn	2.69	53
		Anti	1.00	53
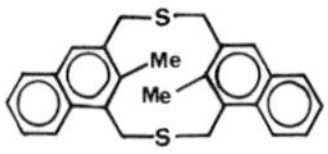	**76**	Syn	2.62	53
		Anti	0.98	53

of **75** be more shielded than H-1_{ax},H-3_{ax} of **76,** which they are, because only one of the latter can be shielded by the opposite ring at any one time. This also requires that H-5,H-16 of **75** be closer on the average to the deshielding S-2,S-13 than their counterparts in **76** H-5,H-20, are to S-2, and they thus should be deshielded as they are: δ 7.97 versus δ 7.84.

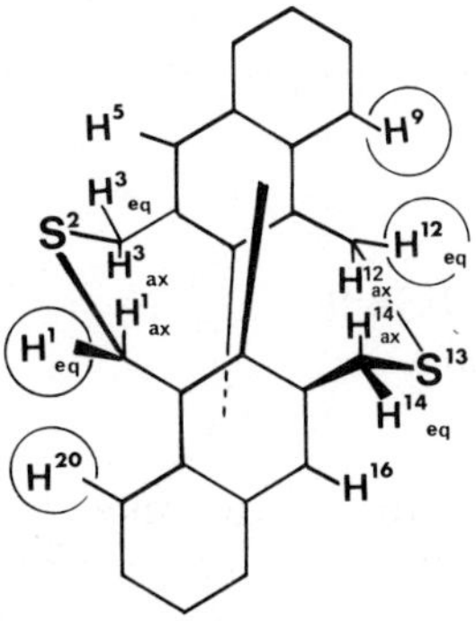

anti-**75** 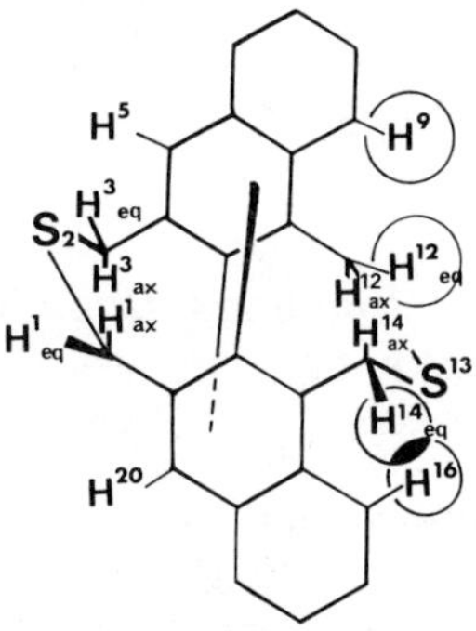 anti-**76**

Similarly, H-9,H-20 of **75** are (on the average) farther from S-13,S-2 than are H-9,H-16 (from S-13 of **76** and therefore should be shielded as they are: δ 8.12 versus δ 8.30.

When the internal substituents of a [3.3]phane such as **77** are H, the molecule may be mobile and hence is discussed in Section III. Vögtle[54]

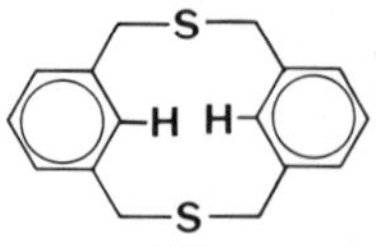

77

made extensive studies of the syn–anti conversions in other dithia[3.3]metacyclophanes, especially in relation to the size of the substituents.[55] Mobile examples are discussed in Section III, but **78A–78C** exist as anti conformers, and **78D–78J** exist as syn conformers, which in the

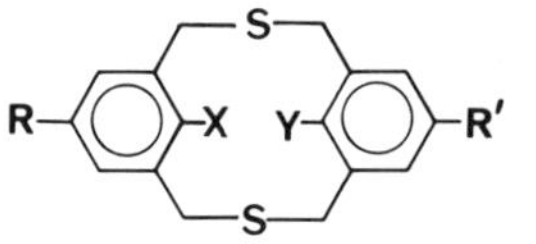

78

	X	Y	R	R′
A	Me	H	NO_2	H
B	Me	H	NH_2	H
C	NH_2	H	H	H
D	NO_2	H	H	H
E	OMe	H	H	NO_2
F	OMe	H	H	NH_2
G	OMe	NO_2	H	H
H	OMe	NH_2	H	H
I	NO_2	NO_2	H	H
J	NH_2	NH_2	H	H

case of **78H–78J** form anti conformers on heating to 150°C. It is interesting that when the NO_2 group of *syn*-**78E** is reduced, *syn*-**78F**, which is stable at room temperature, is formed. On heating to 150°C, however, *syn*-**78F** is cleanly converted to the thermodynamically more stable anti conformer.[54] This clearly indicates that at room temperature there is no conformational ring flip.

C. Metaparacyclophanes: Chemical Shift Reference Points

[2.2]Metaparacyclophanes with H as the internal substituent are mobile and are discussed in Section III. However, when the metabenzeno bridge is replaced with an azulene unit as in **79**[56] and **80**[57] at room temperature, a fixed conformation is adopted because different H_a and H_b resonances can be seen at δ 5.81 and 7.36 for **79** and at δ 5.20 and 6.98 for **80** (see also

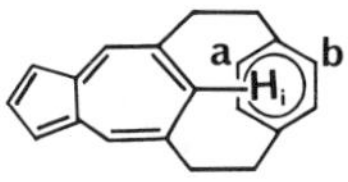

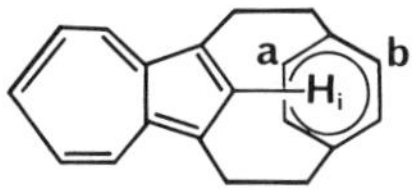

79 80

Section III,C,4). In both cases the internal meta proton H_i is shielded: δ 5.90 for **79,** δ 6.08 for **80.**

When a larger internal group is present (e.g., a CH_3 group as in **81, 82,** and **83**), the molecules are fixed, as evidenced by the multiplet resonances

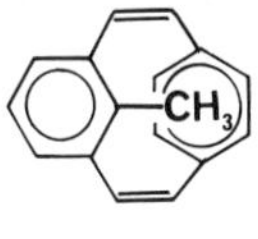

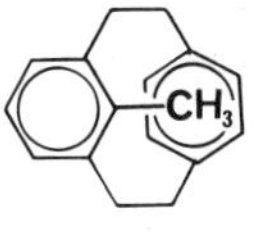

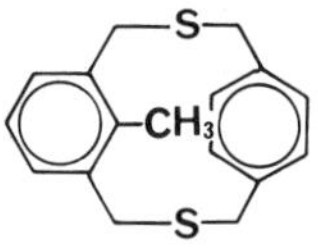

81 82 83

observed for each para ring.[42] In each case the internal CH_3 group is strongly shielded to δ 1.45, 1.78, and 1.9, respectively. In the analogous examples in which the internal substituent is a cyano group, H_a and H_b of each para ring could be assigned as shown in **84** → **86.**[58] This clearly

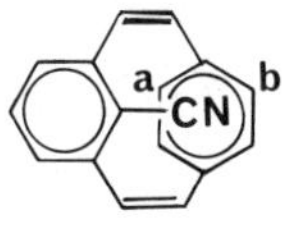

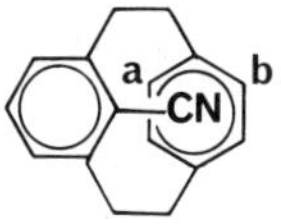

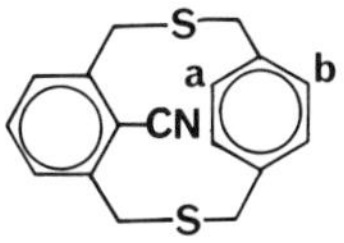

	84	85	86
δH_a	6.15	5.77	6.46
δH_b	7.43	7.35	7.20

indicates conformationally fixed species. In the case of the fluoro compounds, **87** and **88** are fixed (at least up to 150°C), whereas the thiaphane **89** is mobile and is discussed later.

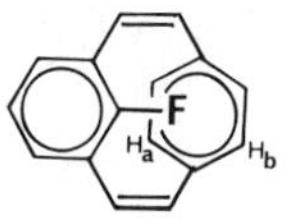

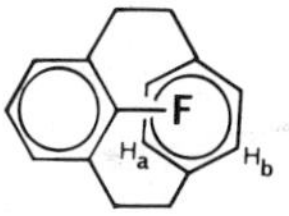

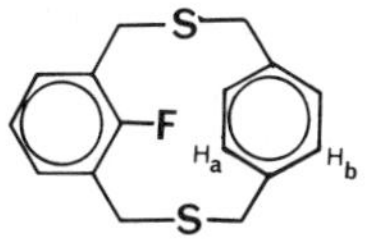

	87	88	89
δH_a	6.26	5.95	6.58
δH_b	7.20	7.15	7.07

Of the heterometaparacyclophanes, **90** and **91** exist as discrete isomers, and hence there is no doubt that these are not conformationally flipping at

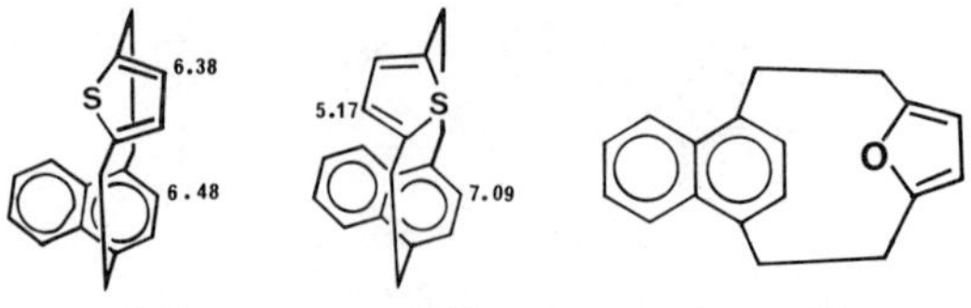

anti-**90** syn-**91** anti-**92**

room temperature.[59] The assignment of stereochemistry is relatively straightforward, because in *syn*-**91** the thiophene protons are markedly shielded by the naphthalene ring, and in *anti*-**90** the 2,3-naphthyl protons are likewise shielded by the thiophene ring. Other examples in which two isomers are not known are assigned by analogy. In the case of **92** this is supported by X-ray determination.[60,60a]

D. Paracyclophanes

1. [2.2]Paracyclophanes

Since [2.2]paracyclophanes would not be expected to be conformationally mobile, the relevant chemical shift reference points are seen in **93–95.** It is interesting that, whereas **94** has been known[64] since 1958, its

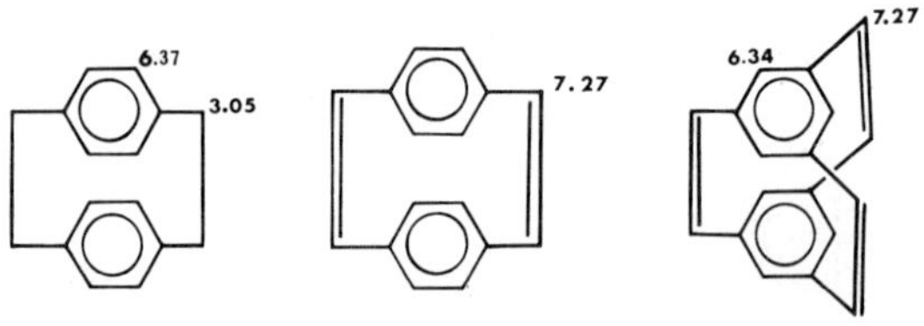

93[61] **94**[62] **95**[63]

^{1}H-nmr spectrum is not completely reported[62]; **95** thus should act as a suitable substitute. The aryl protons appear shielded by the opposite ring, and the vinyl protons appear abnormally deshielded (by the adjacent ring). Data on 1-keto[2.2]paracyclophane **(96)** are also available.[65]

When larger aromatic rings are present, syn and anti isomers can exist (e.g., **97** and **98**). In *anti*-**98** H_c is markedly shielded, with H_a and H_b relatively normal, whereas in *syn*-**97** all aryl hydrogens are somewhat shielded. Similar results were obtained for the corresponding dienes.[62] On melting, *syn*-**97** is known[66] to convert to *anti*-**98,** probably through an open intermediate.

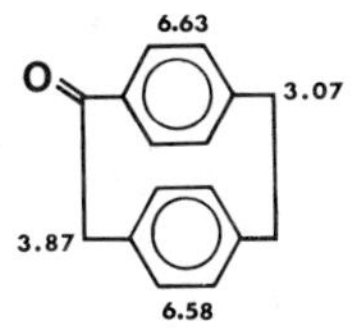

96[65] 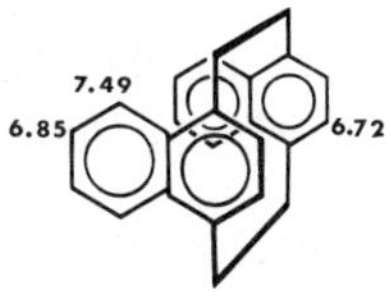 **97**

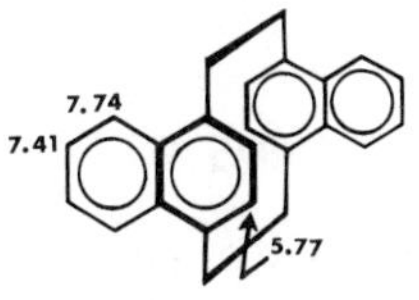

98 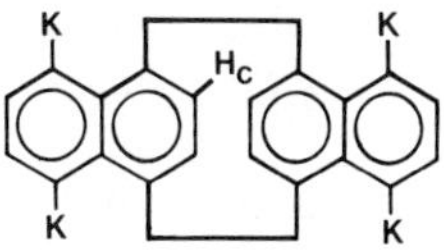 **99** K=OMe

Analogous assignments have been made for more recently prepared methoxy and quinone derivatives[67]; for example, in *syn*-**99** H_c is at δ 6.75, whereas in *anti*-**99** H_c is at δ 5.95.

In the case of the [2.2](1,5)naphthalenophanes in which achiral **(100)** and chiral **(101)** isomers are possible, the original assignment[68] has been

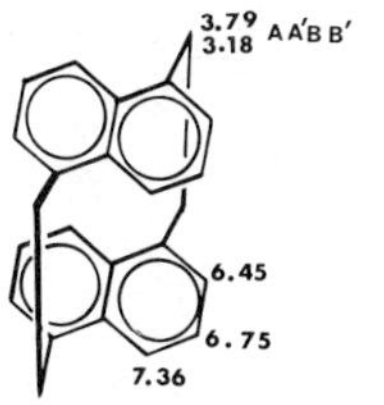

100 (achiral) **101** (chiral)

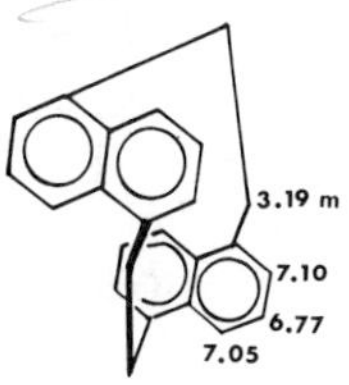

revised[69] on the basis of the 360-MHz spectra and X-ray structure determination. For the corresponding (2,6)phane **(102)** and its diene only the chiral compounds are known[70] and are assigned on the basis of the shielding of H_x to δ 6.45, which is considerable for an α-naphthyl proton. Misumi[71] has prepared a series of anthracenophanes; **103** is an example of

102 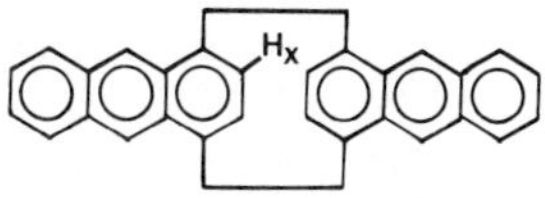 **103**

δH_x syn **6.71**, anti **5.47**

one that exists in syn and anti conformers, as evidenced by the chemical shift of H_X. A number of novel [2.2]paracyclophanes have been prepared. The annulenophanes **104** and **105** have been prepared by Misumi,[72] and

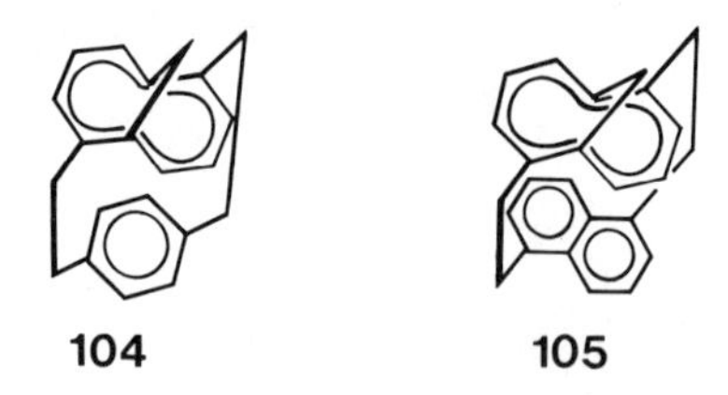

104 **105**

both show more shielded methano bridge protons (δ −1.31 and −1.48, respectively) than the parent (δ −0.52). This may be because of the opposite ring or an increase in macroring planarity or both. The benzene ring protons of **104** are also substantially shielded (δ 5.44).

The azulenophanes **106** and **107** are probably best assigned on the basis of their bridge protons: two (apparent) singlets for **106** and a more complex AA′BB′ for **107.**[73] However, this is supported by X-ray structures

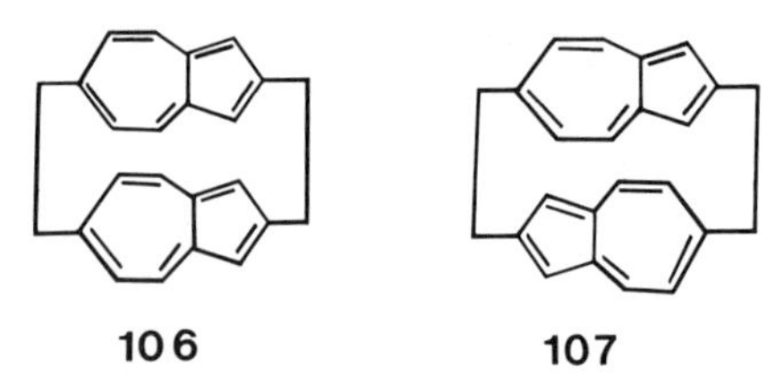

106 **107**

that correct a previous erroneous assignment (cf. **100** and **101**). All the azulene protons are shielded by 0.4 to 0.7 ppm.

The fluorenophanes **108** and **109** and their anions **110** and **111** are known[74] and show H-9 most shielded in the anti conformers **108** and **110.**

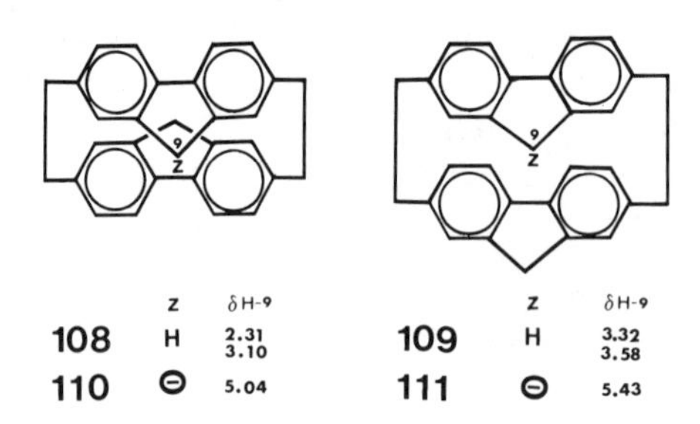

In the phenanthrenophanes **112** and **113,** H_a protons were used to assign[75] stereochemistry: *anti*-**112,** δ 7.43; *syn*-**113,** δ 7.71.

Both the phenanthrobiphenylophane **114** and its diene have been assigned syn stereochemistry on the basis of downfield shifts of the internal

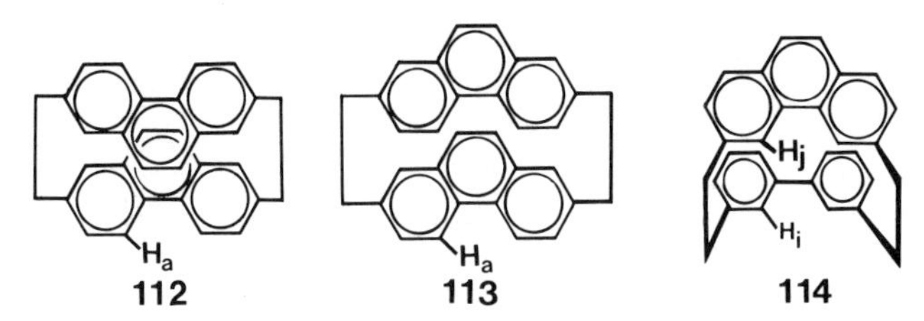

protons H_i and H_j: δ 8.46 and 9.13, respectively, in the diene and δ 7.56 and 8.27, respectively, in the cyclophane. Both exhibit strongly solvent-dependent but temperature-independent spectra, indicating some flexing but no major conformational changes.[76]

Several pyrenophanes have been prepared,[77–79] of which **115**[77] and **116**[78] have simple stereochemistry and show substantial shielding of their aryl

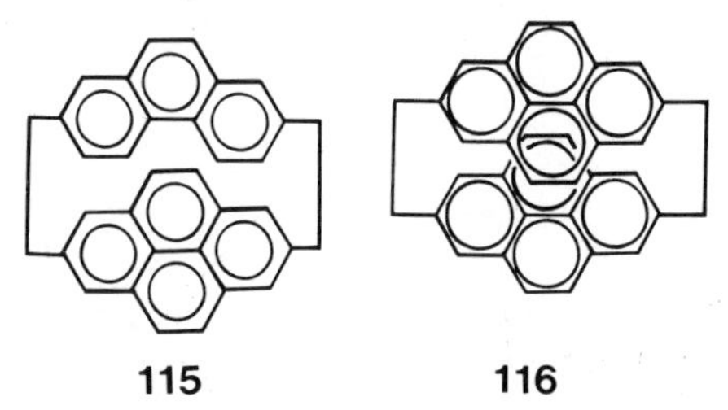

protons (Δδ ≈ 0.50–0.8 ppm). Two stereoisomers of **117** (**A** and **B**) are not interconvertible.[77] The assignment was made on the basis of chemical

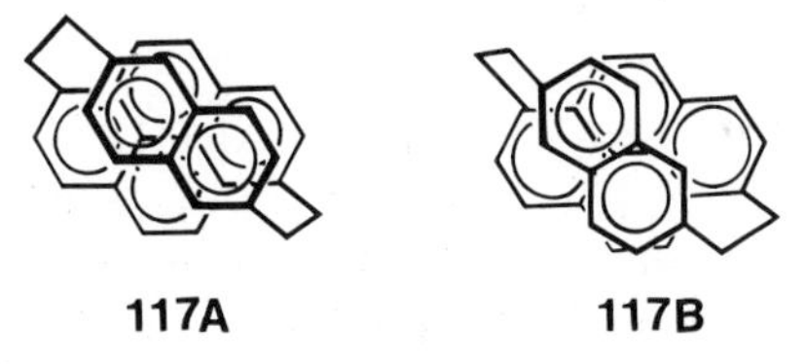

shift differences among the various sets of protons calculated from 360-MHz spectra. Of the larger [2.2]phanes, **118** would appear to have a fixed

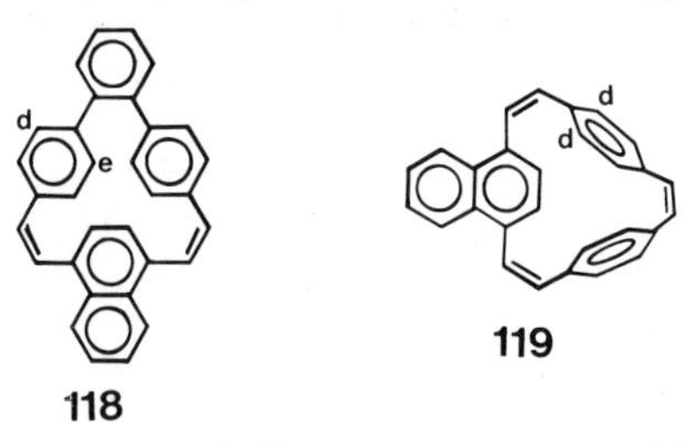

conformation because H_d and H_e appear as a multiplet at δ 6.2–7.2, whereas in the parent **119** H_d appear as a singlet and the molecule thus may have some mobility (see Section III).[80]

Of the [2.2]phanes with substituted bridges, **120–122** are of note, because they have orthogonal rings. It is surprising, however, that very little

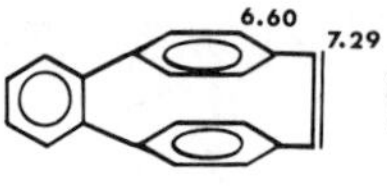

6.60 7.29 6.56 3.11 6.51 3.07

120 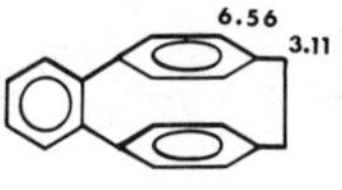 **121** 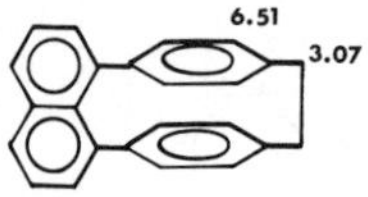 **122**

difference is observed between their ^{1}H-nmr spectra and those of the parents (**93** and **94**), and all are conformationally fixed.[81,82]

The unusual [2.2]paracyclophane derivative **123** was assigned the anti

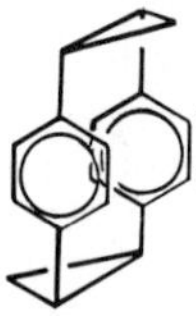

anti-**123** 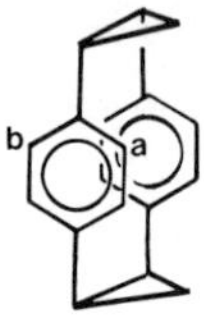 syn-**123**

rather than the syn structure on the basis of the J_{AB} value of 8.0 Hz found[83] for **123,** for which δH_a is 6.74 and δH_b is 6.51.

2. Larger Fixed Paracyclophanes

Because the aromatic signals for [2.4]-, [3.3]-, and [4.4]paracyclophanes are singlets[65] it is not possible to say whether the protons are averaging by ring rotation or bridge movement, or accidentally have the same chemical shifts. This is discussed in Section III. However, when substituents are present on the rings (e.g., in **124, 125**[84] and **126, 127**[85]), discrete isomers exist, indicating that ring rotation is restricted in these [3.3]- and [4.4]phanes. Several thia[3.3]paracyclophanes have been re-

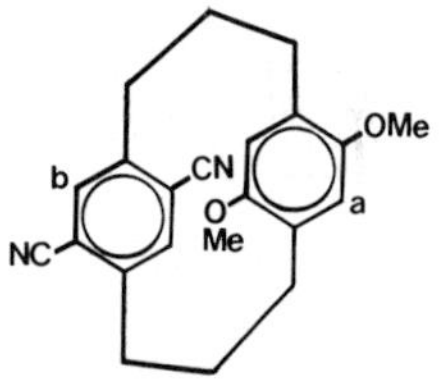

124 **125**

δH_b 6.20 6.50

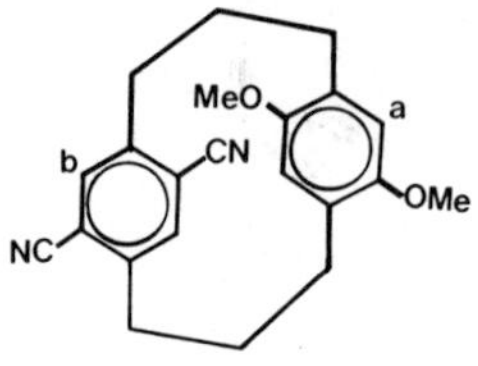

δH_a 7.15 7.40

126 E=OMe **127** E=OMe

ported[62,68] as noninterconvertible syn and anti isomers, for example, *syn*-**128,** δH_c 6.93, and *anti*-**129,** δH_c 6.27; achiral **130,** δH-4 7.61, and chiral

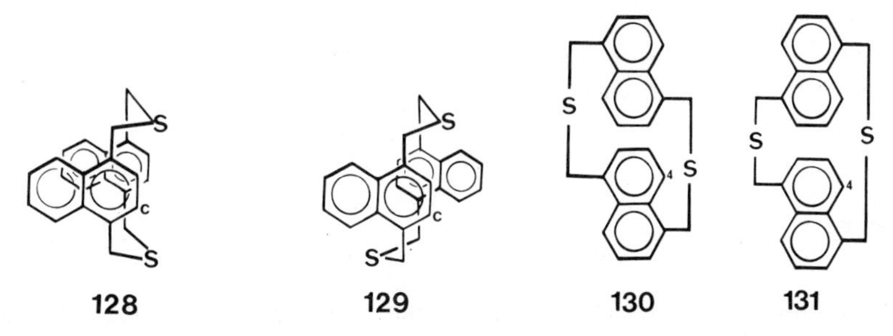

128 **129** **130** **131**

131, δH-4 7.32. However, the thia[4.4]cyclophane **132** exists as easily convertible syn and anti conformers,[86] which after explusion of S give noninterconvertible *syn*- and *anti*-**133,** δH_c 6.93 and 5.99, respectively.

132 **133**

The analogous [3.3](2,6)- and [3.3](1,5)(2,6)naphthalenophanes are also known.[87] As might be expected, the isomeric **134** and **135** are not interconvertible, the less soluble isomer being assigned as the more symmetric.[88]

134 **135**

The shielding of the external aryl proton H_o (δ 6.51) of the derived cyclophane **136** is between that of the metacyclophane **137** (δ 6.83) and

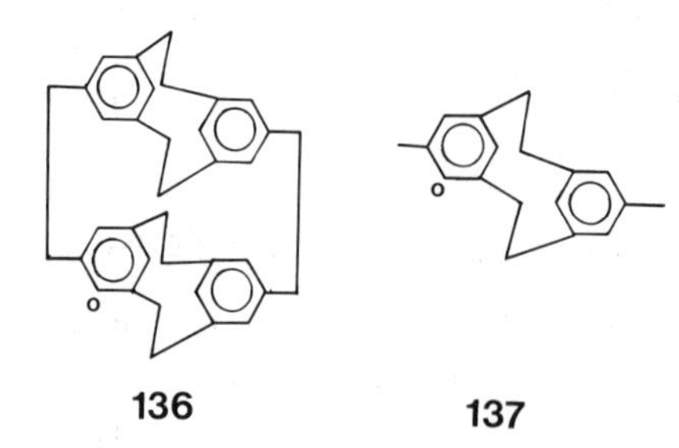

[2.2]paracyclophane (δ 6.46) and indicates the capacity of the metacyclophane bridges to absorb some of the strain.

E. Cage Phanes

Most cage phanes are fixed and, as mentioned in the Introduction, are discussed in Chapter 10. However, because a few mobile cage phanes are described in Section III, pertinent ^{1}H-nmr reference data, represented by **138** to **144**, are noted here.

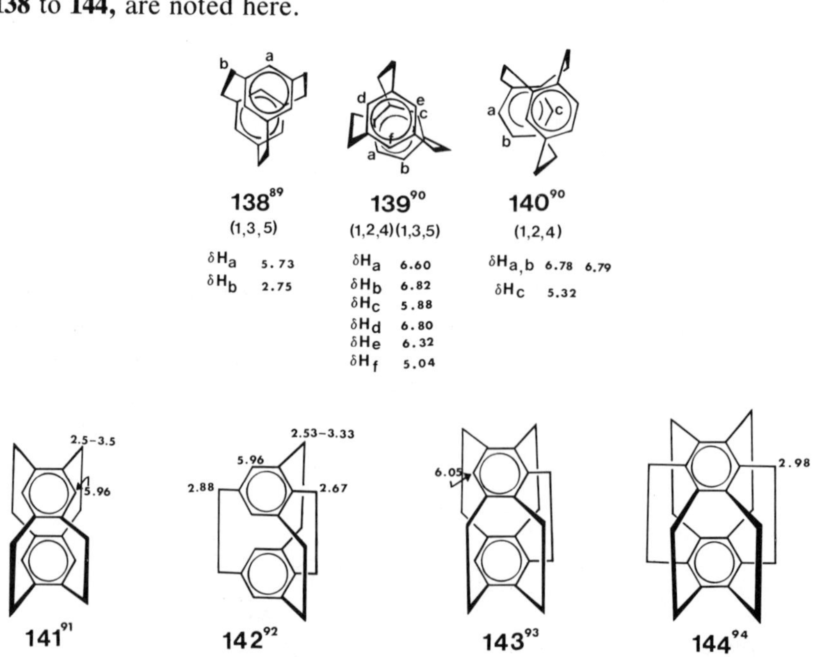

III. MOBILE PHANES

A. Introduction: Variable-Temperature Nuclear Magnetic Resonance Spectroscopy

Conformational changes are now quite commonly studied by variable-temperature nmr (vtnmr) spectroscopy. A list of earlier references can be found elsewhere.[60a] An excellent review by Binsch and Kessler[95] supplements the very complete chapter entitled "Investigation of the Kinetics of Conformational Changes by NMR" by Sutherland,[96] which also contains a section on cyclophanes. Both discuss in detail how exchange rates and activation parameters can be determined by both lineshape analysis and T_c methods and also outline some of the pitfalls, particularly in the determination of $\Delta H^{\ddagger}$ and $\Delta S^{\ddagger}$. Reasonably accurate (± 1 kJ/mol) $\Delta G^{\ddagger}$ values, however, can be obtained by either method, even when the error in temperature estimation is as high as 3K. Indeed, we have studied cyclophanes by both methods and have found that the values of $\Delta G_c^{\ddagger}$ so obtained agree within the limits of experimental error.

The most commonly used equation derives $\Delta G_c^{\ddagger}$ at the coalescence temperature T_c using only T_c (in degrees Kelvin) and the low-temperature separation $\Delta\nu$ (in hertz) of the two exchanging proton signals, both of which are easily measured. The value of $\Delta G_c^{\ddagger}$ is calculated as follows[97]:

$$\Delta G_c^{\ddagger} \text{ (kJ/mol)} = 2.303 \times 8.31434^{*} T_c (10.319 - \log_{10} k_c + \log_{10} T_c)$$

*(for kcal/mol use 1.9872)

where $k_c = (\pi/\sqrt{2})\Delta\nu$. This strictly applies only to equally populated singlets, in which the signal width is small relative to $\Delta\nu$, although it is often used under other circumstances. For a coupled system, for example, an AB, the correct rate at coalescence is

$$k_c = (\pi/\sqrt{2})[(\nu_A + \nu_B)^2 + 6J^2]^{1/2}.$$

When a lineshape analysis is carried out, matching computer-simulated spectra with actual experimental data, a series of exchange rates k are found for temperatures T. The Eyring equation,

$$k = cTe^{-(\Delta H^{\ddagger}/RT)}e^{(\Delta S^{\ddagger}/R)}$$

where $c = k/h$, and k = Boltzmann's constant, and h is Planck's constant, is usually used by plotting $\ln(k/T)$ versus $1/T$ to derive $\Delta H^{\ddagger}$ and $\Delta S^{\ddagger}$ and hence $\Delta G_T^{\ddagger}$ at any required temperature. The advantage of using this method is that different compounds can be compared at standard temperatures. Proton magnetic resonance is suitable only to the study of exchanges in which $\Delta G^{\ddagger}$ falls in the approximate range 20–100 kJ/mol. Nevertheless, as the following sections show, a large number of cyclophane motions fall into this range and hence a great many results are now available.

B. Cyclophanes with One Aromatic Ring

1. Metacyclophanes

Interestingly, the early studies on [7]metacyclophane and [8]metacyclophane derivatives by Mitchell and Boekelheide[98] and Vögtle,[99] respectively (1969), were carried out on the dithia derivatives **145** and **146.** The

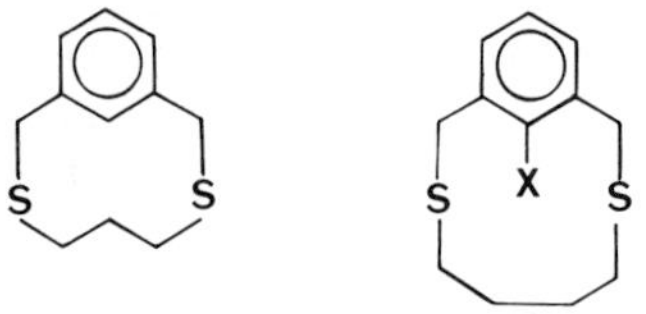

145 **146** X=H **147** X=F

advantage of using these compounds was that the spectra were much easier to interpret. The benzyl protons of **146** were a singlet down to −70°C, and hence at these temperatures no freezing out of conformers **146A** and **146B** occurred, indicating a ring inversion barrier of <38 kJ/

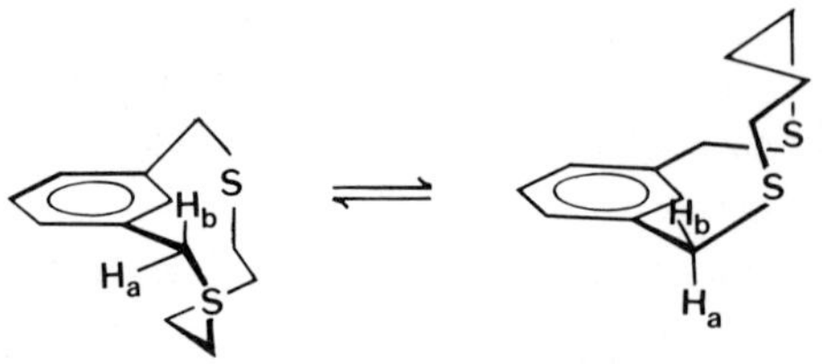

146A **146B**

mol. However, when the internal X group was larger than H, the barrier increased, as Vögtle[100] had shown in 1968; thus, whereas at 35°C the benzyl protons of **147** (X = F) were an AB with H_a at δ 4.10 and H_b at δ 3.13,[99] at temperatures above 185°C these coalesced to a singlet, indicating $\Delta G_c^{\ddagger} = 95$ kJ/mol. For the larger X (Cl or Br) T_c was greater than 200°C. Because the chemical shifts of the other bridge protons have not been reported, little can be said about the conformations of these compounds. The case of the smaller **145,** however, is better documented.[98] At room temperature $H_{c,c'}$ rapidly exchange and appear as a multiplet at δ 0.45, consistent with **145A** ⇌ **145B** ⇌ **145C** ⇌ **145D.** Likewise, the ben-

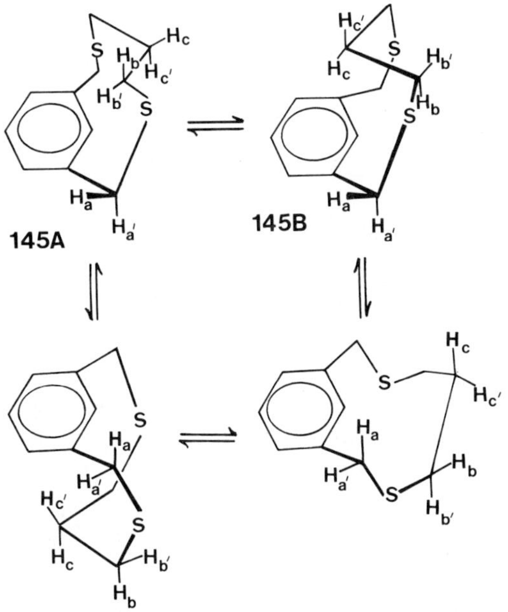

145A 145B 145D 145C

zylic protons $H_{a,a'}$ and the bridge protons $H_{b,b'}$ also exchange, appearing as peaks at δ 3.83 and 2.59, respectively. As the temperature is lowered each set of peaks passes through a different coalescence point (−50°C for $H_{c,c'}$ and −75°C for $H_{aa'}$). Since two different high-field peaks are observed at δ −0.21 and −1.36, it suggests that two different low-temperature conformers are present (e.g., **145A** and **145B**). This is also consistent with the frozen benzylic signals. The high-field peaks would then correspond to H_c of **145B** and $H_{b'}$ of **145A.** The approximate $\Delta G_c^{\ddagger}$ values are 35–40 kJ/mol. However, further comment on these molecules must await 250-MHz spectra, work on which is in progress.

The parent metacyclophanes have not been studied until more recently.[101] In [7]metacyclophane **148** (X = CH), $H_{d,d'}$ appear averaged at

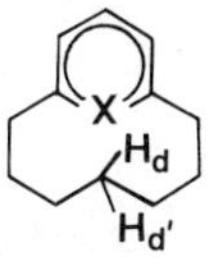

148 X = CH
149 X = N

room temperature at δ −0.18 and, below −28°C, H_d reappears at δ −1.33 (corresponding to H_c in **145B**); this corresponds to $\Delta G^{\ddagger}_{-28}) = 48$ kJ/mol. This barrier is larger than that observed for **146** ($\Delta G^{\ddagger}_{-50} = 43$ kJ/mol)[98] and [7](2,6)pyridinophane (**149:** X = N), where $\Delta G^{\ddagger}_{-75.5}$ was found[102] to be 38 kJ/mol, as might reasonably be expected. The introduction of a large substituent as in **150** freezes a conformation in which H_d is found at δ −1.86 (cf. **145B**), which is not changed at 200°C. The case of the [6]metacyclophanes is also interesting. In the bromo compound **151** (Y = Br) the

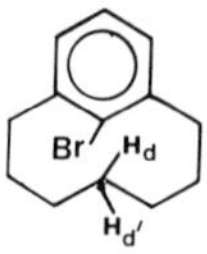

150

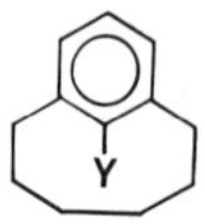

151 Y = Br
152 Y = H

nonequivalent benzyl protons indicate that the hexamethylene chain is frozen on one side of the benzene ring (as in the case of **145A** and **145B**). In this case, however, a pair of enantiomeric conformers (**151A** and **151B**)

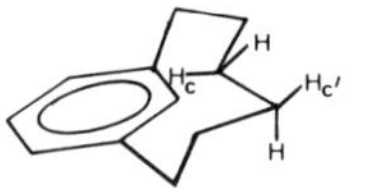

151A 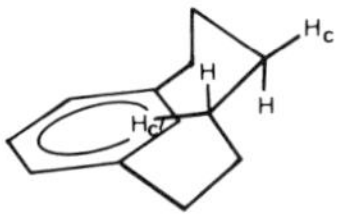 **151B**

is possible, as indicated by the fact that at room temperature $H_{c,c'}$ average to δ −0.03, whereas at low temperature H_c appears at δ −1.68. From $T_c = -4.5$°C, $\Delta G^{\ddagger}_c$ for this pseudorotation is estimated to be 52 kJ/mol. Replacement of Br by I increases the barrier to 53 kJ/mol, and replacement by H to give the parent **152** (Y = H) decreases the barrier to 46 kJ/mol. In the latter case, however, the benzylic signals are still nonequivalent, indicating that the methylene chain is frozen on one side of the ring, in contrast to the dithia analog **145.** However, above 76.5°C the benzylic protons become equivalent, as do $H_{cc',dd'}$ (δ 0.79), suggesting that a process analogous to **145A** ⇌ **145C** now occurs having $\Delta G^{\ddagger}_{76.5} = 73$ kJ/mol.

[5]Metacyclophane **153** was prepared[103] in 1977 and has a rigid fixed conformation, as suggested by its ^{1}H-nmr spectrum, which indicates that H_b are shielded at δ 0.04–0.54. Inspection of a model of **153** indicates a

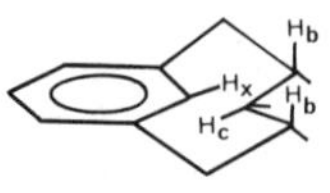

153

TABLE IV
T_c and $\Delta G_c^\ddagger$ Data for Compound 154 [Z = S; Y = $(CH_2)_n$]

X	n	T_c (°C)	$\Delta G_c^\ddagger$ (kJ/mol)	References
SO_2CH_3	12	80	71	104b
CO_2CH_3	10	195	99	104b
SCH_3	10	85	73	104b
OCH_3	10	−55	~44	104a
OCH_3	9	90	75	104a
CH_3	9	−30	50	104a
CH_3	8	60	69	104a
NO_2	8	27	64	104b
Cl	8	—	49	104a
F	5	—	63	104a
H	3	—	43	55

very close proximity of H_c and H_X, and hence both are probably sterically deshielded. Indeed, H_X appears at the remarkable position of δ 7.85; unfortunately, H_c cannot be clearly determined. It is interesting that the protons analogous to H_X in **145, 148,** and **152** are also deshielded from their normal positions (δ 7.66, 7.43, and 7.21, respectively).

Vögtle[104,104a] studied a series of compounds of type **154** for which a conformational process of type **155A** ⇌ **155B** can be conveniently fol-

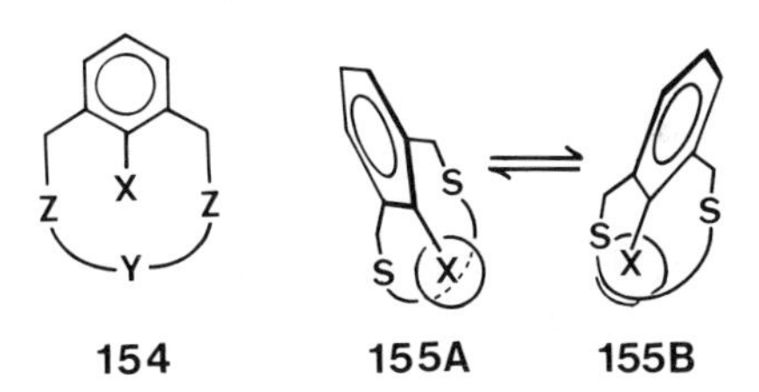

lowed by examination of the benzylic protons. He was thus able to estimate the size of various substituents from his $\Delta G_c^\ddagger$ values and found[55]

$SO_2CH_3 > CO_2CH_3 > SCH_3 > I > OCH_3 > Br > CN > CH_3$
$> Cl, NO_2 > NH_2 > OH > F > H >$ lone pair

Typical examples of the $\Delta G_c^\ddagger$ values that Vögtle found for **154** [Z = S; Y = $(CH_2)_n$] are given in Table IV. He has reviewed his complete set of data.[55]

2. Paracyclophanes

With medium-sized chains (e.g., [10]paracyclophane) the chain of methylene groups creates an arc over the benzene ring, as in **156,** and hence conveniently maps the magnetic ring current. However, as has been shown[105] for **156,** there are advantages to the use of high-field instruments, which make assignments less prone to error. For **156** the new assignments (based on 220-MHz spectra) are shown below.

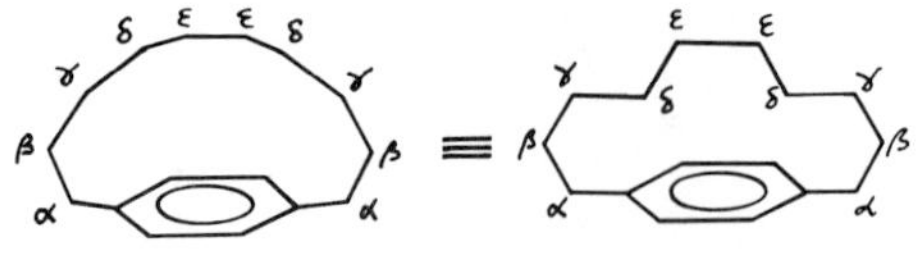

156

Protons	δ (ppm)
Ar	7.04
α	2.62
β	1.54
γ	1.08
δ	0.73
ε	0.51

At this field strength each multiplet is well separated and hence can be confirmed by decoupling experiments. The chemical shifts observed correlated well with ring current calculations, based on a calculated preferred conformation. The ^{1}H-nmr spectrum of [11]paracyclophane-1,10-diene **(157)** has a singlet at δ 7.03 for the four aromatic protons. Since multiplets

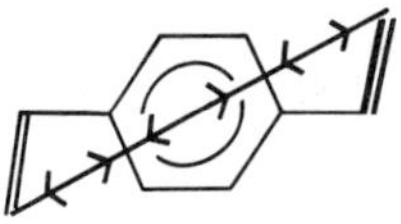

157 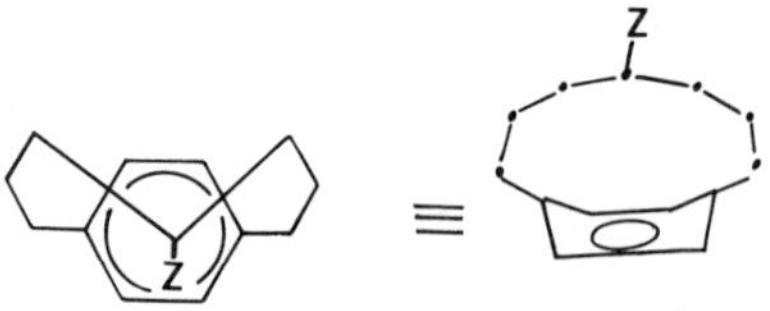 **158** Z=COOH

at δ 0.66 and 1.10 are observed for part of the methylene protons, the conformation of **157** must be such that the chain is disposed symmetrically with respect to the aromatic ring, for example, as shown. The compound is probably fairly mobile with respect to [7]paracyclophane,[106] which would be expected to be much more rigid. 3-Carboxyl[7]paracyclophane **(158)** is known,[107] and its structure has been determined by X-ray crystallography. Although the ^{1}H-nmr spectrum showed the expected high-field multiplets (e.g., two 1H doublets at δ − 1.4), the real interpretation of solution conformations must await a high-field spectrum. It is interesting that an article[108] on the synthesis of [7]paracyclophane **(159):** Y = H) reported that **160** (Y = CF_3) had an upfield proton at δ −

2.4, whereas those of **159** itself are at δ − 0.4! No doubt many more spectral surprises are ahead as investigations into these molecules continue.

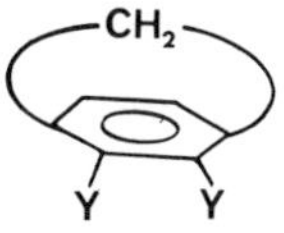

The ^{1}H-nmr spectrum of [6]paracyclophane **(161)** was reported[109] in 1974 to show temperature dependence. At room temperature the aryl protons appear as a singlet at δ 7.17, the benzylic protons appear as a triplet at δ 2.49, and the methylene protons appear as multiplets at δ 1.15 and 0.33. At −80°C the benzylic absorptions become an AB quartet and the higher-field peak due to the methylene protons is split into two (one of which is at δ −0.6). This suggests that the molecule is undergoing a change similar to **151A** and **151B,** in which possible C-3 and C-4 are pseudorotating, which must average out at least the benzylic and high-field multiplets. Again, a high-field study would be valuable. A report[110] on the analogous naphthalenophanes indicates that the [8]-, [9]-, and [10]naphthalenophanes are "frozen," showing benzylic resonances at δ ≈ 2.5 and 3.5, and typical high-field protons up to δ −0.70, whereas for the [14]phane **162,** the benzylic protons appear as an averaged triplet at δ 3.02. The authors interpret this as being a conformational flip of the kind **162A** ⇌ **162B.** Unfortunately, no vtnmr results are reported.

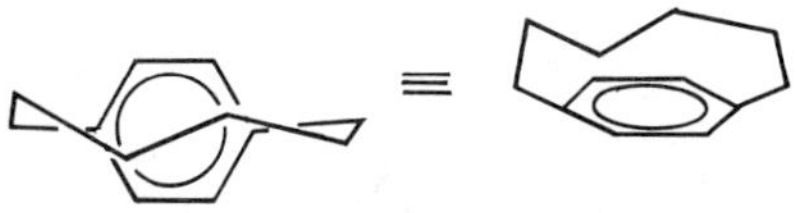

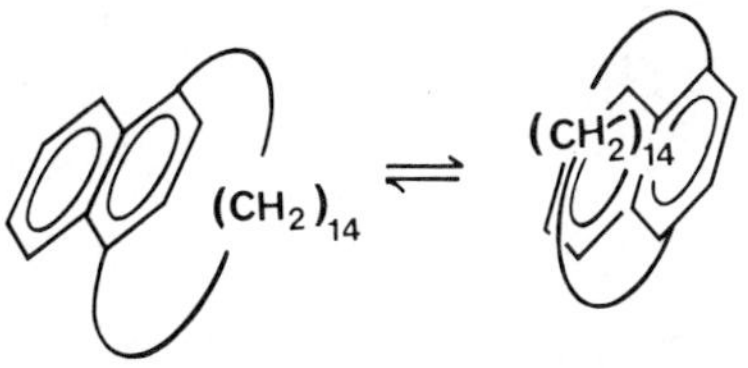

C. Cyclophanes with Two Aromatic Rings

1. Metacyclophanes

Very few small metacyclophanes are mobile (see Section II,A). The furanophane **163** is mobile ($\Delta G^{\ddagger}_{63}$ = 70 kJ/mol),[30] whereas the analogous oxathia **(35)** and dithia **(34)** compounds (Section II,A,2,c) are not. Each benzyl proton signal of the pyrrolofuranophane **164** is also different (ABCD), and this compound is thus immobile.[111]

	X	Y
34	S	S
35	O	S
163	O	O
164	O	NH

	X	Y
165	S	S
166	O	S

In the [3.3] series[112] even the dithia compound **165** is mobile above 105°C. The more stable *anti*-**165** shows the thiophene protons at δ 6.37, whereas *syn*-**165** shows these more shielded at δ 6.00. The isomers are formed initially in a 20 : 1 ratio, and on heating a solution the two peaks above coalesce at 105°C. It is still not possible to state conclusively, however, whether the isomerization involved is a syn ⇌ anti type or whether it is anti ⇌ planar ⇌ anti; high-field analysis of the bridge proton multiplets might be instructive. The analogous oxathia compound did not show similar behavior, even at −90°C.

The six-membered-ring compound [2.2]pyridinophane **(167)** was one of the earliest mobile metacyclophanes to be studied[30]; the benzyl protons at δ 3.28 and 2.55 coalesce at 13.5°C, corresponding to $\Delta G_c^\ddagger$ = 62 kJ/mol.

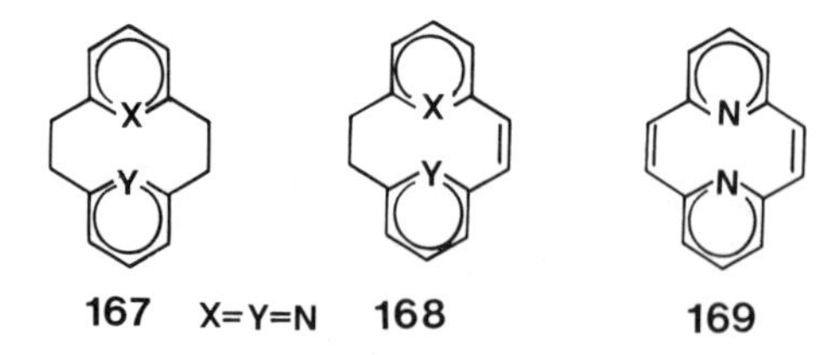

167 X=Y=N **168** **169**

Only more recently, however, has Cooke[113] studied the analogous monoene **168.** At low temperatures the methylene bridge protons are AA′BB′, whereas at room temperature they are a singlet at δ 2.87; from T_c = −43°C, a $\Delta G_c^\ddagger$ value of 46 kJ/mol is obtained. Thus, replacement of one —CH_2CH_2— bridge of **167** with a —CH═CH— bridge substantially lowers the energy barrier to conformation flipping (as in the case of the metaparacyclophanes; see Section III,C,4). Cooke pointed out that in light of this the diene **169,** reported by Boekelheide,[114] may also be mobile, and the invariance of its ^{1}H-nmr spectrum may be explained by the value of $\Delta G_c^\ddagger$, which is too small to observe by vtnmr. He also noted that the compound **168** (X = N; Y = CH) is not yet known and may show a substantially lower barrier than analogous **167** (X = N; Y = CH), where $\Delta G_c^\ddagger > 96$ kJ/mol.[31]

As discussed in Section II,A all cyclophanes of type **167** where X = C—Z have rigid phane ring orientation (normally anti). For a large group Z, restricted rotation of Z may be possible. Vögtle[25] showed this to be the case for *anti*-**22.** At room temperature the two ortho ring protons are a

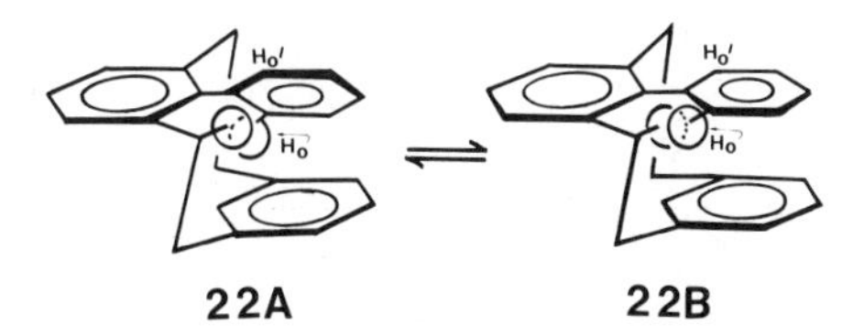

broad singlet at δ 5.9. At low temperatures these become two sharp absorptions at δ 6.55 and 5.29 ($H_{o'}$ and H_o, respectively), with $T_c = 0°C$ and $\Delta G_c^\ddagger = 54$ kJ/mol. Vögtle pointed out that this barrier is between that of biphenyl (<18 kJ/mol) and 2,2′-dimethylbiphenyl (76 kJ/mol), which would seem reasonable. However, it does seem very unlikely to us that complete rotation of the phenyl substituent could occur. We believe that only a partial rotation past the methylene bridges occurs, as is shown for **22A** ⇌ **22B** and as we have found[51] for the analogous thiacyclophanes **73.** We have also found similar results[115] for the phenylmethyl compound **170,** where H_o,$H_{o'}$ appear as a broad signal at δ 6.45 but separate at low temperatures by 0.51 ppm. Most surprisingly, however, we have found T_c for **170** at −59°C, $\Delta G_c^\ddagger = 43.5$ kJ/mol, which is considerably lower than

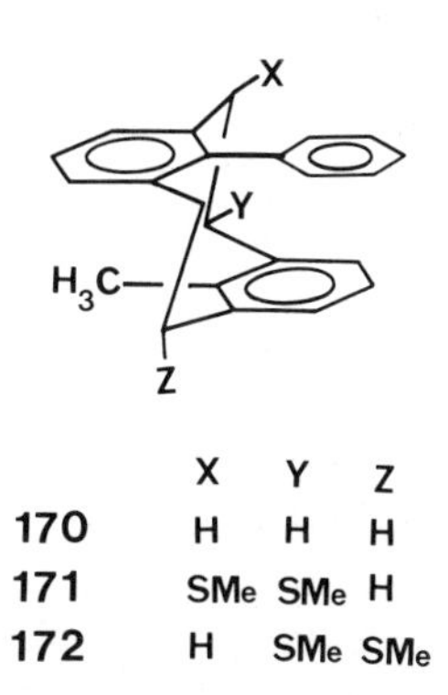

	X	Y	Z
170	H	H	H
171	SMe	SMe	H
172	H	SMe	SMe

that found for **22** (54 kJ/mol). The intermediates in the synthesis that leads to **170** with substituents on the bridges, namely, **171** (X = Y = SMe; Z = H) and **172** (Y = Z = SMe; X = H), have $\Delta G_c^\ddagger = 43$ and 40 kJ/mol, respectively. Clearly, the barrier is very sensitive to geometry. We mentioned in Section II,A,2,a that **22** probably has geometry **B;** the geometry of **170** is probably closer to **A.** Indeed, the internal methyl protons of **170** appear at δ 0.79, actually somewhat less shielded than those of **2** (δ 0.56), which is consistent with a shift from geometry **A** toward geometry **C** (X = ϕ). This change in angle between the phane rings undoubtedly changes the compression between the methylene bridge and ortho ring protons and hence the barrier.

Substituted [3.2]metacyclophanes **173** have been studied by Griffin.[116] He postulated a conformational process of the type **173A** ⇌ **173B,** since

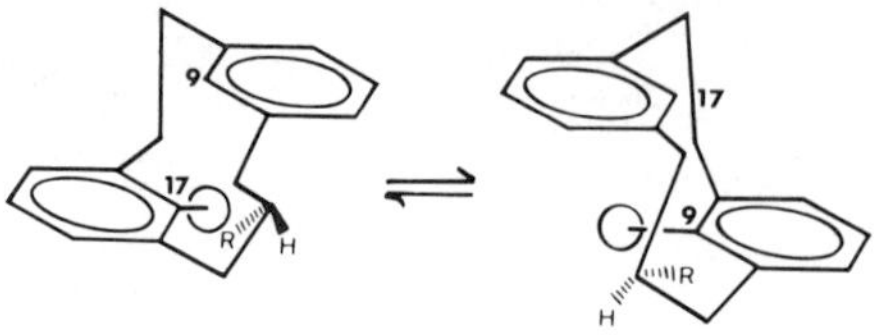

173A **173B**

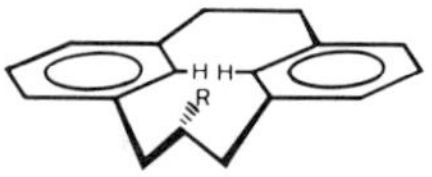

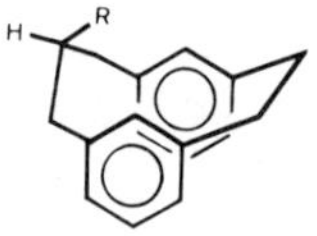

173C **173D**

H-9, H-17 are a singlet at higher temperatures and separate at lower temperatures (e.g., for R = CO_2H, T_c = 53.9°C). He found that $\Delta G^{\ddagger}_{60}$ was between 66 and 70 kJ/mol for a variety of substituents. Similar values were obtained by examination of the bridge protons, consistent with **173A** ⇌ **173B.** *A priori* it is not obvious whether the intermediate in the process **173A** ⇌ **173B** is relatively planar **(173C)** or syn **(173D); 173C** would probably be only a transition state, whereas **173D** could be an insolable conformation, that is, **173A** ⇌ **173C** ⇌ **173B** or **173A** ⇌ **173D** ⇌ **173B.** It is not necessary to pass through **173C** to get to **173D** or vice versa. This intriguing question will plague cyclophane chemists for some time, because with very few exceptions (see following section) it is not clear exactly how the conformation inversion occurs.

[3.1]Metacyclophane **174** has been synthesized and has been reported[117] to be conformationally inverting as **174A** ⇌ **174B.** The diphenylmethane —CH_2— protons appear as an AB quartet at δ 4.20 and 3.58 and exchange at 68°C, yielding $\Delta G^{\ddagger}_{c}$ = 70 kJ/mol. This is amazingly similar to the values found in the [3.2] series. Molecular models indicate that the phenyl rings are at a considerable angle such that the —CH_2— protons just referred to are pseudoaxial and equatorial.

174A 174B

[3.3]Metacyclophane **175** shows[118] the benzyl protons as a triplet at δ 2.74 with H_i at δ 6.84. This indicates a syn conformer, analogous to the thiacyclophanes (see the following section), and this is supported by uv data. A vtnmr study is needed.

175

2. THIA[2.2]-, THIA[2.3]-, AND THIA[3.3]METACYCLOPHANES

a. Proton or Lone Pair as Internal Substituents

This is the largest group of compounds that have been studied and hence is separated from the other metacyclophanes reviewed in Section III,C,1.

The introduction of S into the bridge of a metacyclophane should make conformation inversion easier (S rather than C(H)(H)). Vögtle[55] showed that in the case of the [2.2]metacyclophanes **176** it does (Table V).

TABLE V
Inversion Barriers for Compound 176[a]

X	Y	Z	$\Delta G_c^\ddagger$ (kJ/mol)
N	N	CH_2	62
N	N	S	<57
CH	N	CH_2	>96
CH	N	S	86

[a] When X = Y = CH the molecule is conformationally fixed anti (Section II,B,1,a).

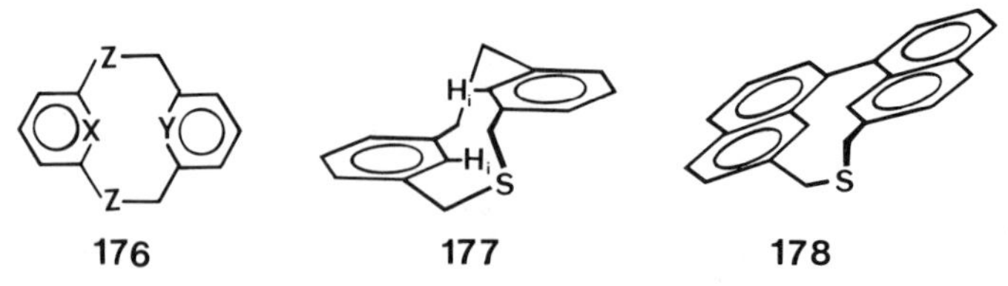

176 **177** **178**

In the [2.3]cyclophanes, Sato[43] showed that **177** is mobile at room temperature, the bridge protons both being singlets; at low temperatures the bridge protons become an AB (δ 3.45 and 3.84) and A_2B_2 (δ 2.22 and 3.08), with the internal proton at δ 5.13: $T_c = -8°C$, $\Delta G_c^{\ddagger} = 55$ kJ/mol (cf. **173,** about 70 kJ/mol). Conversion of the S to a sulfoxide or sulfone increases the barrier to 63 and 68 kJ/mol, respectively. The benzoannelated phane **178** was rigid up to at least 180°C.[26]

The case of 2,11-dithia[3.3]metacyclophane **(179)** is interesting. Sato[119] reported in 1968 that, down to −90°C, **179** showed its benzyl protons as a singlet at δ 3.11 and H_i at δ 6.63. He considered this to be the result of rapid interconversions between anti conformers of **179,** possibly through a "relatively strain free cisoid intermediate."[43] Vögtle,[120] who independently synthesized **179** in 1969, also studied its ^{1}H-vtnmr spectrum and

syn-**179**

likewise considered it to be a rapidly equilibrating mixture of syn and anti conformers. Indeed, for the next decade, **179** was generally written in the anti form, consistent with most other cyclophanes. Because we wished to resolve a melting point anomaly for **179,** we undertook an X-ray structure determination, which showed this molecule to be syn.[121] In fact, reconsideration of the ^{1}H-nmr spectrum[121] conclusively supported the idea that **179** was in the syn conformation in solution also, with very little anti conformer present. The averaging of the benzylic protons can be simply explained by the S-bridge wobble: **H** $\rightleftharpoons$ **I** $\rightleftharpoons$ **J** referred to in Section II,B,1,b. In the ^{13}C-nmr spectrum this motion possibly begins to freeze out at −100°C. This example is illustrative, however, in that despite a chemical shift for H_i of δ 6.6, which would be like no other internal anti proton,[121] we did not reassign the structure for so long!

With this knowledge, the syn structure should also be considered for most other dithia[3.3]metacyclophanes with only internal H substituents; examples and the δ values for H_i are given in Table VI (structures **180–193**).

TABLE VI
Dithia[3.3]cyclophanes That Should Be Reassigned Syn Stereochemistries with δ Values for H_i

Structure	Compound number	δH_i (ppm)	References
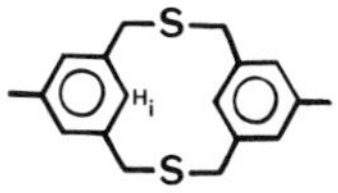	**180**	6.58	122
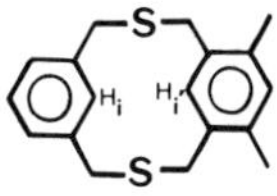	**181**	6.69 6.08 ($H_{i'}$)	34
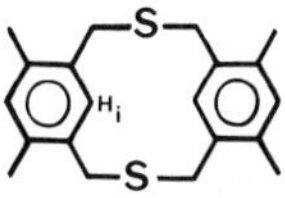	**182**	6.58	43
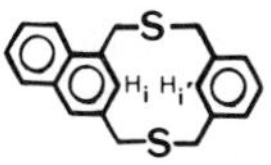	**183**	6.73 6.95 ($H_{i'}$)	123
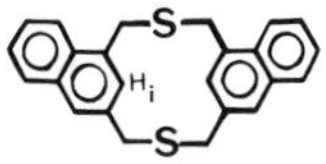	**184**	6.95	52
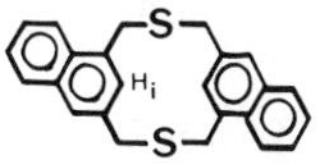	**185**	6.95	52
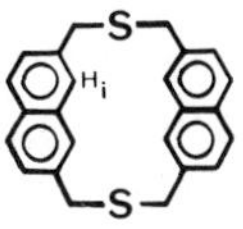	**186**	7.13	39
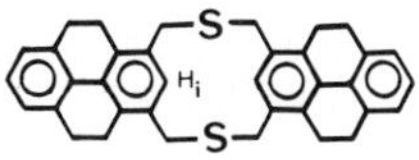	**187**	—	124
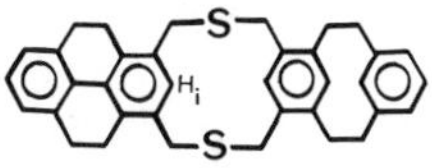	**188**	—	124
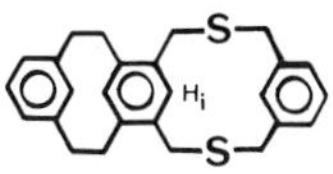	**189**	—	34a
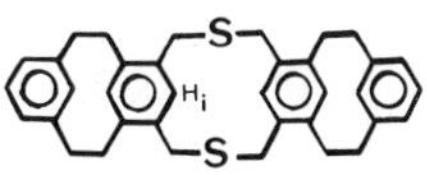	**190**	—	34a

(Continued)

TABLE VI (*Continued*)

Structure	Compound number	δH_i (ppm)	References
	191	6.60	34
	192	H_i 7.61	28
	193	H_i 6.96	125

The extra methyl substituents of **182** raise the barrier for the S-containing bridge inversion discussed previously, causing coalescence at higher temperature ($T_c = -41$°C) than for **179,** giving $\Delta G_c^\ddagger$ **(182)** = 50 kJ/mol.[43] This is probably caused by buttressing of the *o*-methyl substituent with the equatorial benzylic hydrogens. Conversion of the sulfide bridge to sulfoxide or sulfone in **179** did not change the stereochemistry (i.e., it remained syn),[43] although whether a similar effect is present in **182** could not be observed for solubility reasons.

The azulenophane **193** is interesting in that the authors claim that the singlet for the bridges and the upfield shift of H_i (0.52 ppm) relative to the model 1,3-dimethylazulene, which is twice as great as for the outer aryl protons, indicate *anti*-**193** ⇌ *syn*-**193,** and not a conformation change (e.g., **H** ⇌ **I** ⇌ **J**) of the 12 ring; this writer disagrees. The examination of models and the X-ray structure of **179** suggest that in a *syn*-thiacyclophane the aromatic rings are inclined outward at an angle as in **K** and are not parallel as in **L.** The H_o would thus not be expected to be as

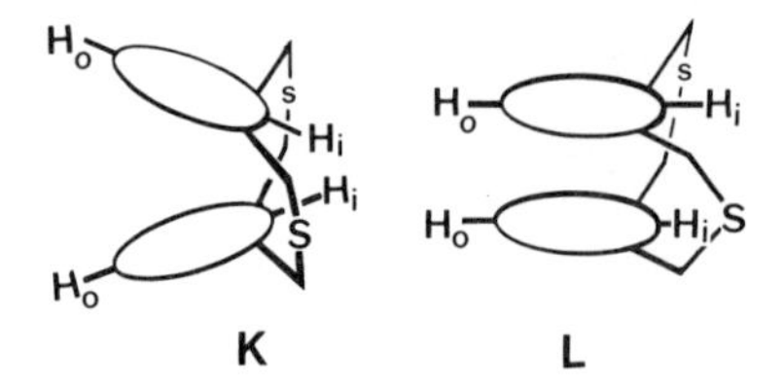

shielded by the opposite ring as H_i. The results (a freezing out of the bridge —CH_2— protons at −120°C) are thus quite compatible with **H** ⇌ **I** ⇌ **J** for the syn isomer.

Replacement of one —CH_2SCH_2— bridge of **179** with a benzene ring leads to **194.** This compound, reported by Vögtle,[21] shows H_i at δ 6.08 and the bridge —CH_2— protons as a singlet at δ 3.59 at room temperature. On cooling, the latter signal separates into an AB quartet with $T_c = -7°C$ and $\Delta G_c^\ddagger = 55$ kJ/mol. This is consistent with an anti structure for **194** showing a bridge rotation as the conformational process **194A** ⇌ **194B** (cf. H_i for *anti*-**1**-1-ene, δ 5.62). The corresponding sulfone **195** has $T_c = 51°C$ and

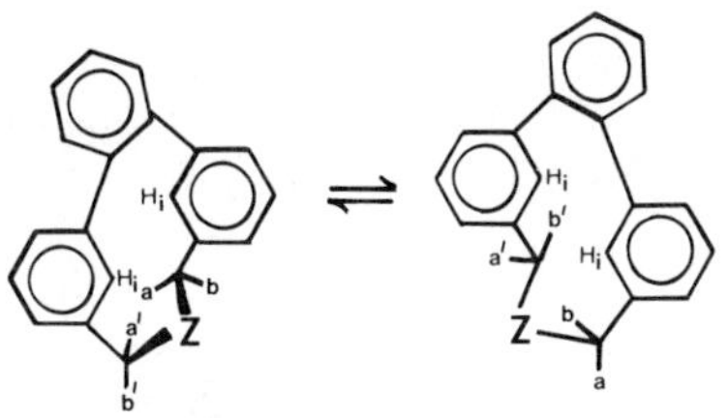

194A Z=S 194B
195A Z=SO_2 195B
196A Z=S, CH_i=N 196B

$\Delta G_c^\ddagger$ 68 kJ/mol,[21] whereas the pyridinophane **196** has $T_c = -26°C$ and $\Delta G_c^\ddagger = 52$ kJ/mol.[126]

The pyridinophane **197** is flexible[120] and probably also adopts a syn conformation, although this cannot be conclusively stated without an X-ray structure. The sulfone **198** is also still flexible,[127] but the bis-*N*-oxide sulfone **199** is fixed, judging by the AX pattern for the bridge methylene protons at δ 4.92 and 6.42.[127] Thus, whether compounds **197–199** are syn or anti remains to be demonstrated!

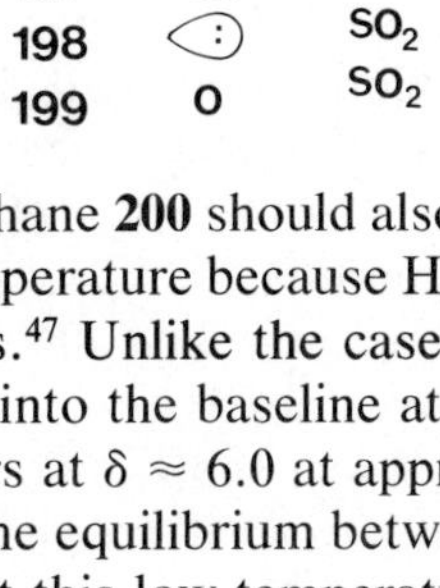

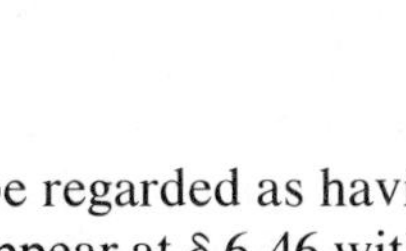

	X	Z
197	(lone pair)	S
198	(lone pair)	SO_2
199	O	SO_2

200

The diselenaphane **200** should also be regarded as having the syn structure at room temperature because H_i appear at δ 6.46 with a singlet for the benzylic protons.[47] Unlike the case of **179,** however, on cooling, the H_i signal collapses into the baseline at approximately −96°C. A small peak (~0.6 H) appears at δ ≈ 6.0 at approximately −115°C. This could be an indication that the equilibrium between *syn*-**200** (70%) and *anti*-**200** (30%) is freezing out at this low temperature. All the peaks, however, are still broad singlets, indicating that the benzylic hydrogens are still exchanging. If the peak at δ 6.0 is due to H_i of *anti*-**200** (not an unreasonable chemical shift), then the room temperature position of H_i would indicate that at this temperature the fast-exchanging mixture is about 1 : 1 syn/anti, and hence, because more syn is present at low temperature, it is the more stable. If the peak at δ 6.0 is H_i of **200** (and there is some doubt because at this low temperature much material is suspended and the solution is almost frozen), then this is the first real evidence for a syn ⇌ anti equilibrium in these systems. This may be a reflection of the longer C—Se bond length (1.98 Å) relative to C—S (1.82 Å), changing the stability of the anti conformer with respect to its syn conformer for **200** versus **179.**

b. Other Internal Substituents

Next in size to H is F, and hence **201** might also be expected to be syn. Vögtle[120] reported that the bridges of **201** are an AB up to T_c = 185°C, and

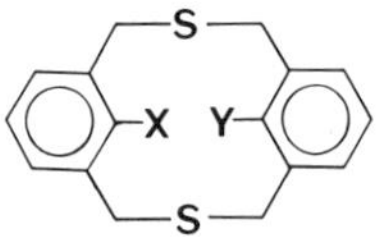

	X	Y
201	H	F
202	F	F
203	Cl	Cl
204	CH_3	F
205-207 (see text)		

hence $\Delta G_c^\ddagger$ = 95 kJ/mol. Unfortunately, he did not report the chemical shift of H_i; thus, whether **201** is syn or anti remains to be determined. The difluoro compound **202** was reported both by Vögtle[120] and by Boekelheide,[29] who (by dipole moment studies) showed that it is syn. Vögtle reported that the AB bridge signals coalesce at 157°C, giving $\Delta G_c^\ddagger$ = 88 kJ/mol. Whether a syn ⇌ anti equilibrium is occurring in these compounds, or whether the **H** ⇌ **I** ⇌ **J** bridge wobble is occurring cannot be said for sure. The analogous dichloro compound **203** is conformationally fixed, because two sets of AB multiplets occur.[120] These could correspond either to syn or anti isomers or to two of **H, I,** or **J** types.

The methylfluoro compound **204** is more instructive. Both syn and anti isomers can be isolated at room temperature and distinguished by the chemical shifts of the methyl group: *syn*-CH_3, δ 2.41; *anti*-CH_3, δ 1.49.[29,104] It is interesting that the syn isomer shows a doublet for the CH_3 signal (J_{FH} = 2.5 Hz), which apparently is a "through space" coupling that is absent in the anti compound. On heating, the CH_3 and the $—CH_2S—$ signals coalesce at ~105°C. The CH_3 signal reappears at $\delta \approx 2.0$, the average of the syn/anti chemical shifts.[104] This would suggest an approximate $\Delta G_c^\ddagger$ value of 78 kJ/mol, somewhat smaller than those for **201** and **202.** Vögtle also reported that syn and anti conformers exist for **205** (X = CH_3; Y = Cl, Br, OCH_3) and **206** (X = OCH_3; Y = OCH_3).[104]

It is interesting that in 1975 Vögtle reported[104a] a number of compounds **(207A–207F)** in which X = H_i and Y is the relatively large group shown in Table VII. From the chemical shift of H_i (Table VII) compounds **207A–207E** must all clearly be assigned syn stereochemistry. The case of **207F,** however, is far from clear. The syn isomer **208** would be expected to have H_i shielded by the adjacent phenyl substituent, whereas the anti isomer **209** would have H_i shielded by the adjacent phane ring. However, we

208 X = H_i 209
210 X = CH_3 211
212 X = phenyl 213

have carried out an X-ray structure determination on this compound, and it is the syn isomer **208.**[128] Thus, in contrast to **61** (**207:** Y = CH_3), when Y is a large group the syn isomer is clearly preferred; in the anti isomer, Y would be compressed by the proximate aromatic ring.

In **208** the rotation of the phenyl substituent presents an interesting problem. Whereas in **22A** $\rightleftharpoons$ **22B** (Section II,C,1) H_o and $H_{o'}$ can be seen clearly; unfortunately this is not the case for **208.** Vögtle reported that the

TABLE VII
Chemical Shifts of H_i for 207A–207F (X = H_i)

Compound	Y	δ (ppm)
A	NO_2	7.33
B	COOH	7.26
C	CO_2Me	7.26
D	CO_2-*t*-Bu	7.45
E	SO_2Me	6.16
F	ϕ	5.60

system is not temperature dependent.[25,54] However, this may depend on the field strength of the nmr spectrometer. We have observed[115] that, at 90 MHz, H_o can be seen at δ 6.70 (35°C) and $H_{o'}$ at δ ≈ 6.85, with H_i at δ 5.49. As the temperature is *lowered* both H_o and H_i steadily move *upfield,* reaching, at −100°C, δ 5.8 and 4.6, respectively. At high temperatures both H_o and $H_{o'}$ appear in the aromatic multiplet. We interpret this behavior as being analogous to **22A** ⇌ **22B,** which can be depicted by **208A** ⇌ **208B,** except that at room temperature the system is "frozen" in one

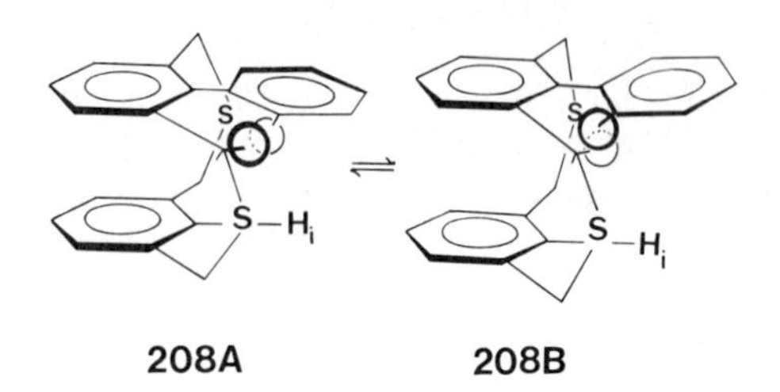

conformation (e.g., **208A**). We believe that as the temperature is lowered the substituent ring rotates, increasing the angle between the biphenyl rings and moving H_o into the shielding region between the two phane rings. This probably also allows the phane angle to increase (see geometry **K**) such that H_i also becomes progessively more shielded. Unfortunately, at 90 MHz when the temperature is raised H_o and $H_{o'}$ have very similar chemical shifts and cannot be differentiated in the aromatic multiplet. Studies are underway, however, at 250 MHz.

For the disulfone of **208,** Vögtle[25] reported that H_o and $H_{o'}$ can be seen at δ 6.10 and 6.70 (H_i at δ 5.9), with T_c being 167°C and hence $\Delta G_c^\ddagger$ being rather large at 91 kJ/mol.

We reported[73] that the analogous phenylmethyl compounds exist in syn form **210** (δCH_3 2.37) and *anti*-**211** (δCH_3 1.46). In each case a process of type **208A** ⇌ **208B** occurs, where at −60°C H_o and $H_{o'}$ are at δ 8.31 and 6.70 for **210** and δ 6.55 and 7.10 for **211,** respectively (from 250-MHz spectra). The T_c values (90 MHz) were −8 and −11°C, leading to $\Delta G_c^\ddagger$ values of 52 and 54 kJ/mol, respectively.

We also prepared[73,115] the diphenyl compounds *syn*-**212** and *anti*-**213.** We established the structure of *anti*-**213** by X-ray determination.[128] For *anti*-**213,** at −50°C H_o is at δ 6.63 and $H_{o'}$ is at δ 7.30, giving $\Delta G^\ddagger_{-30}$ = 50 kJ/mol. The syn isomer **212** shows $H_{o'}$ at δ 6.62 and H_o at δ 7.93, giving $\Delta G^\ddagger_{-55}$ = 43 kJ/mol.

The similar values of $\Delta G_c^\ddagger$ obtained for **22, 170–172,** and **210–213** (40–54 kJ/mol) confirm the similarity of the process (**22A** ⇌ **22B** type) occurring in each case.

3. Thia[4.4]metacyclophanes or Larger Thiametacyclophanes

When the bridges consist of four or more atoms, planar conformations, as well as the syn and anti conformers, become possible. In the smaller phanes, planar conformers are, at best, intermediates between the other two conformers, because compression of the internal groups would be present. Newkome[129] conclusively demonstrated that the [4.4]phane **214** exists as the planar conformer pair **214A** ⇌ **214B.** The pyridine H-3,H-5 appear as two doublets (J = 8 Hz) at δ 6.28 and 6.32, even at −50°C.

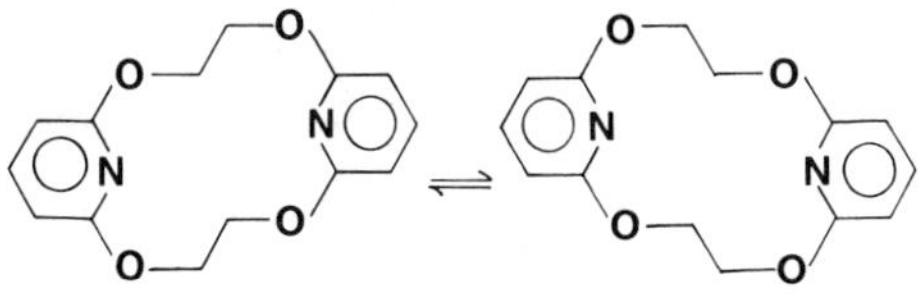

planar **214A** planar **214B**

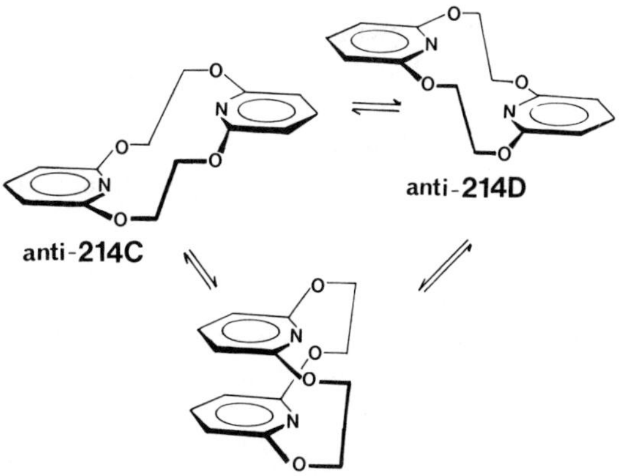

anti-**214C** anti-**214D**

syn-**214E**

Moreover, at −50°C H-4 appears as a single triplet at δ 7.50. Newkome originally[130] interpreted this to be *anti*-**214C** ⇌ *syn*-**214E,** because *anti*-**214C** ⇌ *anti*-**214D** has H-3,H-5 identical by symmetry. However, this requires that the syn and anti conformers be present in equal concentration, which is not very likely over the entire temperature range. He thus proposed (with the support of X-ray data which suggested that —N=C—O—CH_2— is usually close to planar) the process **214A** ⇌ **214B,** which he calls *anti-longitudinal* (as opposed to **C** ⇌ **D,** which is *anti-transverse*).

Newkome proved the occurrence of this isomerization process by synthesis of the bisamide **215.** This shows doublets at δ 7.64 and 6.43 for both

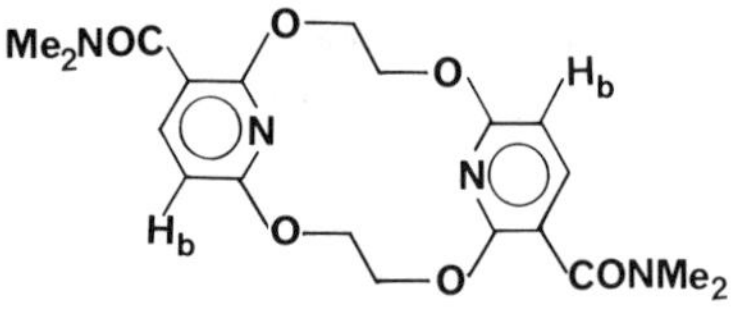

Me_2NOC H_b H_b $CONMe_2$

215A

Me_2NOC H_a H_a $CONMe_2$

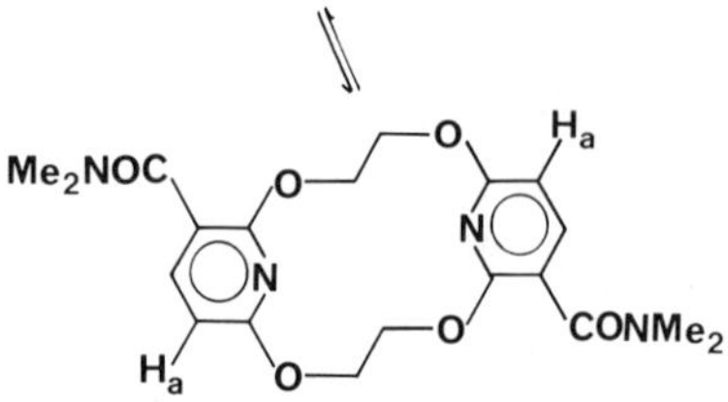

215B

H-4 and H-5. At low temperatures H-5,H-5′(H_a,H_b) appear as two doublets (J = 8 Hz) at δ 6.445 and 6.415 corresponding to the two anti-conformers **215A** ⇌ **215B.** Newkome wrongly calculated $\Delta G_c^{\ddagger}$, because he used, as $\Delta\nu$, the chemical shift difference between H-4 and H-5 (251 Hz) instead of between the exchanging protons H-4 and H-4′ (2 Hz). For T_c = 28°C, $\Delta G_c^{\ddagger}$ is thus ~70 kJ/mol.

Newkome also pointed out that this anti-longitudinal isomerization may be the major process when the bridges of cyclophanes are $—CH_2—X—X—CH_2—$ or heteroaryl $—X—CH_2—$. This may well be the case when X = S, because a disulfide linkage is known[131] to prefer dihedral angles of ~90° between the sulfur lone pairs. Newkome's study of **214** is thus of fundamental importance to the higher thiacyclophanes.

We have reinvestigated[132] the tetrathia[4.4]cyclophane **216** reported by Sato[43] and have found that at low temperatures the bridge protons exhibit AB multiplicity, in contrast to the singlet multiplicity observed at room temperature. Because the external and internal protons remain virtually unchanged over the temperature range +30 to −70°C it seems unlikely that *syn*-**216** ⇌ *anti*-**216.** The chemical shift of H_i is δ 6.63, which is somewhat unlikely for a stepped-anti conformer but possible for a planar

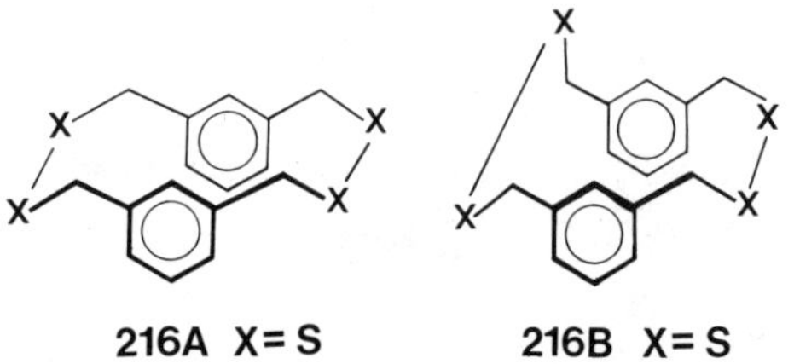

216A X = S **216B X = S**

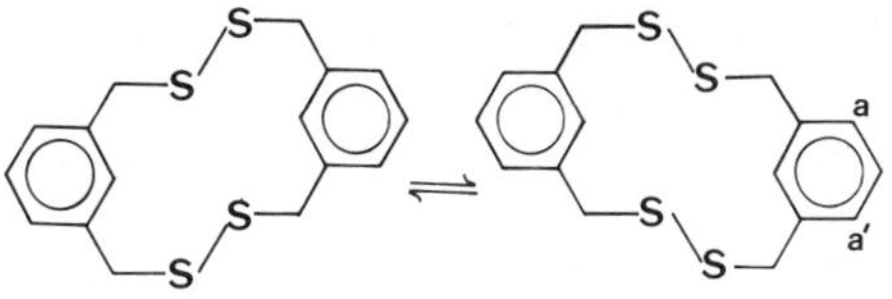

216C ⇌ 216C′

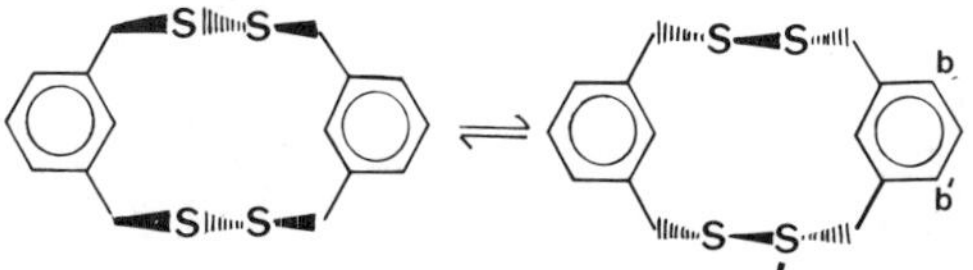

216D ⇌ 216D′

or syn conformer. We originally assigned the syn conformer structure **216A** to the low-temperature form because any nonsymmetric conformer (e.g., **216B**) should show more than one AB pattern for the bridges. However, in light of Newkome's study and the fact that **216A** would have an S—S dihedral angle of 0° (see also **222**), the planar conformers **216C** ⇌ **216C′** are possible. In this instance, however, A,A′ would be different (not observed at 90 MHz; studies at 250 MHz are underway), as are the nearly planar conformers **216D** ⇌ **216D′**, which keep the benzene rings on the same axis (and hence **A** ≡ **A′**). This problem is at present unresolved; we are undertaking a crystal structure determination of **216.** The tetraselenaphane **217** (Z = Se) undergoes a similar process,[132,133] with T_c = −64°C instead of −17°C for **216.** The values of $\Delta G_c^{\ddagger}$ for **216** and **217** are thus 52 and 43 kJ/mol, respectively, with (from lineshape analysis) $\Delta G_{298}^{\ddagger}$ being 53 and 46.5 kJ/mol, respectively. As might be expected the selenaphane is the more floppy.

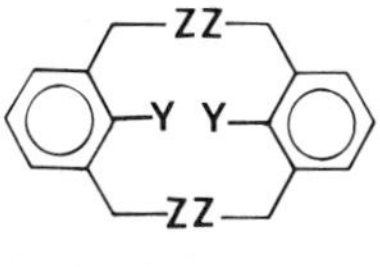

	Z	Y
217	Se	H
218	S	CH_3
219	S	ϕ

Compound **218,** the dimethyl-substituted derivative of **216,** is also known,[134] and the benzylic (δ 3.64), aromatic (δ 7.11), and methyl protons (δ 2.05) all appear as singlets. The chemical shift of the methyl protons is intermediate between those of *anti*-**60** (δ 1.30) and *syn*-**59** (δ 2.54); the tetrasulfide **216** shows H_i at δ 6.63, shielded somewhat from the disulfide *syn*-**179** (δ 6.82), and thus it is not possible to specify the stereochemistry of **218** at present. Variable-temperature nmr studies are required.

The diphenyl derivative **219** exists as two isomers (possibly syn and anti), both of which show singlet absorptions for the benzylic protons, which change on cooling,[25] and one, at least, shows high-field $H_o,H_{o'}$ protons (cf. **212** and **213**). However, high-field spectra probably are required before stereochemistries can be assigned. Pappalardo studied the systems with all-sulfur bridges.[40,41,135] The hexasulfides **220–223** have been synthesized[135] and show typical syn chemical shifts for H_i (perhaps

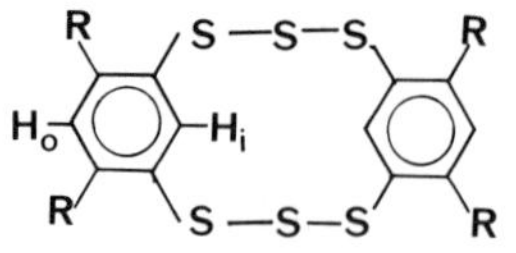

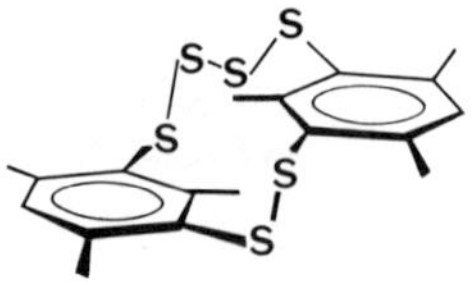

	R	δH_o	δH_i
220	H	7.16	8.37
221	CH_3	7.15	8.26
222	OCH_3	6.74	8.40
223	Cl	-	-

224

deshielded by sulfur). An X-ray structure determination of **222** confirmed its syn stereochemistry.[136] Analysis by vtnmr showed no spectral changes. An anti conformer would not have S—S—S dihedral angles of 90°; it is interesting that no anti conformer is formed (see **216**). The asymmetric hexasulfide **224,** in which one bridge has four sulfur atoms and the other two, has been shown by X-ray analysis to be the anti conformer.[135,136] Here the adjacent S—S atoms have dihedral angles of 74 to 110°. The chemical shift of the inner methyl groups is δ 1.78, and those of the outer are δ 2.71 and 2.53, which are unchanged from −88 to +186°C.

4. Metaparacyclophanes

[2.2]Metaparacyclophane **225** was studied by Cram.[137] The para bridged protons appear at room temperature as a pair of narrow multiplets at δ 5.70 and 6.97. Coalescence of these signals occurs at 140°C, yielding $\Delta G_c^{\ddagger}$ = 87 kJ/mol. The inner proton H_i appears shielded at δ 5.24. The exchange process is represented by **225A** ⇌ **225B.** The bridge protons also are frozen at room temperature but simplify to an AA′BB′ pattern

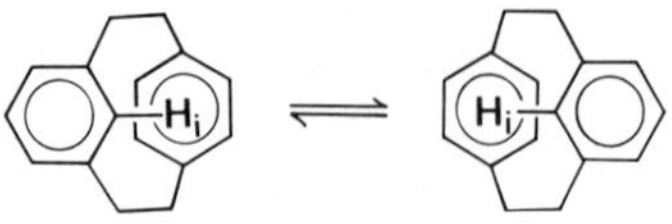

225A ⇌ 225B

at 185°C. The corresponding diene **226,** however, prepared by Boekelheide,[58] exhibits a singlet for H_{para} at δ 6.77 (room temperature),

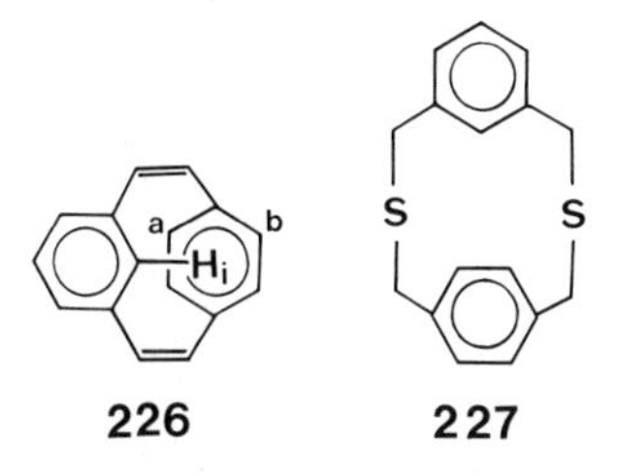

which does not separate into two signals, H_a and H_b (δ 6.31 and 7.22, respectively), until below −96°C. This corresponds to $\Delta G_c^\ddagger$ of only 35 kJ/mol, substantially less than that for **225.** It is perhaps surprising that this should be the case. However, opening the bridge carbon angle to ~120°C (sp^2) from 109° (sp^3) apparently more than outweighs the shorter bond distance of C=C. Thus, H_i in **226** on the average penetrates farther into the π cavity than does H_i for **225** and is more shielded (δ 4.26) at room temperature. The thiacyclophane **227,** with the longer bridges, shows H_i at δ 5.52 and singlets for the para (δ 6.80) and bridge (δ 3.76 and 3.38) protons and is therefore mobile.[58] Removal of H_i would be expected to accelerate the flipping process, and indeed the pyridinophane **228** shows $T_c = -43.5$°C (cf. **225,** 157°C) and $\Delta G_c^\ddagger = 45$ kJ/mol (cf. **225,** 87 kJ/mol).[138] The diene **229** shows H_{para} as a singlet even at −110°C and, by X-ray structure,[139] has the pyridino ring perpendicular to the para ring. If **229** is

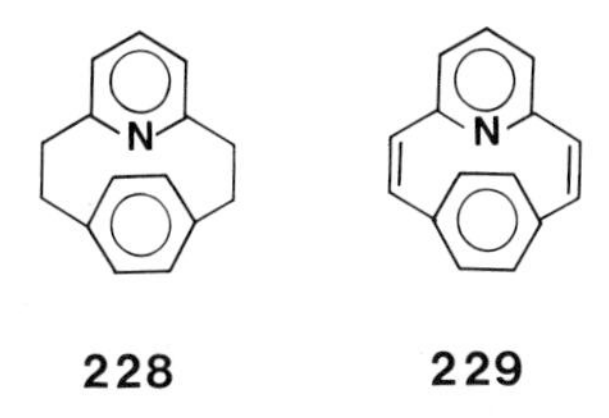

flipping in solution, then it has a very low $\Delta G_c^\ddagger$. In contrast, the crystal of **226** shows the meta ring at 41° to the perpendicular.[140] More recently, Boekelheide has studied several substituted derivatives. Methyl substituents on the para ring of **228** have no effect on $\Delta G_c^\ddagger$ or the conformational process.[141] The HBF_4 salts of both **228** ($T_c \approx 15$°C, solvent dependent) and **229** also appear to undergo rapid conformational flipping.[141] However, the SbF_5 complex of a dimethyl-substituted **229** is definitely not undergoing a flipping process. Clearly, this group is too large to readily pass through the cavity, and hence the pyridino ring is fixed. Substituents on the meta ring do influence the rate of flipping (Table VIII).[142] Similarly, when H_i is

TABLE VIII
Relative Rates of Flipping for Compound 230

X	k_{rel}
Br	1.08
F	1.06
H	1.00
NO_2	0.705
OCH_3	0.595
$COCH_3$	0.556
NH_2	0.189

D, $k_D/k_H = 1.20$, indicating $\Delta\Delta G^{\ddagger} = 468$ J/mol, which suggests that the H-8(D) would have to approach within 2.40 Å of the para ring on flipping.

The fluoro derivatives **87–89** (Section II,C) have also been studied.[58] Only the thiametaparacyclophane **89** is conformationally mobile: T_c = 93°C, $\Delta G_c^{\ddagger} = 76$ kJ/mol. This indicates a rate ratio for **225/89** of $>10^{11}$ at 25°C, which is quite remarkable when the net difference (between Ar—H and Ar—F) is only 0.49Å![58]

230

	X	Z
231	D	O
232	CH_3	O
233	CH_3	S

234 Z = CH
235 Z = N

For the heterometaphanes, the furan rings in **231**[96] and **232**[143] appear mobile [$\Delta G_c^{\ddagger} = 58$ (−40°C) and 50 (−29°C) kJ/mol, respectively], whereas the thiophene ring in **233**[143] is fixed. The five-membered rings in heterophanes **234**[144] and **235**[145] are either fixed or flipping rapidly because no vtnmr changes were observed. A detailed comparison of furan paracyclophanes has appeared.[60]

The azulenophanes **79** and **237** are mobile, in contrast to **236** and **80** (Section II,C). At 70°C H_a,H_b becomes equivalent in **79**, giving $\Delta G_c^{\ddagger} = 68$ kJ/mol,[56] whereas in **237** H_a,H_b separate at −125°C, giving $\Delta G_c^{\ddagger} \approx 30$ kJ/

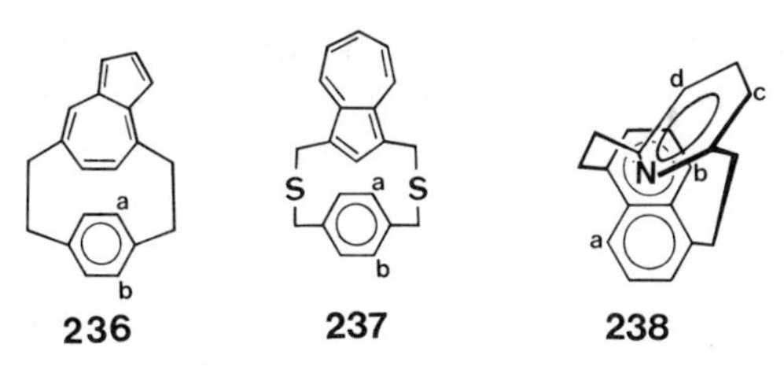

mol.[57] A number of metaparacyclophanes are known with rings larger than benzene.

The metapyridinonaphthalenophane **238** shows T_c for H_a,H_b at −3°C and T_c for H_c,H_d at −15°C. Both yield $\Delta G_c^\ddagger$ = 51 kJ/mol, somewhat larger than the value for **228** (45 kJ/mol). Spectra of the corresponding diene are not temperature dependent, and therefore the aryl rings are probably perpendicular or very rapidly flipping.[146]

The "meta"-naphthalenoparacyclophane **239** appears to be fixed because H_i appear at δ 6.95, whereas H_o are at δ 7.56, and the para ring

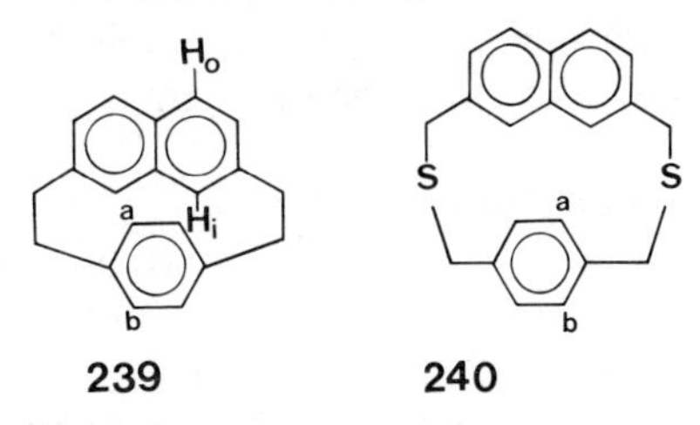

protons H_a,H_b appear at δ 5.11 and 6.95. The analogous diene, unlike previously discussed metaparaphane dienes, also appears fixed because H_a,H_b appear at δ 5.22 and 7.23, respectively.[147] The thiacyclophane **240,** however, has H_a,H_b identical at δ 6.49 and thus is rapidly flipping.

When both rings are naphthalenoid, the system is again mobile; for example, **241** has T_c = 160°C and $\Delta G_c^\ddagger$ = 86 kJ/mol.[148,148a] The diene **242** is

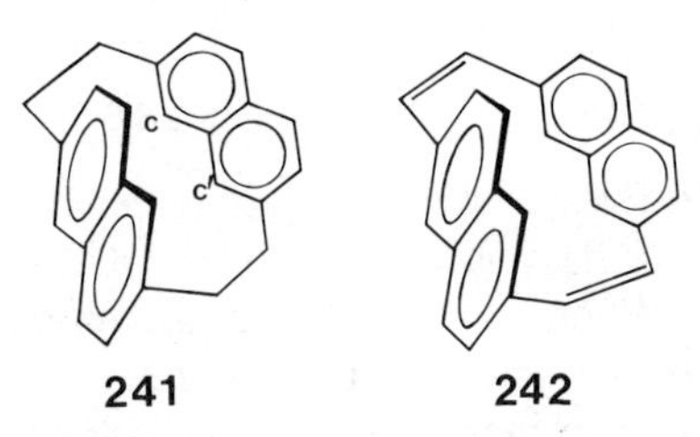

more flexible: T_c = 38°C, $\Delta G_c^\ddagger$ = 62 kJ/mol.[148] The H_c and $H_{c'}$ of **242** appear highly shielded at δ 4.85 and 5.65, respectively. It is interesting that the $\Delta G_c^\ddagger$ values for **241** and **225** are virtually the same, whereas those for **242** and **226** are substantially different.

The "meta"-phenanthroparacyclophane **243** has H_a,H_b at δ 6.42 and H_i at δ 6.93 (~2 ppm shielded compared with the corresponding protons in phenanthrene) and is therefore probably flipping.[149]

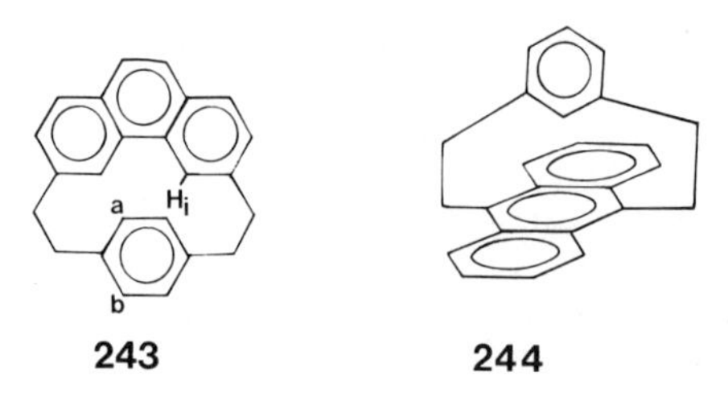

The metacyclo(9,10)anthracenophane **244** has $\Delta G^{\ddagger}_{82}$ = 73 kJ/mol, about 11 kJ/mol less than the parent **225.** This may be because the ground state conformation of **224** is less stable with respect to the transition state for flipping (vertical) than is that of **225,** perhaps because of π–π repulsion.[143]

We shall consider here a number of biphenyl compounds taking the biphenyl unit as a single ring (although they could be considered equally well with the phanes having more than two rings; see Section III,D).

The "meta"-biphenyloparacyclophanes **245** (X = H, S, CH_2CH_2, and CH=CH) have been prepared by Boekelheide.[150] In each case a 4H singlet can be seen for H_a,H_b of the para ring, and thus all are flipping. Only one example appears to be temperature dependent; when X = S and Z = CH=CH, T_c = −10°C and $\Delta G^{\ddagger}_c$ = 50 kJ/mol, H_a and H_b freeze at δ 7.38 and 5.91. Clearly, the conformational process for a biphenylophane is easy except when the biphenyl is tied back by a single atom (e.g., S).

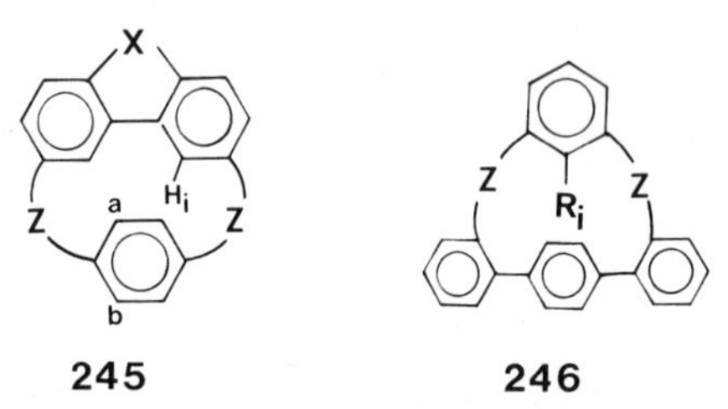

The "meta"-cyclo-"para"-terphenylophanes **246** (Z = CH_2SCH_2, $CH_2SO_2CH_2$, and CH_2CH_2; R_i = H, NO_2, OCH_3, and ϕ) have been studied by Vögtle.[151] When R = H_i the compounds are not temperature dependent, H_i are at δ 5.47, and presumably the molecule is flexible. As R_i increases in size, $\Delta G^{\ddagger}_c$ also increases (Z = CH_2SCH_2):

R_i	T_c (°C)	$\Delta G^{\ddagger}_c$ (kJ/mol)
NO_2	−63	43
OCH_3	35	64
ϕ	64	71

When Z = CH_2CH_2 and R_i = ϕ, the molecule appears not to be flipping. Replacement of the meta ring of **246** by a para ring **(247)** or a biphenyl ring

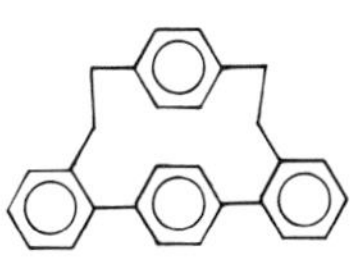

247 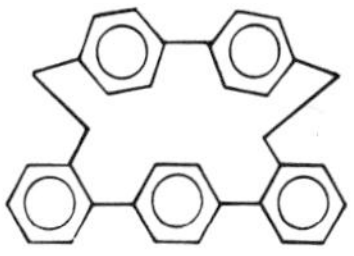 **248**

(248) does lead to vtnmr spectra that indicate for **247** $\Delta G^{\ddagger}_{-12}$ = 56 kJ/mol and for **248** $\Delta G^{\ddagger}_{-111}$ = 34 kJ/mol.

The "meta"-biphenylo-"para"-biphenylophane **249** is also temperature dependent.[152] At −60°C H_i appear shielded at δ 5.84. The 4,4′-biphenyl ("para") protons appear as four sets of doublets. The H_a appears

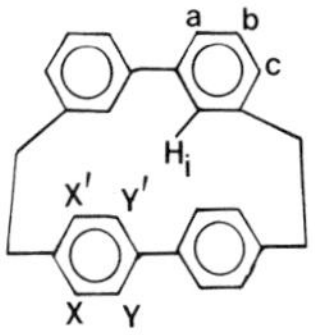

249 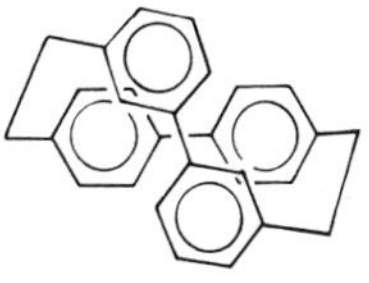 **249A** 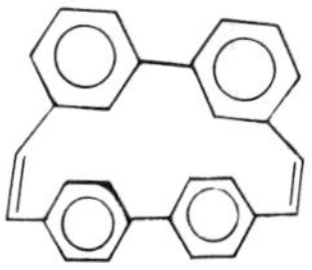 **250**

shielded at δ 6.82, suggesting that the biphenyl is twisted. The authors state that the conformer **249A** best accommodates these data. The T_c was found to be −50°C, and $\Delta G^{\ddagger}_c$ = 43 kJ/mol. The corresponding diene **250** was found to be similarly temperature dependent: T_c = −60°C and $\Delta G^{\ddagger}_c$ = 53 kJ/mol. Although the change in $\Delta G^{\ddagger}_c$ for **249** and **250** was less than that for **225** and **226,** the diene still flips more easily. The aza[3.2]metaparacyclophane **251** has T_c = 60°C and $\Delta G^{\ddagger}_c$ = 56 kJ/mol.[117]

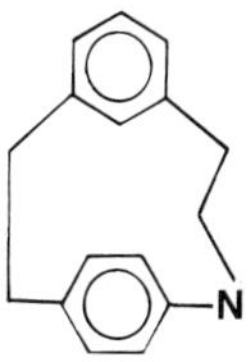

251 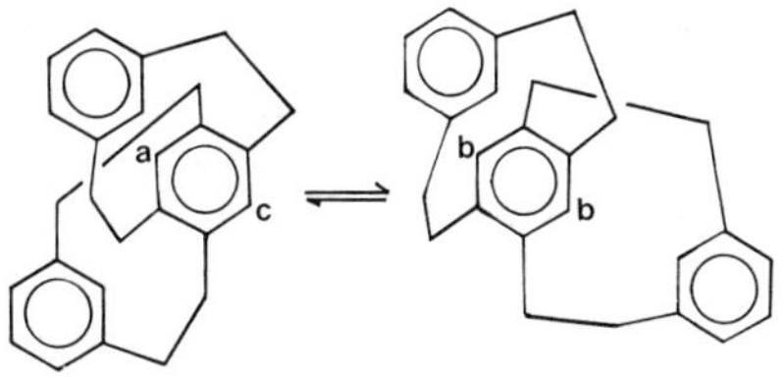 **252A** **252B**

The double-layered metapyridinophane **252** has T_c = −39°C and $\Delta G^{\ddagger}_c$ = 46 kJ/mol, almost identical to the values for **228.**[141] The analogous diene is probably also mobile, because H_{para} appear as a singlet (δ 5.48).[141] At low temperatures two conformers of **252A** and **252B** are formed because the benzene protons appear as three peaks: δH_a 4.41, δH_b 5.82, and δH_c 7.5. These correspond to a 1 : 1 mixture of **252A** and **252B.**

5. Paracyclophanes

Not very much new information has appeared on paracyclophanes since a previous review.[96] Presumably, in all medium-bridge paracyclophanes the bridge can undergo wobbles of the types **H, I,** and **J** found in the metacyclophanes. In **253** this has been studied: For X = S, $T_c = -75°C$ and $\Delta G_c^{\ddagger} = 40$ kJ/mol; for the larger SO_2 = X, $T_c = -65°C$ and

253

$\Delta G_c^{\ddagger} = 45$ kJ/mol.[153] The bridge wobble has also been studied[154] for the substituted [3.3] bridges of **254,** for which two conformers, cis and trans, can exist. These can be distinguished on the basis of the AB coupling constants: For *trans*-**254** $J_{AB} = 7.8$ Hz, whereas for *cis*-**254** $J_{AB} = 1.0$ Hz.

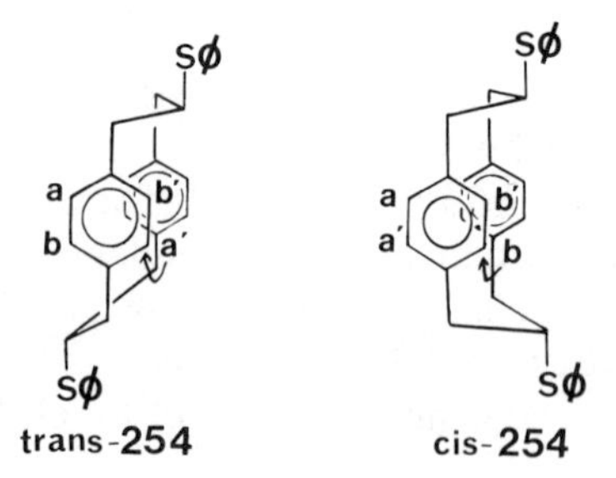

trans-**254** cis-**254**

Apparently, these do not interconvert, unlike the mobile parent [3.3]paracyclophane,[154a] for which $\Delta G_c^{\ddagger} = 49$ kJ/mol.

In the [4.3]paracyclophane series, ring rotation can occur at 160°C, $\Delta G^{\ddagger} \approx 140$ kJ/mol, but this has not been measured by nmr methods (rather, by racemization studies of the carboxylate),[155] whereas [4.4]paracyclophane derivatives have not been resolved,[156] indicating $\Delta G^{\ddagger} < 80$ kJ/mol. Although the aryne ring of **255** cannot be shown to be perpendicular by nmr methods, the formation of **256** implies this geometry for **255.**[157]

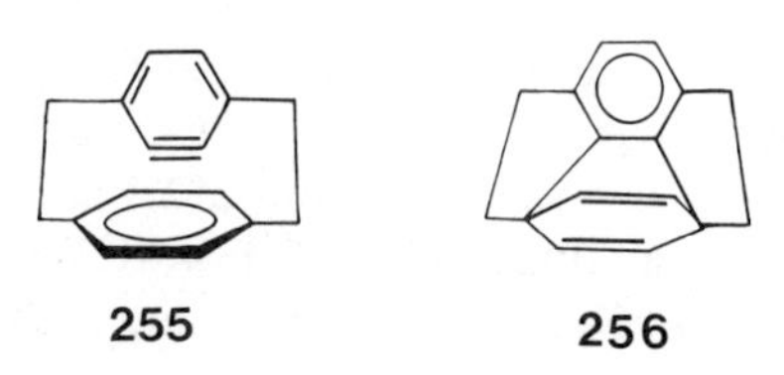

255 **256**

D. Phanes Having More Than Two Rings

1. METACYCLOPHANES

Knowledge of the conformations of higher cyclophanes is scanty. The advent of high-field instruments, however, is beginning to change this situation rapidly, and within the next decade a much clearer picture is likely to emerge.

Vögtle prepared [2.2.0]metacyclophane (a metabiphenylometacyclophane) and [2.0.2.0]metacyclophane (a metabiphenylophane) (**257** and **258**, respectively).[158] The benzylic signals of **257** appear as a sharp singlet

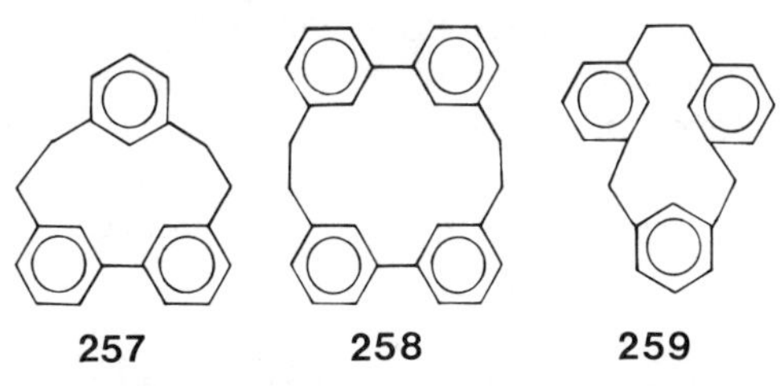

257 **258** **259**

(δ 2.88) at 135°C. (This signal is broad at room temperature and is a multiplet at −20°C.) This clearly indicates that a conformational process is occurring and deserves further study. The bridge protons of **258** appear as a singlet (δ 2.88) down to −45°C. Perhaps lower temperatures would freeze out a conformer of this compound also.

The [2.1.1]cyclophane **259**, reported by Sato[43] to show singlets for the bridge protons at δ 2.78 and 3.60, is also obviously mobile. In his view the shielded aryl protons at δ 6.27 (H_i) and 6.49 ($H_{i'}$) suggest the stepped conformer **259A** rather than the folded **259B**. We have investigated a

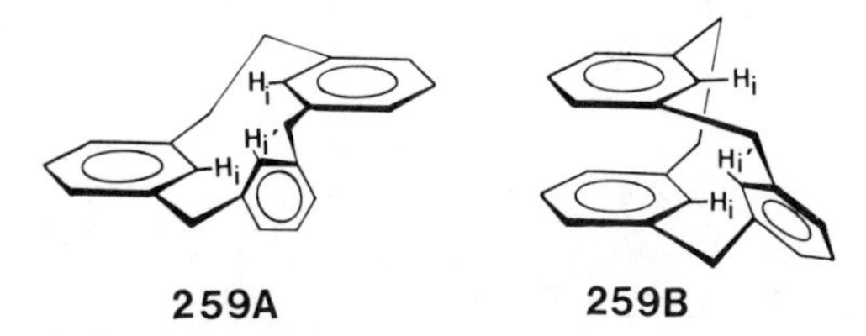

259A **259B**

series of [2.1.2.1]- and [3.1.3.1]metacyclophanes[159,160] in an attempt to synthesize the dihydropyrene **260**. We prepared compounds of the type

X
Y

260 **261**

261, which we assumed would have the traditional anti stereochemistry of [2.2]metacyclophanes. Our assumption proved to be very wrong, because no amount of chemical persuasion would close the central bond between the Ar—CH_2—Ar methylenes. Investigation by vtnmr soon indicated why and demonstrated the correct stereochemistries of such systems. We studied first the parent hydrocarbon **262** (**261:** X = Y = H, H). At room temperature the ^{1}H-nmr signals of **262** are rather broad. On heating to 120°C the internal methyl protons appear as a singlet at δ 1.76, the bridge protons as singlets at δ 2.95 and 3.69, and the aromatic protons as a multiplet at δ 6.5–7. At −20°C a conformer is frozen that shows two different types of methyl protons at δ 2.38 and 1.18, an AB pattern for the —CH_2— bridges at δ ≈ 3.7 and a more complex ABCD(?) pattern for the —CH_2CH_2— bridges. The aryl protons were also resolved into two sets: a shielded AB_2 and a normal AB_2. These spectra have been published.[159] There are five conformers worthy of consideration. The two sets of methyl protons rule out the all-anti or all-syn conformers **262A** and **262B.**

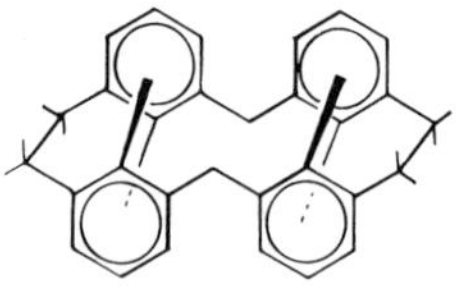

262A 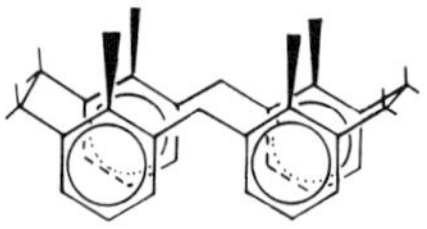 262B

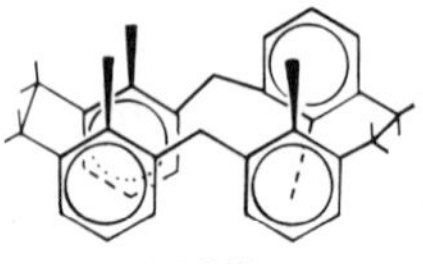

262C 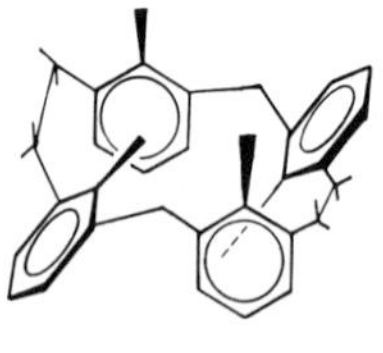 262D

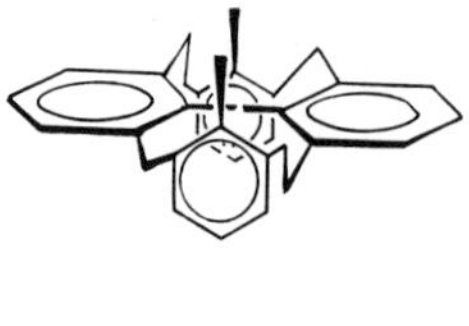

262E 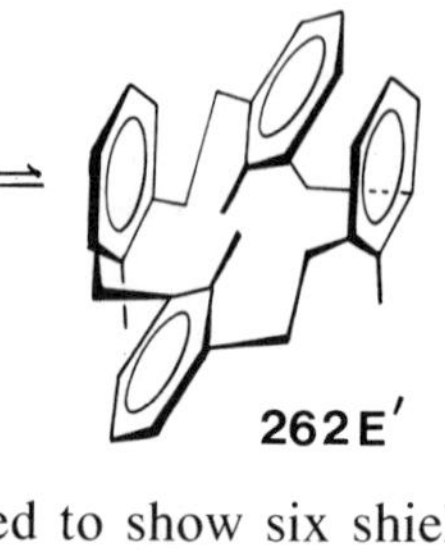 ⇌ 262E′

Compound **262D** would not be expected to show six shielded aromatic protons as **262C** and **262E** would. Those of **262C,** however, should have an ABC multiplicity rather than the observed AB_2 multiplicity. Clearly, the bridges of **262C** and **262E** should be different; however, at 90 MHz the spectra are not sufficiently resolved. The ^{13}C-nmr spectrum solved the problem: At −30°C **262** shows 12 aryl carbons, 2 —CH_2CH_2— carbons, 1 —CH_2— carbon, and 2 —CH_3 carbons, consistent only with the more symmetric **262E,** because **262C** should have 4 —CH_2CH_2 carbons, 2 —CH_2— carbons, and 4 —CH_3 carbons. It is not very likely that all would be degenerate.

Because the high-temperature peaks were found to occur at the average position of the corresponding low-temperature peaks within 0.02 ppm, a true fluxional process **(262E ⇌ 262E′)** is indicated rather than conversion to other conformers. Using the CH_3 group separation of $\Delta\nu$ = 108 Hz and T_c = 60°C, it was found that $\Delta G_c^{\ddagger}$ = 66 kJ/mol.

An identical process was found for the other systems studied **(263A ⇌ 263A′),** and the thermodynamic data are given in Table IX.[160] Several

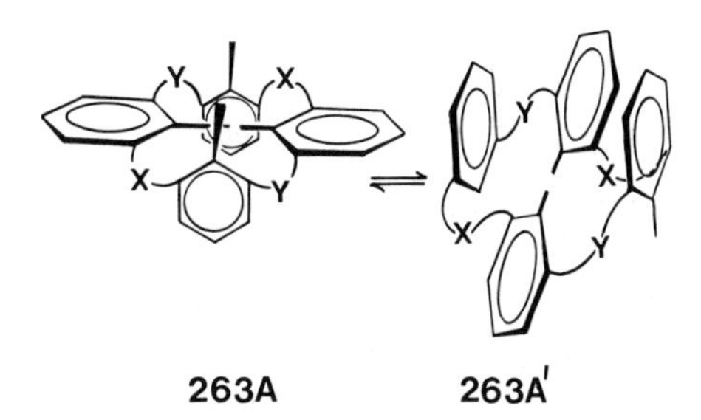

263A **263A′**

interesting points emerge:

1. The same order of flexibility is found for different bridges, that is, CH_2SCH_2 > CH=CH > CH_2CH_2, as with many of the cyclophanes discussed in the previous sections.

TABLE IX

T_c and $\Delta G_c^{\ddagger}$ Data for Cyclophanes 263A ⇌ 263A′ for Varying X and Y

Compound number	X	Y	T_c (°C)[a]	$\Delta G_c^{\ddagger}$ (kJ/mol)
263A	CH_2, CH_2	CH_2SCH_2	−70	39
263B	CH_2, CH_2	CH=CH	15	57
262	CH_2, CH_2	CH_2—CH_2	60	66
263C	CO, CH_2	CH_2SCH_2	−73	39
263D	CO, CH_2	CH=CH	70	68
263E	CO, CH_2	CH_2CH_2	75	68
263F	CO, CO	CH_2SCH_2	Solubility too low	—
263G	CO, CO	CH=CH	>150	>85
263H	CO, CO	CH_2CH_2	90	72

[a] Methyl protons.

2. Changing the $—CH_2—$ bridges from $sp^3(CH_2)$ to $sp^2(C{=}O)$ increases the barrier for each bridge by about 3 kJ/mol in the $—CH_2CH_2—$ series and by about 10 kJ/mol in the $—CH{=}CH—$ series, having almost no effect on the most flexible $—CH_2SCH_2—$ series.
3. A sharp cutoff appears when all the bridges are sp^2 carbon atoms, and the barrier becomes high.

The [2.0.0.0.0]phane **264** with various clamps (K) appears to undergo conformational change of type **264A** ⇌ **264A′**.[161] When K = CH_2CH_2 the

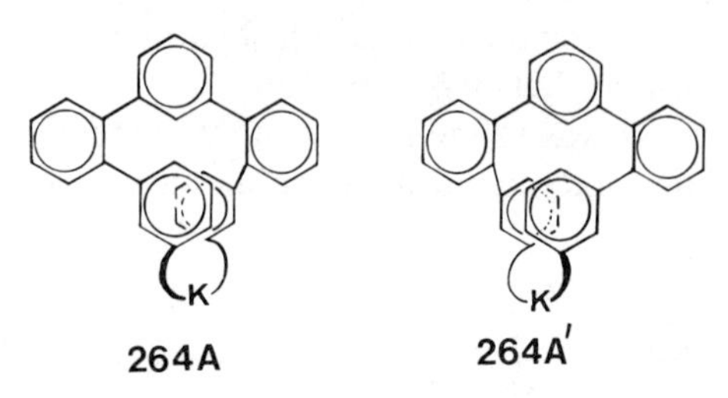

absorption of the benzylic protons changes from a singlet at room temperature to an AB quartet at <-70°C ($\Delta G^{\ddagger}_{-70}$ = 44 kJ/mol).

The [2^n] metacyclophanes* **265** (n = 3–10) have been known for some time.[6b,69,162] Although their ^{1}H-nmr spectra are well documented, the bridge protons all appear as singlets that are not resolved at −60°C. Little can thus be said except that they are mobile. Vögtle,[25] however, studied [2^4]metacyclophane **266** with an internal phenyl substituent, which limits the mobility of the system. Restricted rotation of the phenyl substituent

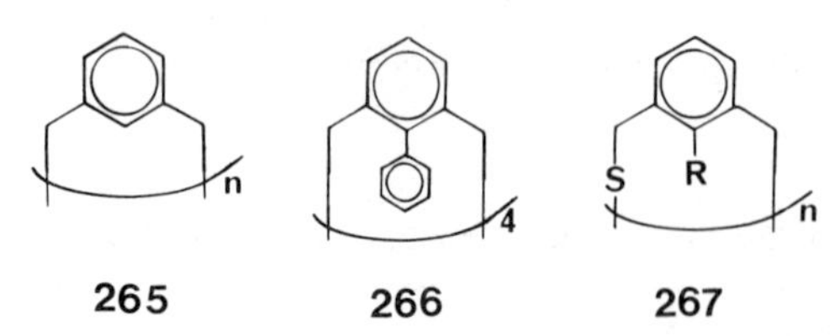

occurs below 45°C because different $H_o, H_{o'}$ signals can be seen. Above 120°C all the bridge protons appear as a singlet, suggesting that at this temperature the substituents can pass over each other (possibly through the center of the ring).

Many thiacyclophane oligomers of type **267** (R = H, CH_3; n = 3, 4) are known[12] but are again mobile to about −60°C.

The [1^4]metacyclophanes **268** give more information.[163] Two conformers, the boat **(268A)** and chair **(268B),** exist. When R = CH_3 and R′ = $COCH_2CH_3$, the chair exhibits a temperature-independent spectrum,

* The superscript n denotes the number of aryl rings in a multiring cyclophane.

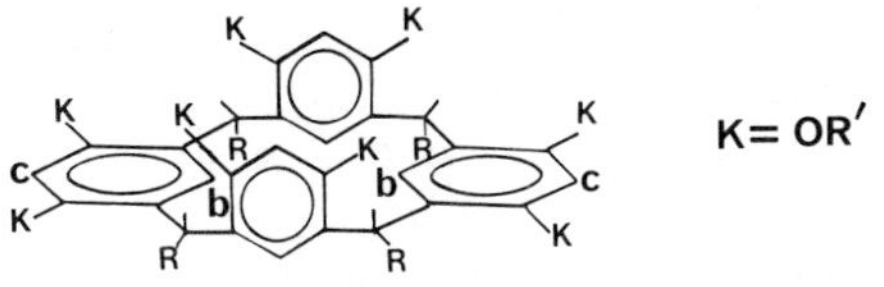

K= OR′

268A

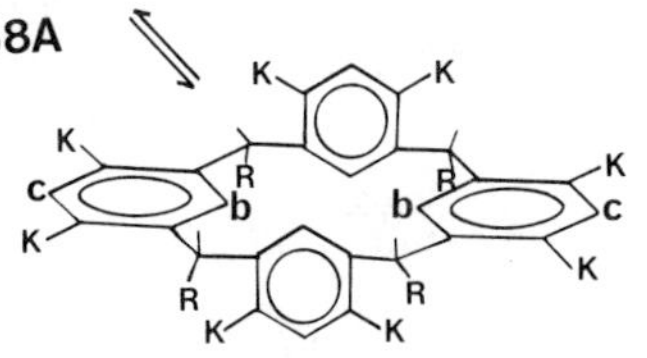

268B

whereas at 28°C the signals for H_b,H_c in the boat form **(268A)** collapse and coalesce, indicating a rotation ($\Delta G_c^{\ddagger}$ = 64 kJ/mol). When R is the larger ϕ, $\Delta G_c^{\ddagger}$ = 74 kJ/mol. The calixarenes **269** provide additional examples of very similar [1^n]metacyclophanes.[164]

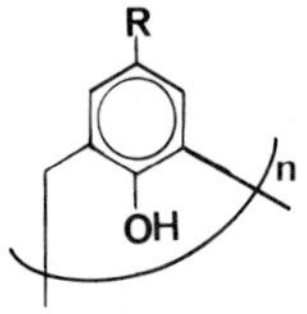

269

2. Paracyclophanes

The [2^n]paracyclophanes **270** (n = 3–8) have been studied by ^{1}H-nmr.[165] Only [2^4]paracyclophane showed a temperature-dependent nmr spectrum

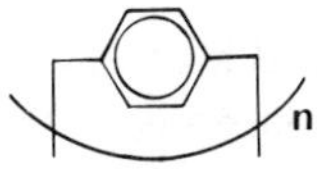

270

in which the methylene bridge protons gave rise to an AB quartet (T_c = −85°C, $\Delta G_c^{\ddagger}$ = 38 kJ/mol). The shift difference between axial and equatorial protons was 0.51 ppm, which suggests that the face conformer **271A** is preferred to the lateral conformer **271B.** However, Wennerström *et al.*[166] have published a rather thorough analysis of the conformations of this and related compounds and has suggested that the data do not distinguish between the D_{2d} and D_2 conformations (**271C** and **271D,** respectively). These authors pointed out that the data obtained by Tabushi[167] on a series

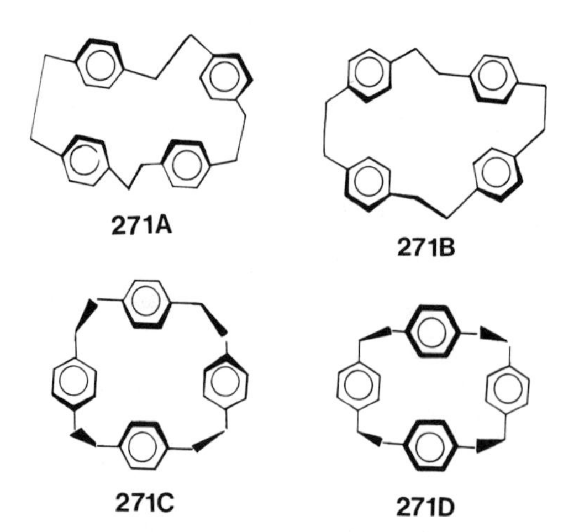

of monosubstituted derivatives of **271,** in which a temperature dependence of protons meta and para to the substituent was observed, better fit the D_2 conformation **(271D).**

The analogous thiophenophane **272** behaves similarly to **271,** with $\Delta G_c^{\ddagger}$ = 34 kJ/mol.[166] Wennerstrom and co-workers have discussed the

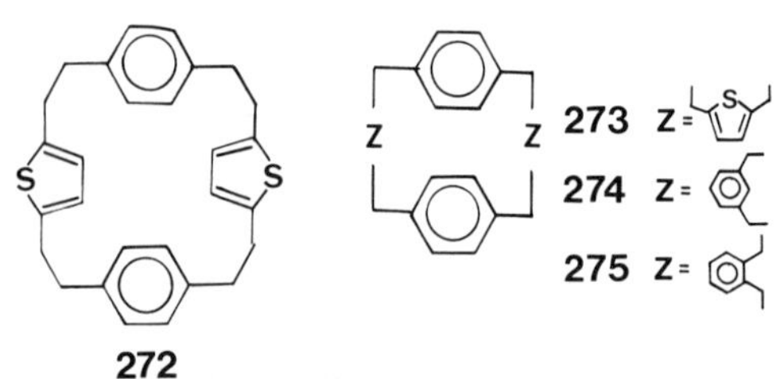

related cyclophanes **273–275** similarly in terms of gauche interactions, to predict the more stable conformations, as well as the tri- and diolefins of compound **271,** and have found barriers for these of ~31 and 29 kJ/mol, respectively. The biphenyl compound **258,** discussed in a previous section, was also reinvestigated,[166] and it was found that the bridge methylene protons reappear at $-110°C$ as two broad peaks, $\Delta\nu$ = 160 Hz. For $T_c = -93°C$, $\Delta G_c^{\ddagger}$ = 35 kJ/mol. This is slightly higher than the barrier for **271** and its olefins. These authors suggested that the interconverting conformers are **258A** and **258B.**

258A 258B

Whereas the tetraene of **271** undergoes rapid ring flipping, the introduction of methyl substituents as in **276** and **277** allows these two conformers

276 **277**

to be observed.[168] At 0°C two separate signals for the methyl protons can be observed at δ 2.10 and 2.05 in a 2 : 1 ratio. At higher temperatures (45°C) these coalesce, corresponding to rotation of one of the substituted rings.

Several [2⁴]naphthalenophanes have also been studied.[169] Compounds **278–280** were too mobile on the nmr time scale to study, but **281** showed

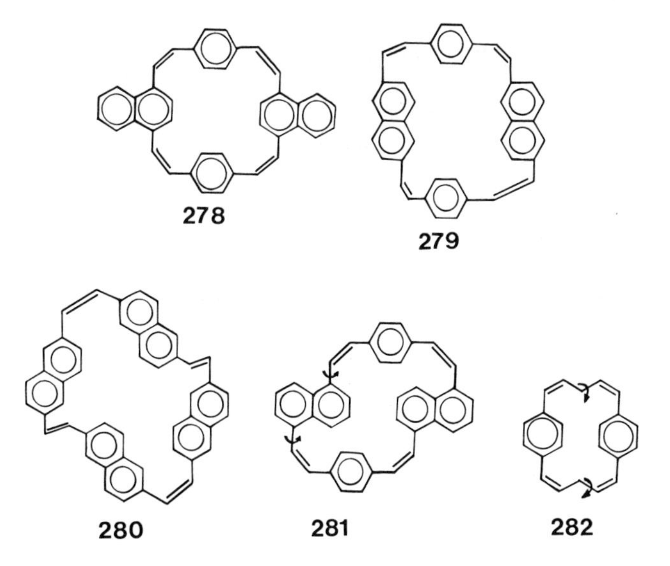

vtnmr spectra. At low temperatures two conformers exist ($T_c \approx 20$°C), which give one conformer at 60°C. This probably corresponds to rotation of one of the naphthalene rings as shown. The saturated cyclophane corresponding to **261** is temperature dependent, with $\Delta G_c^{\ddagger} \approx 40$ kJ/mol, which is very similar to the value for **271.** However, the conformations could not be determined.[169]

The [5.5]paracyclophane **282** resembles **276** and **277** in behavior because rotation of the methylene group of the bridge can occur, with $T_c = -63$°C, $\Delta G_c^{\ddagger} = 37$ kJ/mol.[168]

The perfluorooctathia[2^4]paracyclophane **283** was reported to exhibit temperature-dependent ^{19}F-nmr spectra.[170] The T_c was found to be 55°C, giving $\Delta G_c^{\ddagger} = 75$ kJ/mol. This is higher than the value found for **271** (38 kJ/

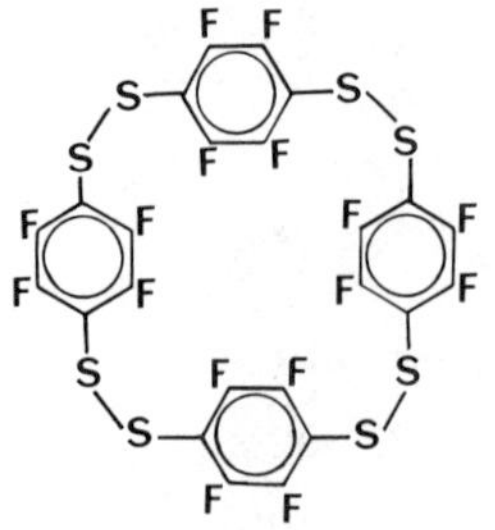

283 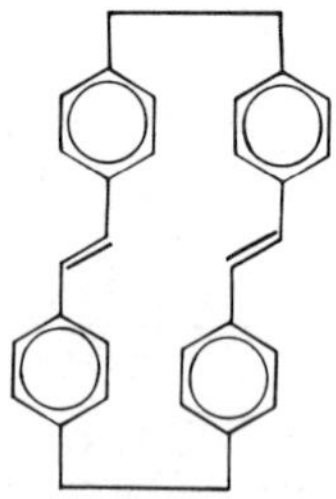 **284**

mol), and hence the fluoro substituents must more than compensate for the longer S—S bridges. In **276–282** the double bonds are all cis. A compound with trans double bonds **(284)** has been prepared.[171] The olefinic protons appear as sharp signals down to −110°C, and hence rotation within this system is fast. The methylene bridge protons, however, broaden below −80°C, indicating a freezing out of gauche forms.

Fewer results are available for the [2^3]paracyclophanes. Compound **119** (Section II,D) must be mobile because H_d appear as a singlet (δ 6.61) whereas, for **118,** H_d,H_e appear as a multiplet (δ 6.2–7.2). Unfortunately, no vtnmr data have been reported.[80]

In thiacyclophanes **285,** as the substituent Y increases in size from H → Cl, the shielding of H_x relative to models decreases from 0.95 to 0.06

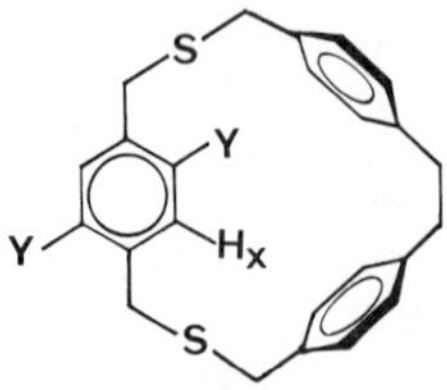

285

ppm. Because rotation is possible, it is assumed that the shielding reflects the average position, with the large substituents being pushed out of the shielding region of the opposite rings.[172] Such systems were also studied under very high pressure to ascertain whether H_x can be pushed into the opposite rings. The effect was small ($\Delta\Delta\delta \approx 3-5$ Hz).[173]

Some very large doughnut-shaped paracyclophanes with eight-membered bridges (e.g., **286** and **287**) have been reported.[174] These seem to

286 **287**

show no restriction of ring rotation. Trimers, 1,4-benzene derivatives, and saturated compounds were also studied.[174]

3. Cage Phanes

This section contains a summary of four reports chosen to represent what might be called *cage phanes*. This writer simply defines these as phanes in which three or more bridges are connected to one or more rings.

The two six-bridge cyclophanes[175] **288** and **289** show barriers ($\Delta G_c^\ddagger = 36$ and 27 kJ/mol, respectively) that are very similar to those found for **271.**

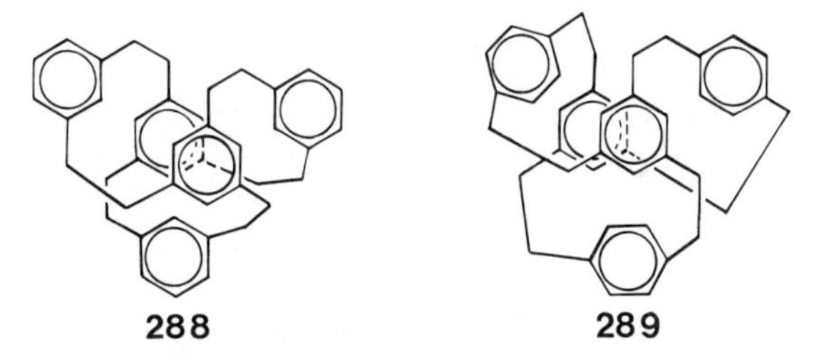

288 **289**

The meta- or para-substituted rings could therefore be considered to rotate about the —CH_2CH_2— bridges. The ^{1}H-nmr spectrum of **288** at room temperature is simple, and the methylene protons appear as an AA′BB′ multiplet (δ 2.68). Below −88°C these protons appear as three peaks, and this observation, along with the strongly shielded meta ring protons (δ 5.73) and trisubstituted ring protons (δ 6.15), is best explained by assuming a compact conformation such as the one of D_3 symmetry shown. For **289** the rather simple room temperature spectrum [singlets at δ 6.5 (para ring protons), δ 6.65 (trisubstituted ring protons), and δ 2.79 (ethylenic protons)] shows broadening at low temperatures (−100°C) only for the aromatic proton signals and gives four signals for the bridge protons. Again, a high-symmetry conformer seems likely. The barriers were determined by lineshape analyses.

The triphenylmethylcyclophanes **290** have been studied[176] for a variety of R groups. The H_m appear at δ 6.4–6.7 and H_c at δ 5.0–5.5. For R = CH_3

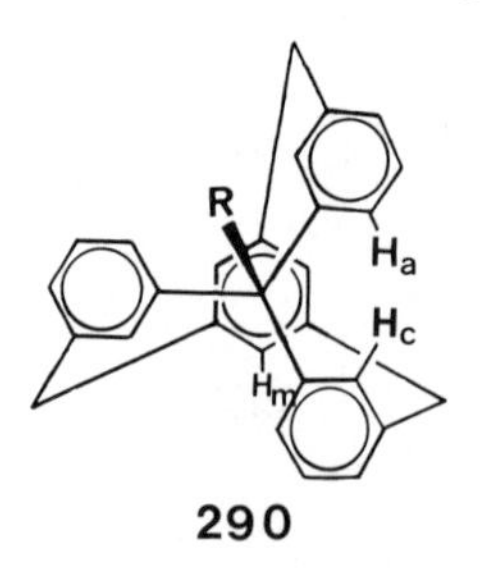

290

the —CH_2CH_2— protons were a sharp singlet at 60°C and a multiplet at lower temperatures. The T_c (40°C) was obtained for the —CD_2CH_2— derivative, giving $\Delta G_c^{\ddagger} = 61$ kJ/mol. The T_c was found to be lower for R = H or OMe (−60°C) and higher for R= Cl (65°C). Possibly, H_a clashes with a large R group on ring flipping.

Finally, the cage phanes **291** and **292** show[177] remarkably simple ^{1}H-nmr spectra; **291** shows only four signals, indicating that it is flexible. The saturated phane **292** shows only three signals, at δ 6.65, 6.48, and 2.79! We await vtnmr results.

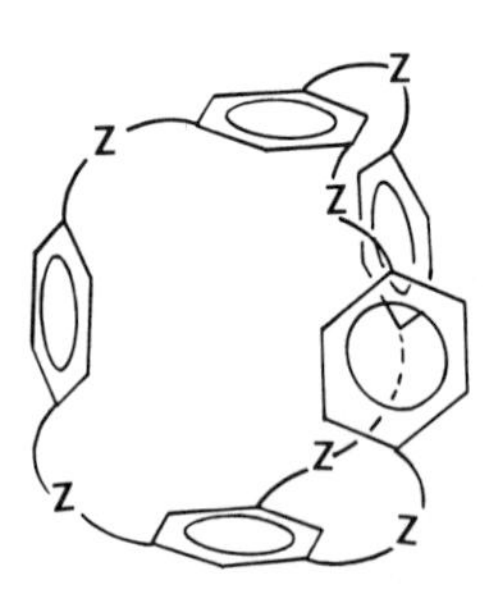

291 Z=CH═CH

292 Z=CH_2—CH_2

IV. NUCLEAR MAGNETIC RESONANCE STUDIES ON NUCLEI OTHER THAN ^{1}H

A. ^{13}C

Very few nmr studies on ^{13}C nuclei have yet appeared in the cyclophane field, possibly in part because ring current effects for carbon are often swamped by other effects caused by geometry changes, steric effects, etc.

However, the basic data for a number of cyclophanes and thiacyclophanes have been published by Sato[178] and our group.[47,121] A few relevant examples are given in Table X.

The inner carbons A in metacyclophanes are ~6 ppm deshielded from those in *m*-xylene. This is believed to be caused by a compression of the sp^2 carbons and adjacent *p* orbitals, which are on the same orbital axis.[178] The methylene bridge carbons suffer an upfield compression shift; this effect is greatest in the paracyclophanes. Ring current effects operate for carbon with about the same magnitude as for protons.[179] This may explain in part the upfield shift of the A carbon in paracyclophanes relative to metacyclophanes. Until ^{13}C-nmr effects are better understood in these molecules, it seems likely that ^{13}C variable temperature nuclear magnetic resonance (^{13}C-vtnmr) may play the more important role, because this often can be used to deduce symmetry changes in the molecule with temperature.

Mann[180] has reviewed ^{13}C-vtnmr. He pointed out that ^{13}C chemical shifts are often five times larger in hertz than ^{1}H chemical shifts. This is not necessarily true for differences in shift, and certainly is not true for shifts associated mainly with ring current effects.[121] However, with this reservation, this means that processes occurring at faster rates can be measured; that is, −100°C in ^{13}C-nmr is equivalent to a much lower temperature in ^{1}H-nmr.

Few cyclophanes have been studied by ^{13}C-vtnmr. We mentioned some previously: **179,**[121] **262,**[159] and **250.**[152] A study[181] of the triazametacyclophane **293** involved detailed ^{13}C-dnmr as well as ^{1}H-nmr studies to

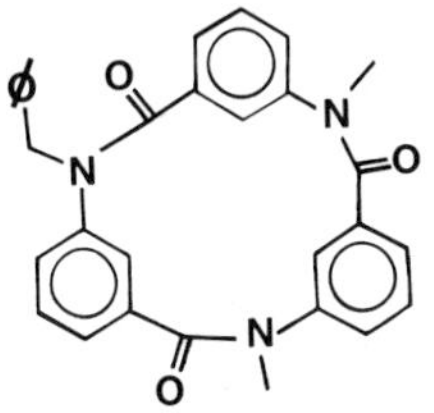

293

show that the crown conformer of **293** is relatively rigid, with a crown ⇌ crown $\Delta G^{\ddagger}$ of 64 kJ/mol and a crown ⇌ saddle $\Delta G^{\ddagger}$ of 62 kJ/mol (the saddle being the more flexible).

B. ^{19}F

Data derived from ^{19}F-nmr studies are even more scarce. In compounds **87, 88,** and **89** the ^{19}F chemical shifts are −86.4, −105.6, and −118 ppm,

TABLE X

[13]C-nmr Data for Selected Cyclophanes[a]: δ Values (ppm) for Selected Carbon Atoms

Structure	A a	B b	C c	D d	E e
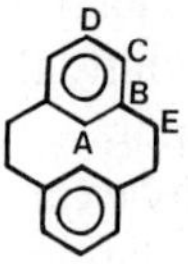	136.3	138.6	125.1	128.6	41.4
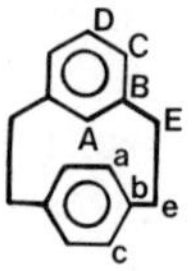	133.0 128.2*	138.5 } 140.6 }	124.7* 129.3*	126.8 —	36.5 } 37.7 }
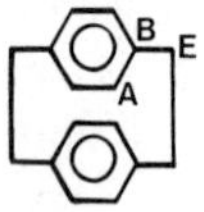	132.8	139.1	—	—	35.7
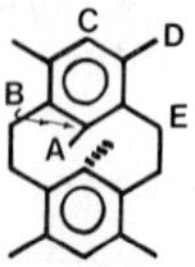	15.6	141.9	127.6	19.1	29.9
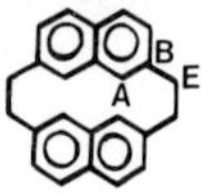	Anti, 130.8 Syn, 127.2	135.9 137.8	— —	— —	38.8 35.8
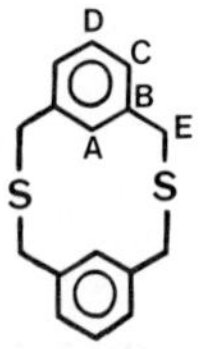	132.1	137.7	127.3	128.7	38.3
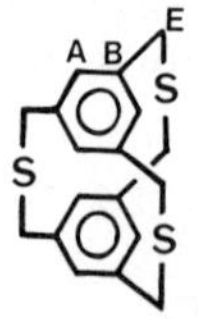	130.4	137.2	—	—	38.7

[a] A brace or asterisk indicates that values are interchangeable.

respectively (relative to $CFCl_3$).[58] For the corresponding proton compounds **226, 225,** and **227** the ^1H chemical shifts are δ 5.24, 4.26, and 5.52, respectively. With the exception of the olefinic compounds **87** and **226** the data are consistent. However, until the anisotropy effect of the C=C bond is documented in these systems, ^{19}F-nmr shifts hardly seem likely to be used much.

The $\Delta G_c^\ddagger$ for **283** discussed above was obtained by ^{19}F-vtnmr.[170] The high-temperature peak at δ −131.2 becomes two peaks at δ −128.8 and −132.7 at low temperature.

C. ^{33}S

Data from ^{33}S-nmr studies are indeed scarce.[182] We have attempted and failed to measure $\phi CH_2SCH_2\phi$ on our Brucker WM-250 instrument in relatively concentrated solutions. This author holds no hope for the thiacyclophanes (yet!).

V. EPILOGUE

Clearly, modern nmr methods have allowed the study of more bizarre and complex cyclophane systems. No doubt, these and other more advanced nmr techniques will be used for many years to come, although X-ray methods will be used more frequently to supplement (and perhaps supplant) the structural and conformational data obtained by nmr. In fact, signs of this state of affairs are already appearing in the literature. A paper

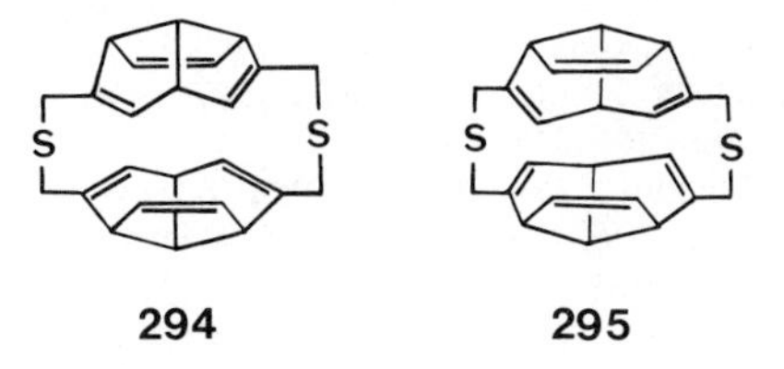

describing the novel *syn*- and *anti*-cyclophane-like structures **294** and **295** has appeared[183] with no nmr but rather only X-ray data given.

REFERENCES

1. B. H. Smith, "Bridged Aromatic Compounds." Academic Press, New York 1964.
2. F. Vögtle and G. Hohner, *Top. Curr. Chem.* **74,** 1 (1978).
3. F. Vögtle and P. Neumann, *Tetrahedron Lett.* p. 5329 (1969); *Tetrahedron* **26,** 5847 (1970).
4. F. Vögtle and P. Neumann, *Angew. Chem., Int. Ed. Engl.* **11,** 73 (1972).
5. D. J. Wilson, V. Boekelheide, and R. W. Griffin, *J. Am. Chem. Soc.* **82,** 6302 (1960).
6. N. L. Allinger, M. A. De Rooge, and R. B. Herman, *J. Am. Chem. Soc.* **83,** 1974 (1961); H. S. Gutowsky and C. Juan, *J. Chem. Phys.* **37,** 120 (1962); M. Fujimoto, T. Sato, and K. Hata, *Bull. Chem. Soc. Jpn.* **40,** 600 (1967); N. L. Allinger, B. J. Gordon, S. E. Hiu, and R. A. Ford, *J. Org. Chem.* **32,** 2272 (1967); R. Flammang, H. R. Figeys, and R. H. Martin, *Tetrahedron* **24,** 1171 (1968).

6a. T. Sato, S. Akabori, M. Kainosho, and K. Hata, *Bull. Chem. Soc. Jpn.* **39,** 856 (1966); **41,** 218 (1968).

6b. K. Burri and W. Jenny, *Helv. Chim. Acta* **50,** 1978 (1967).

7. C. E. Johnson and F. A. Bovey, *J. Chem. Phys.* **29,** 1012 (1958); C. W. Haigh and R. B. Mallion, *Org. Magn. Reson.* **4,** 203 (1972).
8. C. J. Brown, *J. Chem. Soc.* p. 3278 (1953).
9. T. Sato, S. Akabori, S. Muto, and K. Hata, *Tetrahedron* **24,** 5557 (1968).
10. W. S. Lindsay, P. Stokes, L. G. Humber, and V. Boekelheide, *J. Am. Chem. Soc.* **83,** 943 (1961).
11. A. W. Hanson, *Acta Crystallogr.* **15,** 956 (1962).
12. R. H. Mitchell and V. Boekelheide, *Tetrahedron Lett.* p. 1197 (1970); *Chem. Commun.* p. 1555 (1970); *J. Am. Chem. Soc.* **96,** 1547 (1974).
13. D. Kamp and V. Boekelheide, *J. Org. Chem.* **43,** 3470 (1978).
14. H. Blaschke, C. E. Ramey, I. Calder, and V. Boekelheide, *J. Am. Chem. Soc.* **92,** 3675 (1970).
15. A. W. Hanson and M. Rohrl, *Acta Crystallogr., Sect. B* **B28,** 2032 (1972).
16. V. Boekelheide and J. B. Phillips, *J. Am. Chem. Soc.* **89,** 1695 (1967).
17. S. Akabori, T. Sato, and K. Hata, *J. Org. Chem.* **33,** 3277 (1968).
18. R. W. Griffin, R. W. Baughman, and C. E. Ramey, *Tetrahedron* **24,** 5419 (1968).
19. W. Jenny and H. Hotzrichter, *Chimia* **21,** 509 (1967).
20. F. Vögtle, *Angew. Chem., Int. Ed. Engl.* **8,** 274 (1969).
21. E. Hammerschmidt and F. Vögtle, *Chem. Ber.* **113,** 1125 (1980).
22. J. S. H. Yan, M.Sc. Thesis, University of Victoria (1978); R. H. Mitchell, J. S. H. Yan, and T. W. Dingle, *J. Am. Chem. Soc.* **104,** 2551 (1982).
23. E. Langer and H. Lehner, *Tetrahedron Lett.* p. 1357 (1974); D. Krois, E. Langer, and H. Lehner, *Tetrahedron* **36,** 1345 (1980).
24. M. Tashiro and T. Yamato, *Chem. Lett.* p. 595 (1979).
25. K. Bockmann and F. Vögtle, *Chem. Ber.* **114,** 1048 (1981).
26. T. Sato, H. Matsui, and R. Komaki, *J. Chem. Soc., Perkin Trans. 1* p. 2051 (1976).
27. J. R. Davy and J. A. Reiss, *Aust. J. Chem.* **29,** 163 (1976).
28. Y. Nesumi, T. Nakazawa, and I. Murata, *Chem. Lett.* p. 771 (1979).
29. V. Boekelheide and P. H. Anderson, *J. Org. Chem.* **38,** 3928 (1973).
30. I. Gault, B. J. Price, and I. O. Sutherland, *Chem. Commun.* p. 540 (1967).
31. J. R. Fletcher and I. O. Sutherland, *Chem. Commun.* p. 1504 (1969).

32. H. E. Winberg, F. S. Fawcett, W. E. Mochel, and C. W. Theobald, *J. Am. Chem. Soc.* **83,** 1428 (1960).
33. G. R. Newkome, J. D. Sauer, J. M. Roper, and D. C. Hager, *Chem. Rev.* **77,** 513 (1977).
34. T. Umemoto, T. Otsubo, Y. Sakata, and S. Misumi, *Tetrahedron Lett.* p. 593 (1973).
34a. T. Umemoto, T. Otsubo, and S. Misumi, *Tetrahedron Lett.* p. 1753 (1974).
35. H. A. Staab, C. P. Herz, and A. Döhling, *Chem. Ber.* **113,** 233 (1980).
36. H. A. Staab, C. P. Herz, and A. Döhling, *Tetrahedron Lett.* p. 791 (1979); H. Staab and A. Döhling, *ibid.* p. 2019.
37. A. Kasahara, T. Izumi, and H. Umezawa, *Chem. Lett.* p. 1039 (1980).
38. N. Finch and C. W. Gemenden, *J. Org. Chem.* **44,** 2804 (1979).
39. F. Vögtle, R. Schafer, L. Schunder, and P. Neumann, *Justus Liebigs Ann. Chem.* **734,** 102 (1970).
40. F. Bottino, S. Foti, S. Pappalardo, P. Finocchiaro, and M. Ferrugia, *J. Chem. Soc., Perkin Trans. 1* p. 198 (1979).
41. G. Ronisvalle, F. Bottino, and S. Pappalardo, *Org. Magn. Reson.* **14,** 344 (1980).
42. V. Boekelheide and C. H. Tsai, *J. Org. Chem.* **38,** 3931 (1973).
43. T. Sato, M. Wakabayashi, K. Hata, and M. Kainosho, *Tetrahedron* **27,** 2737 (1971).
44. A. W. Hanson and E. W. Macauly, *Acta Crystallogr., Sect. B* **B28,** 2287 (1972).
45. T. Otsubo, D. Stusche, and V. Boekelheide, *J. Org. Chem.* **43,** 3466 (1978).
46. M. Tashiro and T. Yamato, *J. Org. Chem.* **46,** 1543 (1981).
47. R. H. Mitchell, *Can. J. Chem.* **58,** 1398 (1980).
48. R. H. Mitchell and R. Mahadevan, *Tetrahedron Lett.* **22,** 5131 (1981).
49. R. Mahadevan, Ph.D. Thesis, University of Victoria (1981).
50. D. Kamp and V. Boekelheide, *J. Org. Chem.* **43,** 3475 (1978).
51. R. H. Mitchell and W. Anker, *Tetrahedron Lett.* **22,** 5135 (1981).
52. R. H. Mitchell, R. J. Carruthers, and L. Mazuch, *J. Am. Chem. Soc.* **100,** 1007 (1978); R. H. Mitchell, R. J. Carruthers, L. Mazuch, and T. W. Dingle, *ibid.* **104,** 2544 (1982).
53. R. H. Mitchell, R. V. Williams, and T. W. Dingle, *J. Am. Chem. Soc.* **104,** 2560 (1982).
54. K. Böckmann and F. Vögtle, *Chem. Ber.* **114,** 1065 (1981).
55. H. Förster and F. Vögtle, *Angew. Chem., Int. Ed. Engl.* **16,** 429 (1977).
56. T. Kawashima, T. Otsubo, Y. Sakata, and S. Misumi, *Tetrahedron Lett.* p. 1063 (1978).
57. Y. Fukazawa, M. Aoyagi, and S. Ito, *Tetrahedron Lett.* p. 1067 (1978).
58. V. Boekelheide, P. H. Anderson, and T. A. Hylton, *J. Am. Chem. Soc.* **96,** 1558 (1974).
59. S. Mizogami, N. Osaka, T. Otsubo, Y. Sakata, and S. Misumi, *Tetrahedron Lett.* p. 799 (1974); T. Otsubo, S. Mizogami, N. Osaka, Y. Sakata, and S. Misumi, *Bull. Chem. Soc. Jpn.* **50,** 1859 (1977).
60. C. B. Shana, S. M. Rosenfeld, and P. M. Keehn, *Tetrahedron* **33,** 1081 (1977).
60a. M. Corson, B. M. Foxman, and P. M. Keehn, *Tetrahedron* **34,** 1641 (1978).
61. D. J. Cram, C. K. Dalton, and G. R. Knox, *J. Am. Chem. Soc.* **85,** 1088 (1963).
62. T. Otsubo and V. Boekelheide, *J. Org. Chem.* **42,** 1085 (1977).
63. V. Boekelheide and R. A. Hollins, *J. Am. Chem. Soc.* **92,** 3512 (1970).
64. K. C. Dewhirst and D. J. Cram, *J. Am. Chem. Soc.* **80,** 3115 (1958).
65. D. J. Cram and R. C. Helgeson, *J. Am. Chem. Soc.* **88,** 3515 (1966).
66. H. H. Wasserman and P. M. Keehn, *J. Am. Chem. Soc.* **91,** 2374 (1969).
67. H. A. Staab and C. P. Herz, *Angew Chem., Int. Ed. Engl.* **16,** 392 (1977).
68. M. W. Haenel, *Tetrahedron Lett.* p. 3053 (1974).
69. M. W. Haenel, *Chem. Ber.* **111,** 1789 (1978).

70. M. Haenel and H. A. Staab, *Chem. Ber.* **106,** 2203 (1973).
71. A. Iwama, T. Toyoda, M. Yoshida, T. Otsubo, Y. Sakata, and S. Misumi, *Bull. Chem. Soc. Jpn.* **51,** 2988 (1978).
72. M. Matsumoto, T. Otsubo, Y. Sakata, and S. Misumi, *Tetrahedron Lett.* p. 4425 (1977).
73. N. Kato, H. Matsunaga, S. Oeda, Y. Fukazawa, and S. Ito, *Tetrahedron Lett.* p. 2419 (1979).
74. M. Haenel, *Tetrahedron Lett.* p. 1273 (1977).
75. H. A. Staab and M. Haenel, *Chem. Ber.* **106,** 2190 (1973).
76. D. N. Leach and J. A. Reiss, *Aust. J. Chem.* **32,** 361 (1979).
77. R. G. Kirrstetter and H. A. Staab, *Liebigs Ann. Chem.* p. 899 (1979).
78. H. A. Staab and R. G. Kirrstetter, *Liebigs Ann. Chem.* p. 886 (1979).
79. T. Kawashima, T. Otsubo, Y. Sakata, and S. Misumi, *Tetrahedron Lett.* p. 5115 (1978).
80. T. Otsubo, R. Gray, and V. Boekelheide, *J. Am. Chem. Soc.* **100,** 2449 (1978).
81. N. Jacobson and V. Boekelheide, *Angew. Chem., Int. Ed. Engl.* **17,** 46 (1978).
82. W. Bieber and F. Vögtle, *Chem. Ber.* **112,** 1919 (1979).
83. E. A. Truesdale and R. S. Hutton, *J. Am. Chem. Soc.* **101,** 6475 (1979).
84. M. Haenel, A. Flatow, V. Taglieber, and H. A. Staab, *Tetrahedron Lett.* p. 1733 (1977).
85. H. A. Staab and C. P. Herz, *Angew. Chem., Int. Ed. Engl.* **16,** 799 (1977).
86. M. Yoshinaga, T. Otsubo, Y. Sakata, and S. Misumi, *Bull. Chem. Soc. Jpn.* **52,** 3759 (1979).
87. N. E. Blank and M. W. Haenel, *Chem. Ber.* **114,** 1520 (1981).
88. R. H. Mitchell, R. J. Carruthers, and J. C. M. Zwinkels, *Tetrahedron Lett.* p. 2585 (1976).
89. V. Boekelheide and R. A. Hollins, *J. Am. Chem. Soc.* **95,** 3201 (1973).
90. M. Nakazaki, K. Yamamoto, and Y. Miura, *J. Org. Chem.* **43,** 1041 (1978).
91. R. Gray and V. Boekelheide, *J. Am. Chem. Soc.* **101,** 2128 (1979).
92. W. Gilb, K. Menke, and H. Hopf, *Angew. Chem., Int. Ed. Engl.* **16,** 191 (1977).
93. P. F. T. Schirch and V. Boekelheide, *J. Am. Chem. Soc.* **101,** 3125 (1979).
94. Y. Sekine and V. Boekelheide, *J. Am. Chem. Soc.* **103,** 1777 (1981).
95. G. Binsch and H. Kessler, *Angew. Chem., Int. Ed. Engl.* **19,** 411 (1980).
96. I. O. Sutherland, *in* "Annual Reports on NMR Spectroscopy" (E. F. Mooney, ed.), Vol. 4, pp. 71–235. Academic Press, New York, 1971.
97. I. C. Calder and P. J. Garratt, *J. Chem. Soc. B* p. 660 (1967).
98. R. H. Mitchell and V. Boekelheide, *J. Heterocycl. Chem.* **6,** 981 (1969).
99. F. Vögtle, *Chem. Ber.* **102,** 1784 (1969).
100. F. Vögtle, *Tetrahedron Lett.* p. 5221 (1968).
101. S. Hirano, H. Hara, T. Hiyama, S. Fujita, and H. Nozaki, *Tetrahedron* **31,** 2219 (1975).
102. S. Fujita and H. Nozaki, *Bull. Chem. Soc. Jpn.* **44,** 2827 (1971).
103. J. W. van Straten, W. H. de Wolf, and F. Bickelhaupt, *Tetrahedron Lett.* p. 4667 (1977).
104. F. Vögtle and P. Neumann, *Tetrahedron* p. 5299 (1970).
104a. F. Vögtle, J. Grütze, R. Natscher, W. Wieder, E. Weber, and R. Grun, *Chem. Ber.* **108,** 1694 (1975).
105. A. Agarwal, J. A. Barnes, J. L. Fletcher, M. J. McGlinchey, and B. G. Sayer, *Can. J. Chem.* **55,** 2575 (1975).
106. E. Doomes and R. M. Beard, *Tetrahedron Lett.* p. 1243 (1976).
107. N. L. Allinger, T. J. Walter, and M. G. Newton, *J. Am. Chem. Soc.* **96,** 4588 (1974).

108. P. G. Gassman, T. F. Bailey, and R. C. Hoye, *J. Org. Chem.* **45,** 2923 (1980).
109. V. V. Kane, A. D. Wolf, and M. Jones, *J. Am. Chem. Soc.* **96,** 2643 (1974).
110. K. B. Wiberg and M. J. O'Donnell, *J. Am. Chem. Soc.* **101,** 6660 (1979).
111. S. M. Rosenfeld and P. M. Keehn, *Chem. Commun.* p. 119 (1974).
112. Y. Miyahara, T. Inazu, and T. Yoshino, *Chem. Lett.* p. 397 (1980).
113. M. P. Cooke, *J. Org. Chem.* **46,** 1747 (1981).
114. V. Boekelheide and J. A. Lawson, *J. Chem. Soc. D* p. 1558 (1970).
115. W. Anker, Ph.D. Thesis, University of Victoria (1982).
116. R. W. Griffin and R. A. Coburn, *J. Am. Chem. Soc.* **89,** 4638 (1967).
117. C. Lin, P. Singh, M. Maddox, and E. F. Ullman, *J. Am. Chem. Soc.* **102,** 3261 (1980).
118. T. Otsubo, M. Kitasawa, and S. Misumi, *Bull. Chem. Soc. Jpn.* **52,** 1515 (1979).
119. T. Sato, M. Wakabayashi, M. Kainosho, and K. Hata, *Tetrahedron Lett.* p. 4185 (1968).
120. F. Vögtle and L. Schunder, *Chem. Ber.* **102,** 2677 (1969).
121. W. Anker, G. W. Bushnell, and R. H. Mitchell, *Can. J. Chem.* **57,** 3080 (1979).
122. R. H. Mitchell and R. J. Carruthers, *Can. J. Chem.* **52,** 3054 (1974).
123. R. H. Mitchell and R. J. Carruthers, *Tetrahedron Lett.* p. 4331 (1975); R. H. Mitchell, R. J. Carruthers, L. Mazuch, and T. W. Dingle, *J. Am. Chem. Soc.* **104,** 2544 (1982).
124. T. Umemoto, T. Kawashima, Y. Sakata, and S. Misumi, *Chem. Lett.* p. 837 (1975).
125. Y. Fukazawa, M. Aoyagi, and S. Ito, *Tetrahedron Lett.* p. 1055 (1979).
126. G. R. Newkome, J. M. Roper, and J. M. Robinson, *J. Org. Chem.* **45,** 4380 (1980).
127. H. J. J. B. Martel, S. McMahon, and M. Rasmussen, *Aust. J. Chem.* **32,** 1241 (1979).
128. G. W. Bushnell, K. Beveridge, W. Anker, and R. H. Mitchell, *Can. J. Chem.* **60,** 362 (1982).
129. G. R. Newkome and T. Kawato, *J. Am. Chem. Soc.* **101,** 7088 (1979).
130. G. R. Newkome, A. Nayak, G. L. McGlure, F. Danesh-Khoshboo, and J. Broussard-Simpson, *J. Org. Chem.* **42,** 1500 (1977).
131. S. J. Abrahams, *Q. Rev., Chem. Soc.* **10,** 407 (1956).
132. K. R. Dixon and R. H. Mitchell, *Can. J. Chem.* (in press) (1983).
133. R. H. Mitchell, *Can. J. Chem.* **54,** 238 (1976).
134. V. Boekelheide and J. L. Mondt, *Tetrahedron Lett.* p. 1203 (1970).
135. F. Bottino and S. Pappalardo, *Tetrahedron* **36,** 3095 (1980).
136. F. Bottino, S. Foti, S. Pappalardo, and N. B. Pahor, *Tetrahedron Lett.* p. 1171 (1979).
137. D. T. Hefelfinger and D. J. Cram, *J. Am. Chem. Soc.* **92,** 1073 (1970); D. J. Cram, R. C. Helgeson, D. Lock, and L. A. Singer, *ibid.* **88,** 1324 (1966).
138. V. Boekelheide, K. Galusko, and K. S. Szeto, *J. Am. Chem. Soc.* **96,** 1578 (1974).
139. L. H. Weaver and B. W. Matthews, *J. Am. Chem. Soc.* **96,** 1581 (1974).
140. A. W. Hanson, *Acta Crystallogr., Sect. B* **B27,** 197 (1971).
141. I. D. Reingold, W. Schmidt, and V. Boekelheide, *J. Am. Chem. Soc.* **101,** 2121 (1979).
142. S. A. Sherrod, R. L. Da Costa, R. A. Barnes, and V. Boekelheide, *J. Am. Chem. Soc.* **96,** 1565 (1974).
143. S. Mizogami, T. Otsubo, Y. Sakata, and S. Misumi, *Tetrahedron Lett.* p. 2791 (1974).
144. C. Wong and W. W. Paudler, *J. Org. Chem.* **39,** 2570 (1974).
145. M. D. Bezoari and W. W. Paudler, *J. Org. Chem.* **45,** 4584 (1980).
146. M. W. Haenel, *Tetrahedron Lett.* p. 4007 (1978).
147. J. R. Davey, M. N. Iskander, and J. R. Reiss, *Tetrahedron Lett.* p. 4085 (1978); *Aust. J. Chem.* **32,** 1067 (1979).
148. M. N. Iskander and J. A. Reiss, *Tetrahedron* **34,** 2343 (1978).
148a. V. Boekelheide and C. H. Tsai, *Tetrahedron* **32,** 423 (1976).
149. J. T. Craig, B. Halton, and S. F. Lo, *Aust. J. Chem.* **28,** 913 (1975).

150. R. B. Duvernet, O. Wennerström, J. Lawson, T. Otsubo, and V. Boekelheide, *J. Am. Chem. Soc.* **100,** 2457 (1978).
151. K. Böckmann and F. Vögtle, *Liebigs Ann. Chem.* p. 467 (1981).
152. D. N. Leach and J. A. Reiss, *Aust. J. Chem.* p. 33, 823 (1980).
153. W. Bieber and F. Vögtle, *Angew. Chem., Int. Ed. Engl.* **16,** 175 (1977).
154. M. W. Haenel and A. Flatlow, *Chem. Ber.* **112,** 249 (1979).
154a. F. A. L. Anet and M. A. Brown, *J. Am. Chem. Soc.* **91,** 2389 (1969).
155. H. J. Reich and D. J. Cram, *J. Am. Chem. Soc.* **91,** 3517 (1969).
156. See, e.g., D. J. Cram *et al., J. Am. Chem. Soc.* **80,** 3094, 3126 (1958).
157. D. T. Longone and J. A. Gladysz, *Tetrahedron Lett.* p. 4559 (1976).
158. F. Vögtle, *Justus Liebigs Ann. Chem.* **728,** 17 (1969).
159. R. H. Mitchell and Y. H. Lai, *Tetrahedron Lett.* **21,** 2633 (1980).
160. Y. H. Lai, Ph.D. Thesis, University of Victoria (1980).
161. F. Vögtle and E. Hammerschmidt, *Angew. Chem., Int. Ed. Engl.* **17,** 268 (1978).
162. R. Paioni and W. Jenny, *Helv. Chim. Acta* **52,** 2041 (1969).
163. A. G. S. Högberg, *J. Am. Chem. Soc.* **102,** 6046 (1980); *J. Org. Chem.* **45,** 4498 (1980).
164. C. D. Gutsche, B. Dhawan, K. H. No, and R. Muthukrishnan, *J. Am. Chem. Soc.* **103,** 3782 (1981).
165. I. Tabushi, H. Yamada, and Y. Kuroda, *J. Org. Chem.* **40,** 1946 (1975).
166. T. Olsson, D. Tanner, B. Thulin, O. Wennerström, and T. Liljefors, *Tetrahedron* **37,** 3473 (1981).
167. I. Tabushi and H. Yamada, *Tetrahedron* **33,** 1101 (1977).
168. T. Olsson, D. Tanner, B. Thulin, and O. Wennerström, *Tetrahedron* **37,** 3491 (1981).
169. D. Tanner, B. Thulin, and O. Wennerström, *Acta Chem. Scand., Ser. B* **B33,** 443 (1979).
170. M. S. Raasch, *J. Org. Chem.* **44,** 2629 (1979).
171. D. Tanner and O. Wennerström, *Tetrahedron Lett.* **22,** 2313 (1981).
172. F. Imashiro, M. Oda, T. Iida, Z. Yoshida, and I. Tabushi, *Tetrahedron Lett.* p. 371 (1976).
173. F. Imashiro, A. Saika, H. Yamada, and A. Sera, *Chem. Lett.* p. 247 (1981).
174. E. T. Jarvi and H. W. Whitlock, *J. Am. Chem. Soc.* **102,** 657 (1980); B. J. Whitlock, E. T. Jarvi, and H. W. Whitlock, *J. Org. Chem.* **46,** 1832 (1981); S. P. Adams and H. W. Whitlock, *ibid.* p. 3474.
175. T. Olsson, D. Tanner, B. Thulin, and O. Wennerström, *Tetrahedron* **37,** 3485 (1981).
176. M. Nakazaki, K. Yamamoto, and T. Toya, *J. Org. Chem.* **45,** 2553 (1980); **46,** 1611 (1981).
177. H. E. Högberg, B. Thulin, and O. Wennerström, *Tetrahedron Lett.* p. 931 (1977).
178. T. Takemura and T. Sato, *Can. J. Chem.* **54,** 3412 (1976); *J. Chem. Soc., Perkin Trans. 2* p. 1195 (1976); *Chem. Commun.* p. 97 (1974).
179. R. Duvernet and V. Boekelheide, *Proc. Natl. Acad. Sci. U.S.A.* **71,** 2961 (1974).
180. B. E. Mann, *Prog. Nucl. Magn. Reson. Spectrosc.* **11,** 95 (1977).
181. F. E. Elhardi, W. D. Ollis, J. F. Stoddart, D. J. Williams, and K. A. Woode, *Tetrahedron Lett.* **21,** 4215 (1980).
182. R. Faure, E. J. Vincent, J. M. Ruiz, and L. Léna, *Org. Magn. Reson.* **15,** 401 (1981).
183. W. P. Roberts and G. Shohan, *Tetrahedron Lett.* **22,** 4895 (1981).
184. A. W. Hanson, *Acta Crystallogr. Sect. B,* **31,** 2352 (1975).

CHAPTER 5

[*n*]Cyclophanes

STUART M. ROSENFELD AND K. ANN CHOE

Department of Chemistry
Smith College
Northhampton, Massachusetts

ISBN 0-12-403001-7

I. INTRODUCTION

[*n*]Cyclophanes are arguably among the simplest of the cyclophanes and therefore occupy a special place in the realm of cyclophane chemistry. The very simplicity of these molecular systems affords the cyclophane chemist opportunities to isolate and probe individual aspects of the chemical and physical properties of these bridged aromatic compounds. A list of questions amenable to study would certainly include the following:

1. What is the relationship between planarity of aromatic rings and aromaticity? How does nonplanarity manifest itself in the chemistry and spectroscopy of the aryl ring?
2. What effect does transannular interaction have on the chemistry of the aryl ring and on the chemistry of the aliphatic bridge?
3. How are conformational motions of an aliphatic chain affected by a proximate aryl ring? How is the conformational mobility of the aryl ring within the macrocycle affected by the bridge length and the ring substitution?
4. Do the unique structural features of these molecules make them attractive synthons for the preparation of other molecules of interest?

These systems also provide models for improving theoretical methods and for exploring the ways in which molecules distribute strain energy. One purpose of this chapter is to describe the approaches made to some of these questions in the period between the exhaustive review by Smith[1] and the present (through 1981).

The synthesis and chemical and physical properties of [*n*]cyclophanes (**1**) are discussed here. Coverage includes systems with benzene, naphthalene, and anthracene rings as well as the heteroaromatics pyridine, furan, pyrrole, and thiophene. The relatively few reports of higher condensed aromatics and the less common heteroaromatics are also discussed. Molecules with saturated, unsaturated, and otherwise functionalized all-carbon bridges receive attention, as do those with heteroatoms in

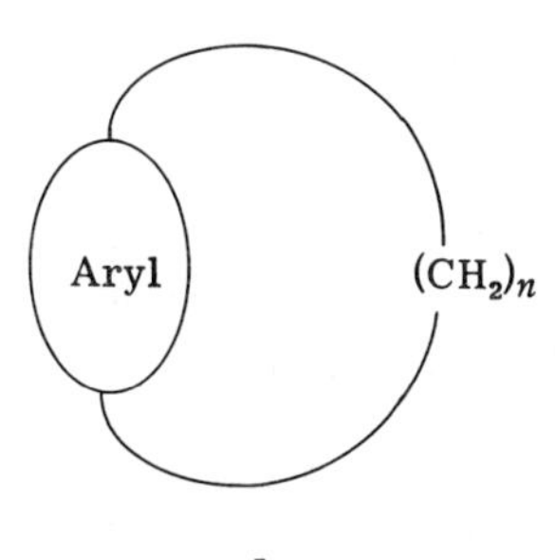

1

the bridging chain. There is no coverage of ortho- or peri-bridged systems. Macrocyclic polyethers and related compounds are generally excluded because they are more appropriately covered in other chapters. Neither aromatics with transition metals incorporated in the aromatic moiety, for example, [*n*]ferrocenophanes **(2)**,[2] nor [*n*]cyclophanes with interlocking bridges, for example, **3**[3] (a catenanophane!), are discussed.

2 **3**

In preference to the more awkward IUPAC nomenclature, we have employed exclusively the system that has come into common use.[1,4,5] Specifically, [*n*] denotes bridge length (or [*m*][*n*] for two bridges), and meta or para (or numbers for nonbenzenoid compounds) describes points of connection. Numbers are used to identify the position(s) of heteroatoms in the bridge. For example, **4** is given the name 2,13-dithia[14]paracyclophane, and **5** is referred to as 4-keto[8](1,4)anthracenophane.

4 **5**

II. SYNTHESIS OF [*n*]PARACYCLOPHANES

Historically, the synthesis of [*n*]paracyclophanes has been approached through the reactions commonly used for the preparation of medium- and large-ring alicyclic and heterocyclic compounds. Included among the methods described by Smith[1] are acyloin condensation, Friedel–Crafts reaction, pyrolysis of diacids, cyclization by amide formation, halo-amine cyclization, halo-ether cyclization, Ziegler cyclization, and oxidative coupling of acetylenes and mercaptans. At the time *Bridged Aromatic Compounds*[1] was written [8]paracyclophane was the smallest [*n*]paracyclophane described in the literature, although Allinger *et al.*[6] had speculated that [7]paracyclophane would be about as strained as cyclopropane and therefore amenable to synthesis. It was in the syntheses of these most highly strained target molecules that new methods were clearly required. In this section emphasis is placed on the methods most commonly used during the past 20 years, typically those of most general applicability and those that address the special problems involved in the preparation of strained compounds. The Smith book[1] should be consulted for a comprehensive treatment of the earlier literature.

A. Intramolecular Ring Closure Methods

Intramolecular ring closure methods have historically been the most significant approach to [*n*]paracyclophane derivatives. For the larger values of *n* no problems unique to paracyclophanes exist. These methods generally fail, however, for compounds with less than nine atoms in the bridge. The acyloin cyclization (Eq. 1) has been applied successfully to the preparation of [9]-,[7,8] [10]-,[9] [12]-,[10] and [14]paracyclophanes[11] and various bridge-functionalized derivatives, often with yields of 70 to 75% for the ring closure step:

$$p\text{-}C_6H_4[(CH_2)_xCO_2CH_3]_2 \xrightarrow[\text{xylene}]{\text{Na}} p\text{-}C_6H_4\,(CH_2)_x\text{—}C(=O)\text{—}CHOH\text{—}(CH_2)_x \quad (1)$$

In fact, the first well-characterized [*n*]paracyclophane was prepared by this method in 1950 by Wiesner and co-workers.[9] [*n*](1,4)Naphthalenophanes with $n = 10$ and 14 have been synthesized by acyloin condensation of the appropriate diesters.[12] However, the reaction fails for [9](1,4)naphthalenophane. A modified acyloin condensation has been attempted for the preparation of an [8](9,10)anthracenophane derivative and, although spectral data suggested formation of the desired acyloin, the compound could not be isolated.[13]

High-dilution Friedel–Crafts acylation occurs at the para position when $x = 8$ or larger (Eq. 2) and has provided [*n*]paracyclophane derivatives with $n = 9$, 10, 11, 12, 13, 14, and 16.[14–16]

$(CH_2)_xCOCl$ → $(CH_2)_x$ (2)

Intramolecular cyclization via Eglinton oxidative coupling of bisethynyl compounds (Eq. 3) yields [*n*]paracyclophadiynes:

$(CH_2)_xC{\equiv}CH$, $(CH_2)_yC{\equiv}CH$ $\xrightarrow[\text{pyridine}]{Cu(OAc)_2}$ $(CH_2)_x$, $(CH_2)_y$, C≡C–C≡C (3)

$x = y = 3$ (21.3%)
$x = 2$; $y = 4$ (41.1%)
$x = 3$; $y = 4$ (37.2%)
$x = y = 4$ (46.6%)
$x = y = 5$ (26.1%)

Misumi and co-workers used the method to synthesize [10]- [11]-, [12]- and [14]paracyclophadiynes[17,18] and later the [*n*](9,10)anthracenophadiynes **6** and **7**.[19] This group also prepared [10](9,10)anthracenophane by catalytic hydrogenation of [10](9,10)anthracenopha-4,6-diyne.[20]

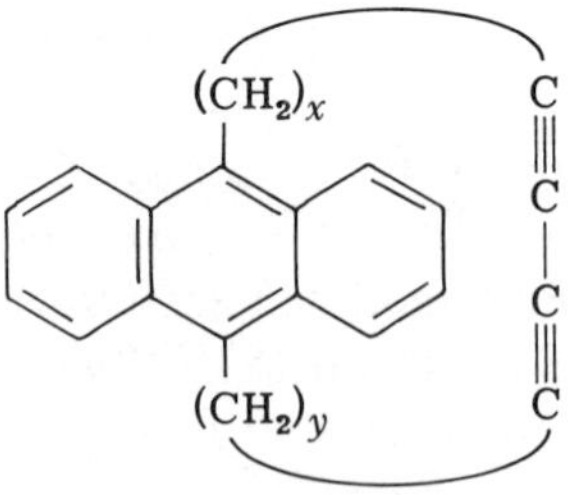

$(CH_2)_x$ C≡C–C≡C $(CH_2)_y$

6 $x = y = 3$
7 $x = y = 4$

B. Intermolecular Coupling Methods

In 1978 Otsubo and Misumi[21] reported the preparation of [8]-, [9]-, [10]-, [11]-, [12]-, and [14]paracyclophanes by the intermolecular coupling of dihalides and dithiols. Subsequent oxidation of the disulfides to disulfones followed by pyrolysis yielded the desired compounds (Eq. 4) in 30 to 46% overall yield:

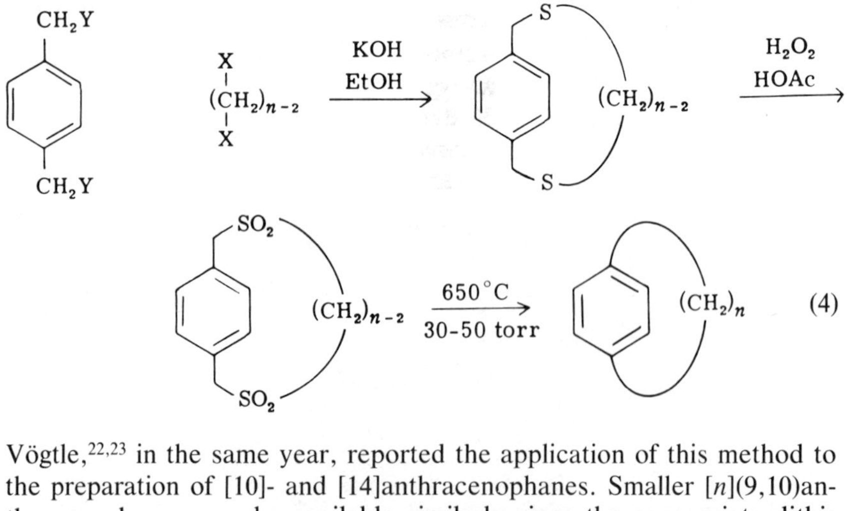

CH_2Y / CH_2Y + X–$(CH_2)_{n-2}$–X $\xrightarrow{\text{KOH, EtOH}}$ S, S, $(CH_2)_{n-2}$ $\xrightarrow{H_2O_2,\ \text{HOAc}}$

SO_2, SO_2, $(CH_2)_{n-2}$ $\xrightarrow[\text{30-50 torr}]{650^\circ\text{C}}$ $(CH_2)_n$ (4)

Vögtle,[22,23] in the same year, reported the application of this method to the preparation of [10]- and [14]anthracenophanes. Smaller [*n*](9,10)anthracenophanes may be available similarly since the appropriate dithia precursors for [8]-, [7]-, and [6](9,10)anthracenophanes have been made.[24] Partial reduction of a dithia[3.3]paracyclophane prepared in this fashion afforded a 2,9-dithia[10]paracyclophane.[25] [8](3,7)Tropolonophane **(8),** a potential precursor to a tropylium cyclophane, was prepared similarly via the appropriate dithia compound, which was oxidized and subsequently

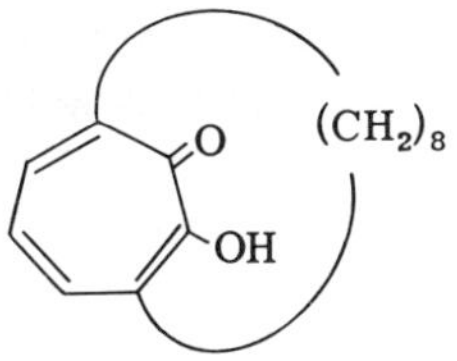

8

pyrolized.[26] The use of this method in the preparation of functionalized dithia[*n*](2,5)pyridinophanes ($n = 8, 10, 12$)[27,28] as catalytically potent pyridoxal model compounds[29,30] has been reported.

[2.2]Cyclophanes formed by "cross-breeding" cycloaddition (Hofmann elimination) have served as precursors of [*n*]paracyclophanes. The first reported synthesis of [8]paracyclophane, by Cram and Knox,[31,32] was accomplished by the opening of the furan ring of [2.2]paracyclo-(2,5)furanophane (Eq. 5):

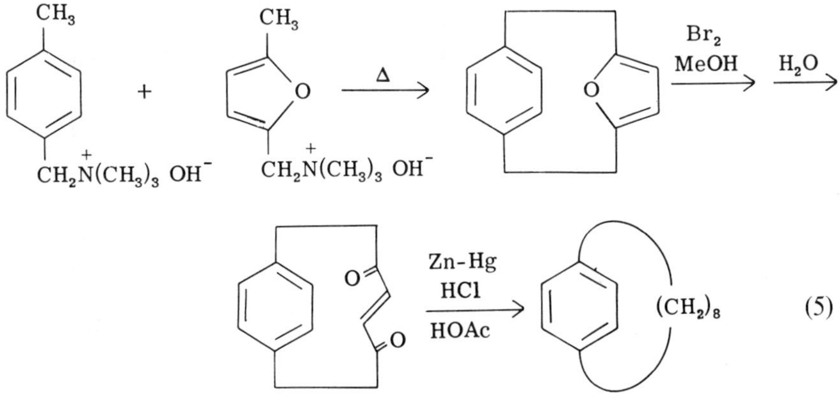

(5)

The synthesis of [8](1,4)naphthalenophane was accomplished in an analogous fashion,[12] as was the preparation of 3,6-diketo[8](9,10)anthracenophane, although the latter was not fully characterized.[13]

Nakazaki and co-workers[33,34] prepared the unusual doubly bridged and chiral [8][8]- and [8][10]paracyclophanes as described in Eq. (6):

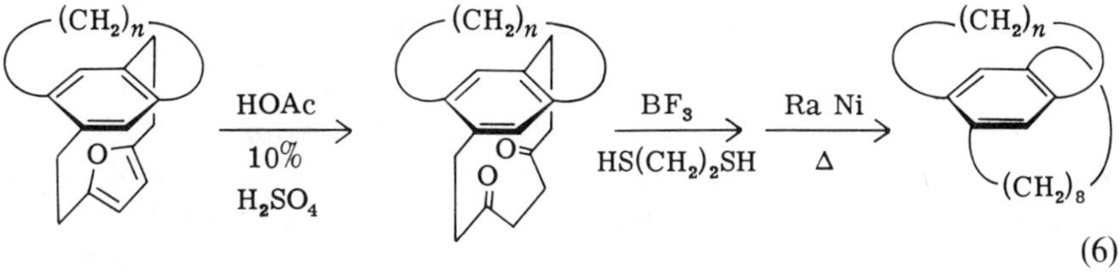

(6)

Optical resolution of precursors for this synthesis allowed the isolation of the individual optical isomers (+)-[8][8]- and (−)-[8][10]paracyclophanes, which were found to have specific rotations of +5.4° and −6.3°, respectively. Absolute configurations were established by relation to compounds of known configuration and were confirmed by analysis of CD spectra using the method developed by Schlögl *et al.*[35] for optically active cyclophanes.

Methods involving amide or ester formation continue to be useful in the preparation of [*n*]paracyclophanes with heteroatoms in the bridge. Oki and Sakamoto[36] reported that reaction of 1,4-phenylenediacetyl chloride with a series of aliphatic diamines under high-dilution conditions provides diamido[*n*]paracyclophanes in low yield (Eq. 7):

CH_2COCl, R, R, CH_2COCl + $H_2N{-}(CH_2)_n{-}NH_2$ ⟶ CH_2, C=O, NH, R, $(CH_2)_n$, R, CH_2, NH, C=O (7)

R = H; n = 6-10
R = Me; n = 5-10
R = OMe; n = 5-10, 12
R = OEt; n = 12

Similarly, these authors used the reaction of aliphatic diols with 1,4-phenylenediacetyl chloride to prepare several of the analogous diesters.[37–39]

C. Intermolecular Coupling with Either Subsequent or Concomitant Aromatization

In a series of papers Tsuji *et al.*[40–42] reported that dispiro[2.2.2.2]deca-4,9-diene **(9)** undergoes thermal cycloaddition with variously substituted butadienes to give [8]- and [10]paracyclophane derivatives (Eq. 8) with yields dependent on the butadiene structure:

$$R^1CH{=}CR^2{-}CR^2{=}CHR^3 + \mathbf{9} \xrightarrow{150\,^\circ C}$$

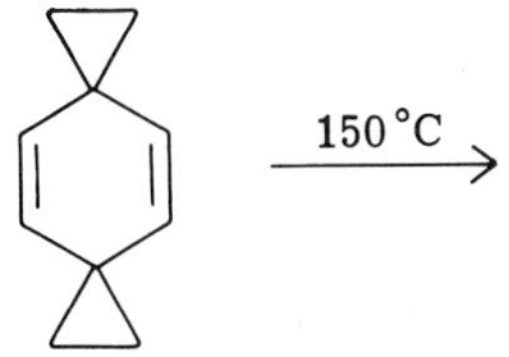

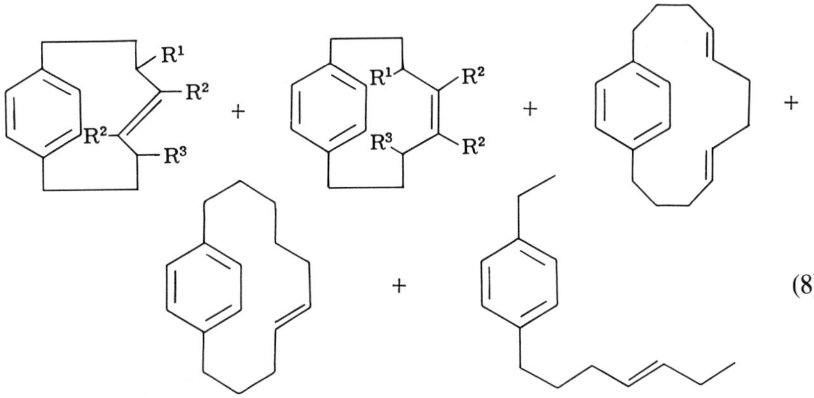

(8)

The production of 1 : 2 cycloadducts and both *cis*- and *trans*-[8]paracyclophenes supports the proposed diradical mechanism (Eq. 9),

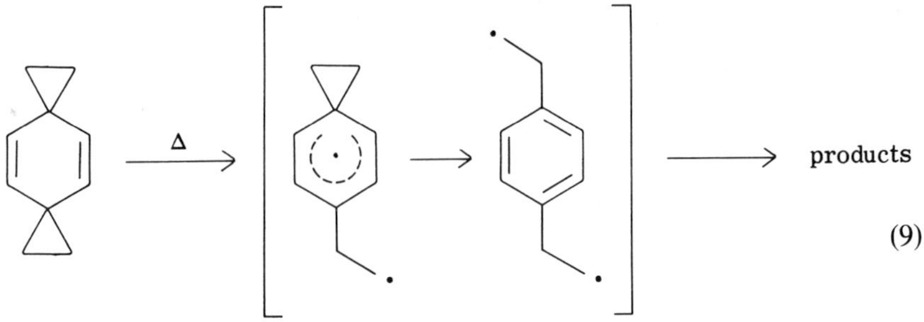

products

(9)

as do CIDNP studies of the reaction.[43–47] An extension of this approach to the preparation of bridged hydroquinones has also been described.[48] Attempts to add **9** to isolated double bonds as an entry into the [6]paracyclophane series failed.[49] Substituted [10]paracyclophadienes have been obtained by thermal cycloaddition of tetracyanoquinodimethane with substituted 1,2-dicyclopropylethylenes[50] (Eq. 10), tetracyclopropylethylene, and 1,2-dicyclopropylstilbenes.[51]

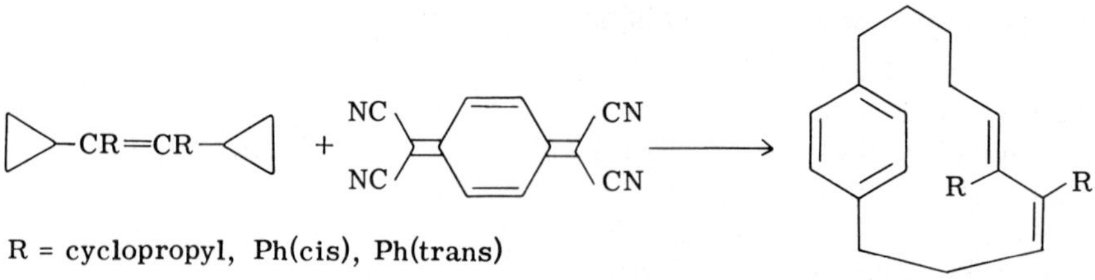

$\triangleright$—CR=CR—$\triangleleft$ + NC, NC, CN, CN → R, R

R = cyclopropyl, Ph(cis), Ph(trans)

(10)

Diels–Alder addition of cyclic dienes with maleic anhydride leads to adducts that can be dehydrogenated to [*n*]paracyclophanes[9] (Eq. 11):

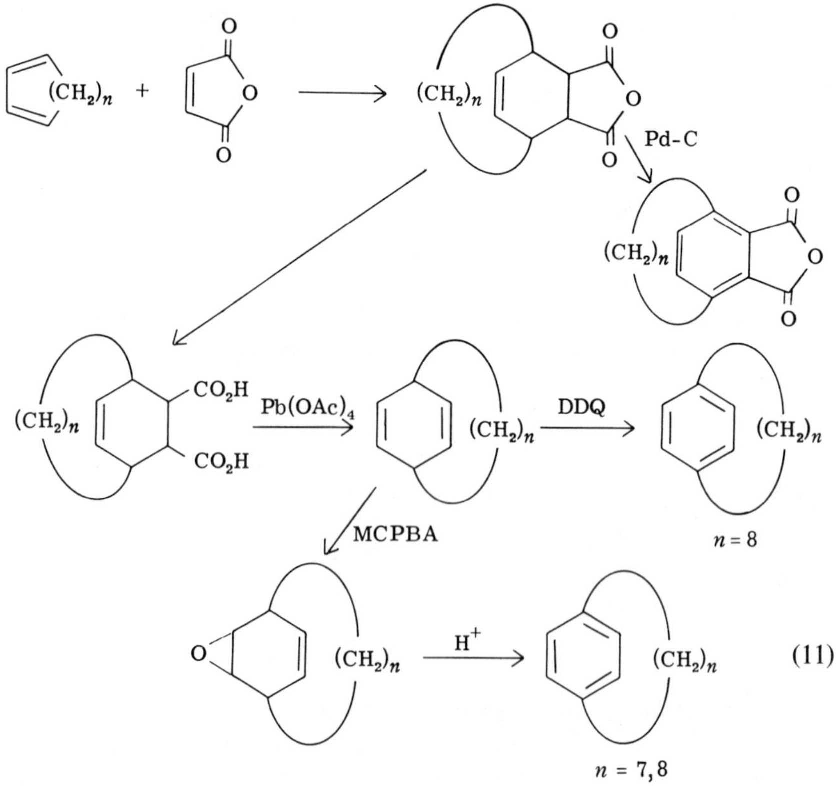

$(CH_2)_n$ + O, O, O → $(CH_2)_n$ O, O, O

Pd-C

$(CH_2)_n$ O, O, O

$(CH_2)_n$ CO_2H, CO_2H $\xrightarrow{Pb(OAc)_4}$ $(CH_2)_n$ $\xrightarrow{DDQ}$ $(CH_2)_n$

$n = 8$

MCPBA

O $(CH_2)_n$ $\xrightarrow{H^+}$ $(CH_2)_n$ (11)

$n = 7,8$

An extension of the method to phthalic anhydride to form a [9]paracyclophane derivative was successful, although the dehydrogenation step was more difficult.[9,52] An application of this general approach to the synthesis of [7]- and [8]paracyclophanes afforded the desired compounds in ~69% overall yield.[53] Ring-substituted [7]- and [8]paracyclophanes (trifluoromethyl substituents) were also reported.

A novel synthesis of an [8]paracyclophane was accomplished by reaction of [8](2,5)thiophenophane with dicyanoacetylene. The reaction may involve the formation of a paddlane with subsequent loss of sulfur (Eq. 12), although this hypothesis has not been explicitly tested.[13]

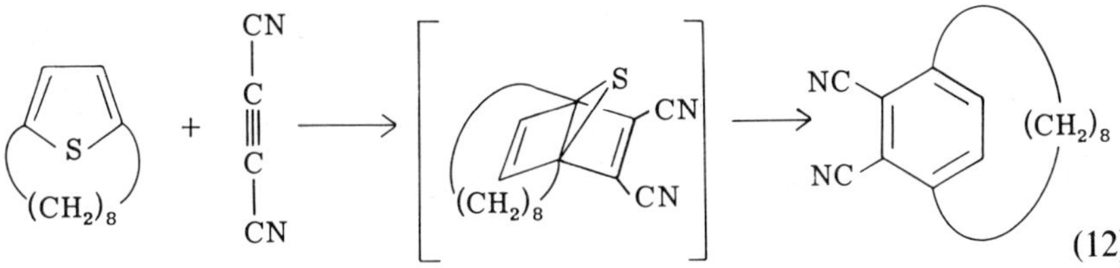

(12)

D. Intramolecular Coupling with Concomitant Aromatization

Jones and co-workers have prepared the smallest [*n*]paracyclophane so far isolated, [6]paracyclophane,[54] as well as [7]paracyclophane,[55] via spirodienones **10,** as in Eq. (13):

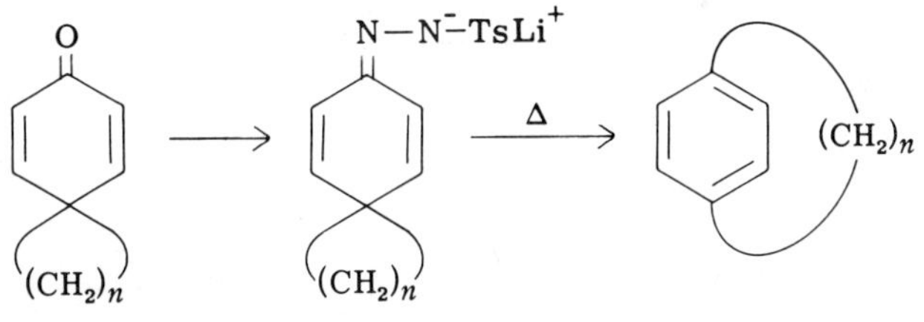

(13)

10 $n = 6, 7$

$n = 6, 7$

A Dewar benzene has been postulated as an intermediate from rearrangement of the initially formed carbene, although a mechanistic study has not yet been reported.

An attempt to prepare [*n*]anthracenophane derivatives by Lewis acid-catalyzed cyclization of spirocyclopropyl alcohols **11** was unsuccessful despite expected exothermicity due to simultaneous relief of cyclopropyl ring strain and aromatization (Eq. 14):

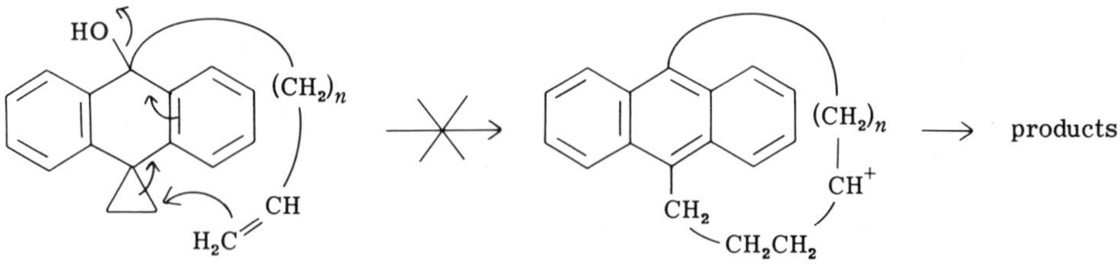

11 (14)

Although indirect evidence suggested cyclization as a minor pathway, the principal products were 9,10-disubstituted anthracenes, derived from opening of the cyclopropyl ring by fortuitous nucleophiles.[56]

E. Ring Contraction of Higher [*n*]Paracyclophanes

The Wolff rearrangement is ideally suited to attempts to ring-contract paracyclophanes, because loss of N_2 makes the contraction step exothermic. Allinger and co-workers[6,57–60] applied this approach successfully to the preparation of [7]- and [8]paracyclophane derivatives (Eq. 15):

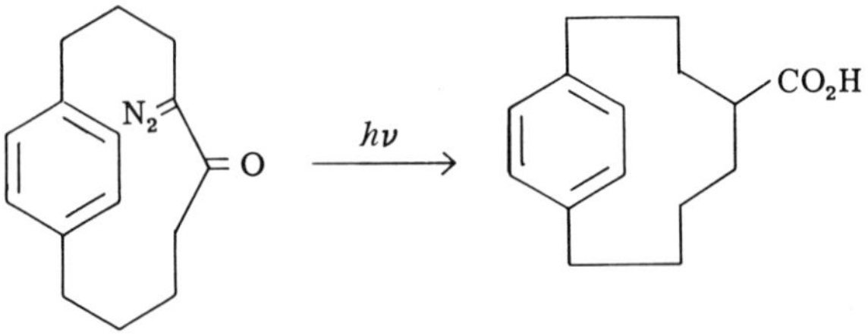

$\xrightarrow{h\nu}$ (15)

An extension to the corresponding [6]paracyclophane failed because it was not possible to isolate the required diazo ketone.[60]

F. Aromatization of Bridged Precursors

Although [6]paracyclophane is the smallest known member of this series, the synthesis of smaller [*n*]paracyclophanes was attempted via aromatization of the corresponding Dewar benzenes. Silver-catalyzed valence isomerization of 1,1′-tri-, 1,1′-tetra-, and 1,1′-pentamethylenebicyclopropenyls gave the Dewar benzenes **12** (Eq. 16):

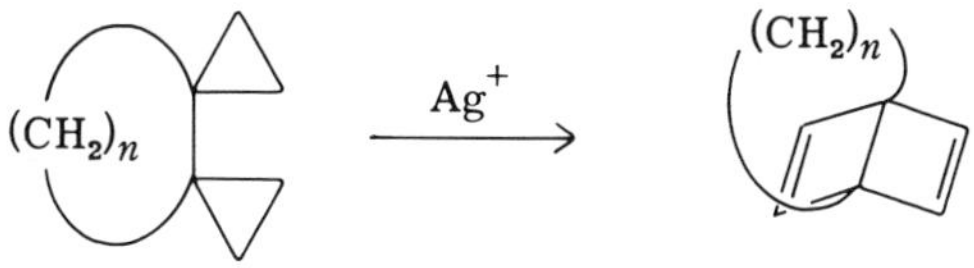

(16)

12

Attempts at aromatization by flow pyrolysis, at 300°C, yielded the products in Eq. (17) due to the greater stability of the Dewar isomer compared with the [n]paracyclophane for these smaller values of n[61–63]:

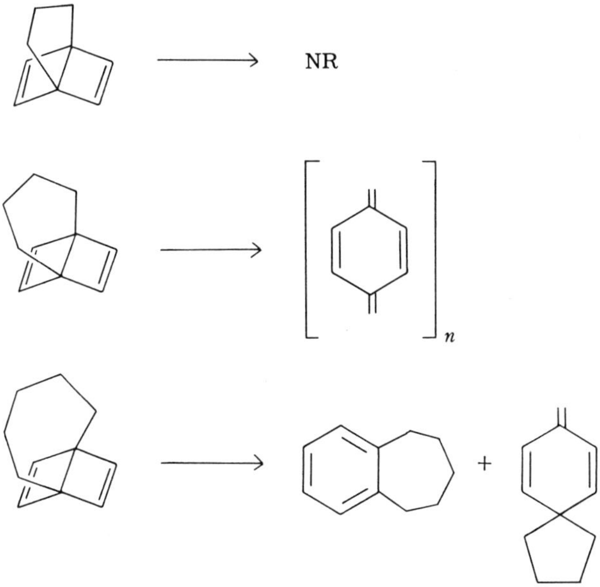

(17)

13

In contrast, [6]paracyclophane is formed in good yield by thermolysis of the corresponding Dewar isomer ($E_A = 19.9 \pm 0.9$ kcal/mol), which is the sole product of photolysis of [6]paracyclophane.[64]

Bickelhaupt *et al.*[63] postulated the intermediacy of [5]paracyclophane in the thermal rearrangement of its Dewar isomer to form **13.** Thermal fragmentation of [5]paracyclophane to a diradical that closes to **13** was suggested as a likely pathway. The possibility of reversibility in these mechanistic steps led these workers to a novel synthetic approach that afforded both [7]- and [8]paracyclophanes in low yield[65] (illustrated for [8]paracyclophane in Eq. 18).

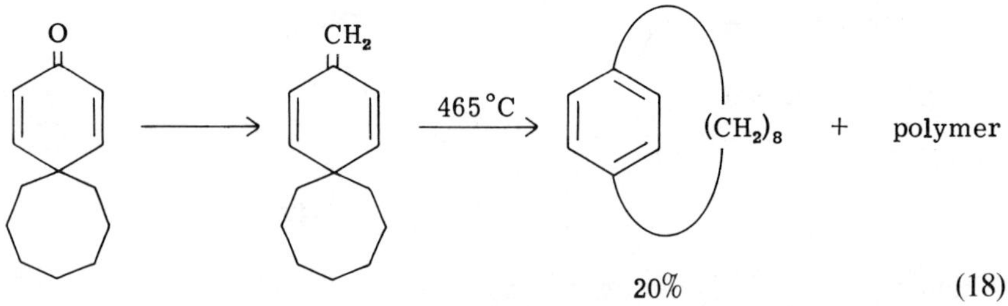

O

CH_2

465 °C

$(CH_2)_8$ + polymer

20%

(18)

Synthesis of the first reported tetrazinophane, [9](3,6)-*s*-tetrazinophane **(14),** was accomplished by the route shown in Eq. (19).[66]

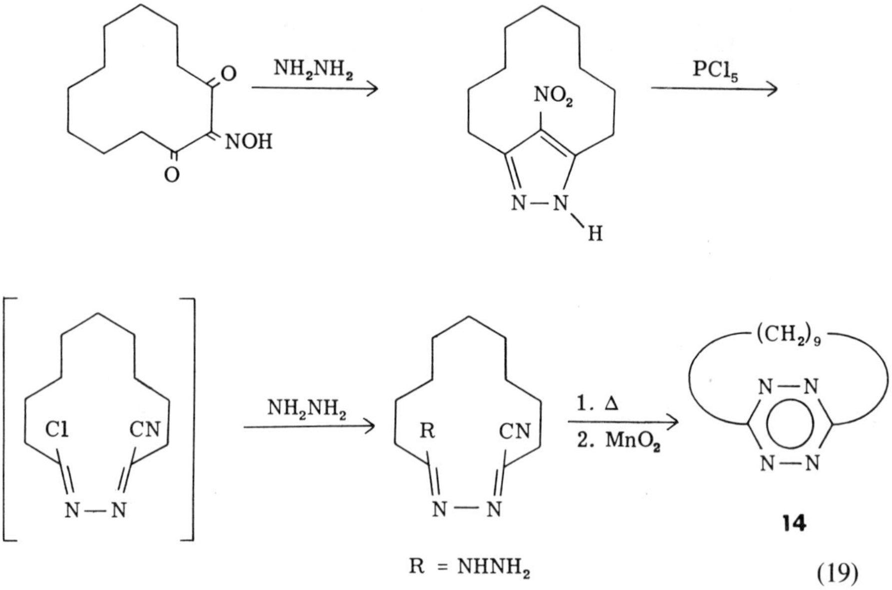

O

NH_2NH_2

NOH

O

NO_2

N—N

H

PCl_5

Cl CN

N—N

NH_2NH_2

R CN

N—N

R = $NHNH_2$

1. Δ

2. MnO_2

$(CH_2)_9$

N—N

N—N

14

(19)

G. Homologation of Smaller [*n*]Paracyclophanes

Wilberg and O'Donnell[12] prepared [9]- and [10](1,4)naphthalenophanes by diazomethane homologation of 3,6-diketo[8](1,4)naphthalenophane followed by Wolff–Kishner reduction. Many additional examples of homologation appear in the earlier literature covered by Smith.[1]

H. Conclusion

It is of interest to briefly examine which of the [*n*]cyclophanes incorporating the common aromatic nuclei and saturated all-carbon bridges have been prepared and which have not. All members of the [*n*]paracyclophane series from $n = 6$ to $n = 16$ with the exception of $n = 15$ are known. Since $n = 6$ is almost certainly the smallest isolable member, this series is essentially complete. In contrast, only the [*n*](1,4)naphthalenophanes with $n = 8$, 9, 10, and 14 and the [*n*](9,10)anthracenophanes with $n = 10$ and 14 are known. Members of the latter two series down to $n = 6$ are likely to be available by methods used for [*n*]paracyclophanes. There is even some likelihood that a [5](9,10)anthracenophane will be isolated since the central ring of anthracene appears to be more flexible than benzene. (For example, Boekelheide and co-workers[67] demonstrated that the meta-bridged benzene ring in **15** undergoes conformational flipping about 45 times as rapidly as the meta-bridged ring in **16.**) Although

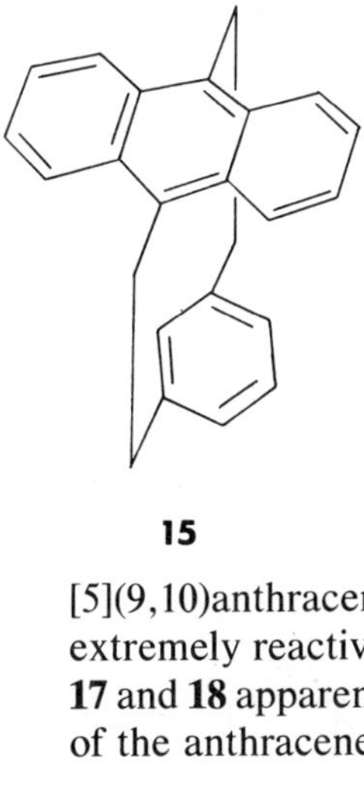

15 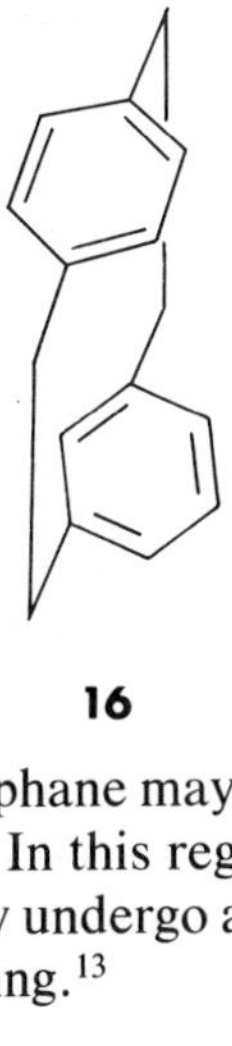 **16** 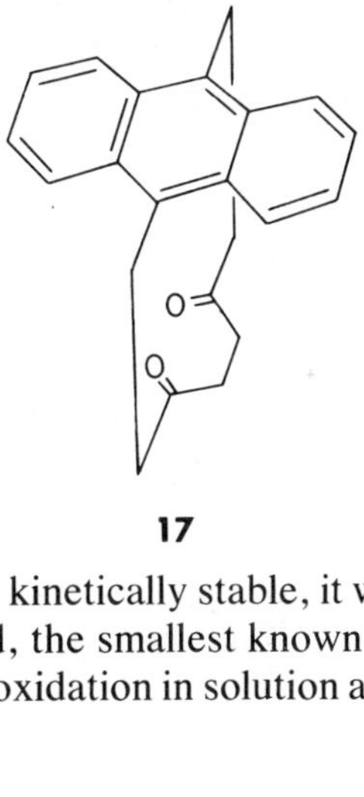 **17** 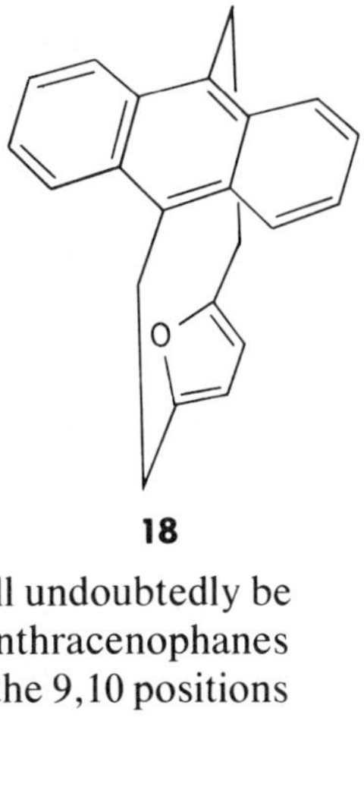 **18**

[5](9,10)anthracenophane may be kinetically stable, it will undoubtedly be extremely reactive. In this regard, the smallest known anthracenophanes **17** and **18** apparently undergo air oxidation in solution at the 9,10 positions of the anthracene ring.[13]

III. SYNTHESIS OF [*n*]METACYCLOPHANES AND OTHER 1,3-BRIDGED AROMATICS

As with [*n*]paracyclophanes, the common ring-forming reactions have proved to be useful in the service of [*n*]metacyclophane synthesis.[1,68]

Again, emphasis is placed on those methods that have continued to be valuable during the past 20 years and on those new approaches developed specifically to augment earlier methods. The synthesis of heterocyclophanes has been reviewed by Fujita and Nozaki.[69]

A. Intramolecular Ring Closure Methods

Acyloin cyclization provided an early entry into the [*n*](2,5)-thiophenophane series[70,71] but has otherwise received little attention. Friedel–Crafts acylation, in contrast, has been applied to a variety of synthetic tasks despite the drawback of more favorable ortho and para cyclization in certain systems. This route is especially useful in the benzene series when the para position is blocked and the chain length is too long for facile ortho cyclization. It has also served in the preparation of [*n*]metacyclophanes with $n > 7$. [*n*](2,5)Thiophenophanes (n = 8–12) have been prepared by Friedel–Crafts acylation, although competing cyclization at the 3 position (for lower values of n) and dimer formation limit the general utility of this approach.[71]

B. Intermolecular Coupling Methods

In 1975 Kumada and co-workers[72] reported the coupling of Grignard reagents with aryl dihalides (catalyzed by a nickel–phosphine complex), affording a variety of [*n*]metacyclophanes in yields from 3 to 33% (Eq. 20):

$$\text{Cl}-\text{C}_5\text{H}_3\text{Y}-\text{Cl} + \text{XMg}-(\text{CH}_2)_n-\text{MgX} \longrightarrow \text{C}_5\text{H}_3\text{Y}(\text{CH}_2)_n \quad (20)$$

Y = N, n = 6-10, 12
Y = CH, n = 8-10, 12

The method was also applied to the preparation of the cyclophane natural product muscopyridine (**19,** previously prepared via a 10-step route[73]) as well as the cyclophane ethers **20** and **21.**

3,6-Diketo[8](2,5)thiophenophane has been prepared by hydrolysis of [2.2](2,5)furano(2,5)thiophenophane formed by the previously discussed (Section II,B) cross-breeding cycloaddition route.[74]

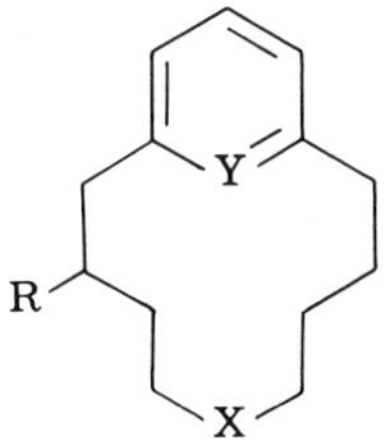

19 Y = N; R = CH_3; X = CH_2CH_2
20 Y = CH; R = H; X = O
21 Y = N; R = H; X = O

Intermolecular coupling of dihalides with dithiols affords dithiametacyclophanes.[21] Mitchell and Boekelheide[75] reported the preparation of 2,6-dithia[7]metacyclophane in this fashion in 43% yield, and Vögtle[76] prepared 2,7-dithia[8]metacyclophane **(22)** in low yield by this route (Eq. 21):

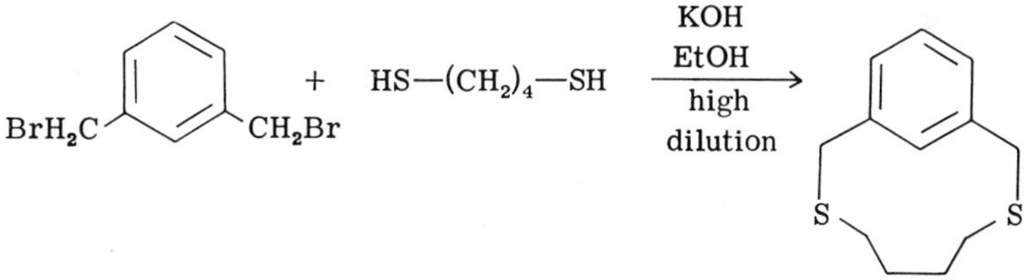

$$BrH_2C\text{-}C_6H_4\text{-}CH_2Br + HS\text{—}(CH_2)_4\text{—}SH \xrightarrow[\text{high dilution}]{\text{KOH, EtOH}} \mathbf{22} \quad (21)$$

Vögtle later extended this work to a series of substituted compounds (**23**[77,78] and **24**[79]), generally formed in good yield. Extensions of this

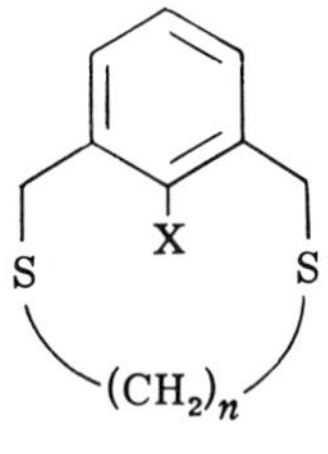

23 X = H; n = 3-9
X = F; n = 3-9
X = Cl; n = 4-9
X = Br; n = 4-9
24 X = CH_3, OCH_3; n = 3-10

method to the preparation of metacyclophane polyethers have been made,[80] but these compounds are treated elsewhere in this volume (see Chapter 12).

A number of examples of cyclization by amine, amide, or ester formation appeared in the earlier literature.[1] [*n*]Metacyclophanes **25** and **26** were prepared in generally low yield[81] by reaction of the appropriate aliphatic diol and aryl diacid (for **25**) or the appropriate *N*,*N*-ditosyldiaminoalkanes and 1,3-di(bromomethyl)benzenes.

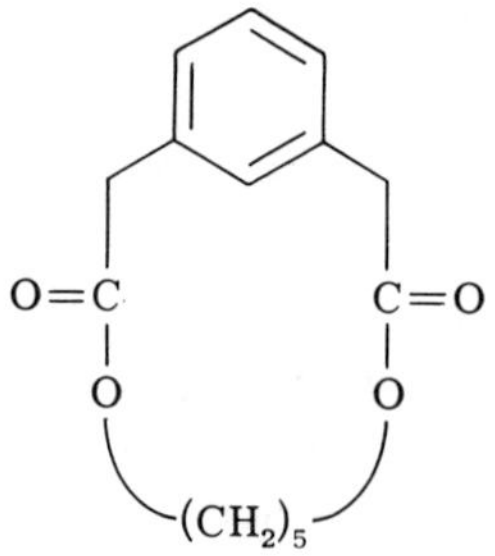

25

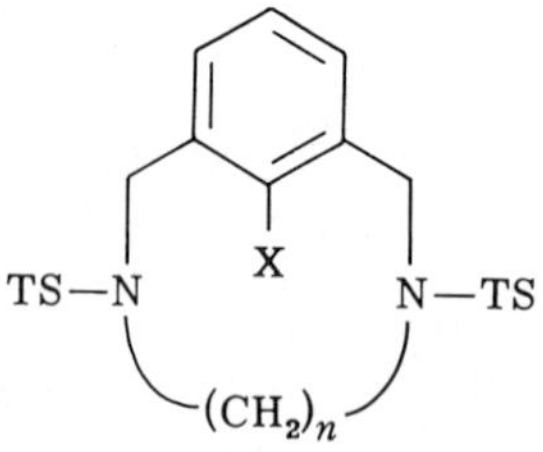

26 X = H, F, Cl; $n = 4$
X = F; $n = 5$
X = H, F, Cl, Br; $n = 6$

C. Formation of the Aryl Ring via Paal–Knorr Type Cyclization

The well-known Paal–Knorr cyclization of diketones was applied by Nozaki[82,83] to the preparation of [8](2,5)furanophane **(27)**, [8](2,5)thiophenophane **(28)**, [8](2,5)pyrrolophane **(29)**, and various N-substituted derivatives of **29** (Eq. 22):

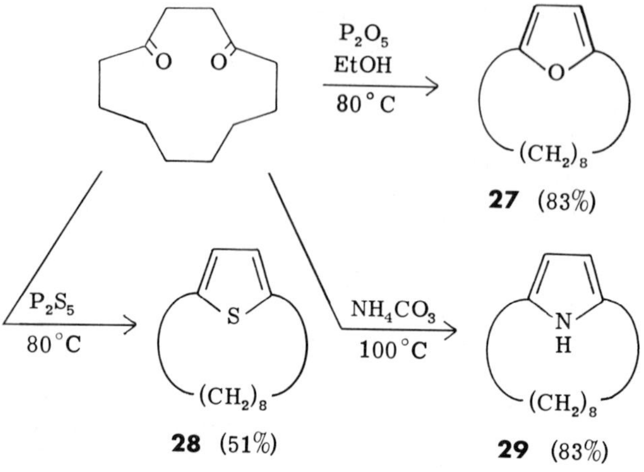

P_2O_5, EtOH, 80 °C → **27** (83%)

P_2S_5, 80°C → **28** (51%)

NH_4CO_3, 100°C → **29** (83%)

(22)

Similarly, Fujita and co-workers[84,85] reported the analogous preparation of 5-methyl[6](2,4)thiophenophane **(30)** and several 5-methyl[8](2,4)pyrrolophane derivatives (**31,** Eq. 23), 5-methyl[*n*]thiophenophanes with $n =$ 6, 7, and 9 (51–70% yields), and 5-methyl[*n*]pyrrolophanes with $n = 7$ and 9 as well as *N*-aryl-substituted pyrrolophanes with $n = 6$ and 9 by this method.

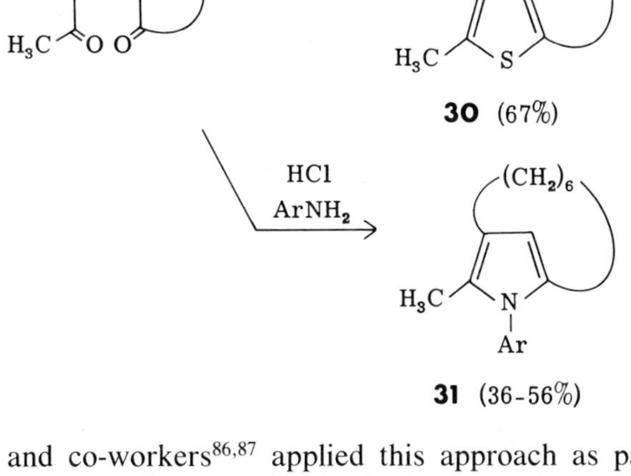

$(CH_2)_6$ H_3C O O $\xrightarrow{P_2S_5}$ $(CH_2)_6$ H_3C S

30 (67%)

$\xrightarrow[ArNH_2]{HCl}$ $(CH_2)_6$ H_3C N Ar (23)

31 (36-56%)

Wasserman and co-workers[86,87] applied this approach as part of their synthesis of the naturally occurring 2,4-bridged pyrrole metacycloprodigiosin **(32).**

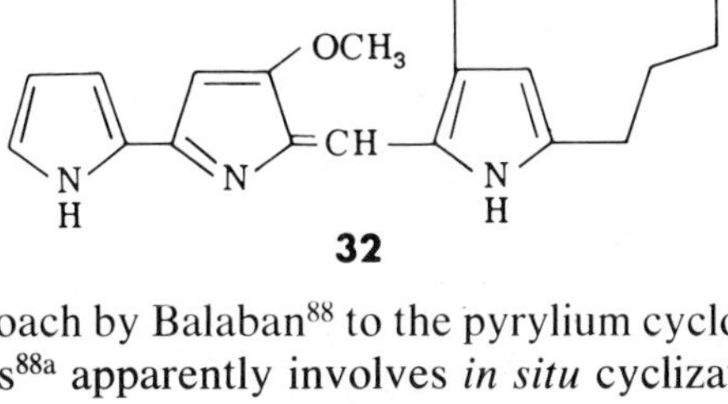

OCH_3 N H N CH N H

32

A related approach by Balaban[88] to the pyrylium cyclophane salt **33** and other compounds[88a] apparently involves *in situ* cyclization of a diketone (Eq. 24):

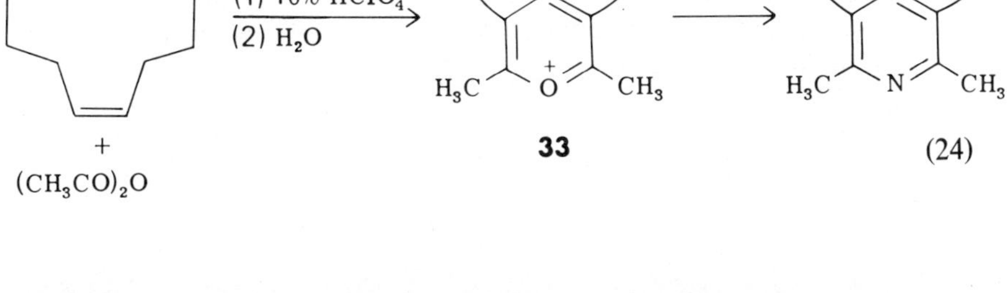

\+ $(CH_3CO)_2O$ $\xrightarrow[(2)\ H_2O]{(1)\ 70\%\ HClO_4}$ $(CH_2)_9$ H_3C O^+ CH_3 $\longrightarrow$ $(CH_2)_9$ H_3C N CH_3

33 (24)

Georgi and Rétey[89] prepared **34** by this route and converted it to the corresponding bridged pyridine in 1% overall yield.

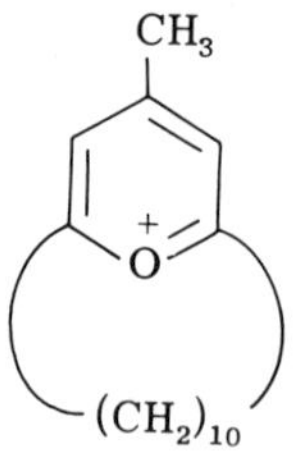

CH_3

$(CH_2)_{10}$

34

A series of related cyclization methods has supplied another entry to the 2,4-bridged systems. Bradamante and co-workers[90] prepared [9](2,4)thiophenophane **(36)** in low yield, as shown in Eq. (25):

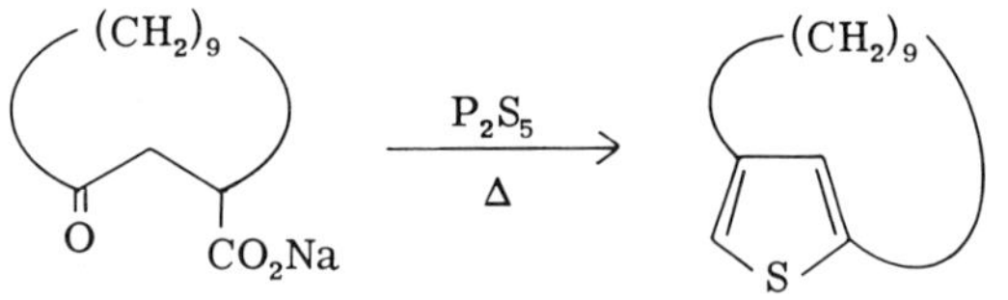

$(CH_2)_9$ O CO_2Na $\xrightarrow[\Delta]{P_2S_5}$ $(CH_2)_9$ S (25)

35 **36**

Formation of the α,β-unsaturated γ-lactone of the parent acid of **35** followed by treatment with diisobutylaluminum hydride provided [9](2,4)furanophane, also in low yield. Isoxazolophane **(37)** and pyrazolophane **(38)** as well as their corresponding hydrochlorides were also reported. Hirano[85] reported the preparation of **38** and other members of

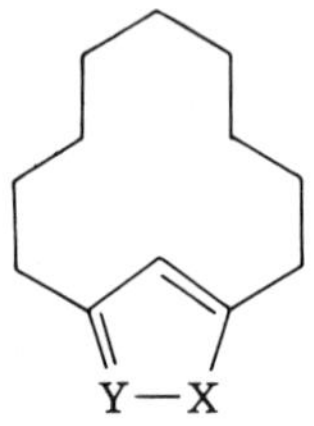

Y—X

37 Y = N; X = O

38 Y = N; X = NH

this series by treatment of α,β-unsaturated ketones with hydrazine or phenylhydrazine (Eq. 26) as done previously by Fujita[91] in the preparation of [7](3,5)pyrazolophane:

$(CH_2)_n$ O $\xrightarrow{RNHNH_2}$ $(CH_2)_n$ N—N R $\xrightarrow{S_8}$ $(CH_2)_n$ N—N R (26)

n = 6, 7, 9
R = H, Ph

[9](3,5)Pyrazolophane, 11-methyl[9](2,4)furanophane,[92] [9](2,5)oxazolophane, and [9](2,4)imidazolophane[93] have also been prepared.

Related methods involving cyclization with subsequent aromatization have been applied by Bradamante, Marchesini, and co-workers[94] in the synthesis of [9](2,4)pyridinophane (Eq. 27) and [9]metacyclophane (Eq. 28).

$(CH_2)_9$ O $CHCO_2Et$ CN $\xrightarrow[H_2O]{OH^-}$ $\xrightarrow[180\text{-}200°C]{Cu}$ $(CH_2)_9$ O CH_2CN $\xrightarrow{LiAlH_4}$

$(CH_2)_9$ OH $(CH_2)_2NH_2$ $\xrightarrow{\text{Jones oxidation}}$ $(CH_2)_9$ N $\xrightarrow[\text{reflux}]{\text{Pd-C, xylene, nitrobenzene}}$ $(CH_2)_9$ N (27)

$(CH_2)_9$ O $CHCOCH_3$ CO_2Et + $(CH_2)_9$ HO O CO_2Et $\xrightarrow{H^+}$ $(CH_2)_9$ O $\xrightarrow{LiAlH_4}$ $\xrightarrow{PBr_3}$

(mixture from previous step)

$(CH_2)_9$ $\xrightarrow{DDQ}$ $(CH_2)_9$ (28)

(mixture of four double-bond isomers)

D. Aromatization of Bridged Precursors

In a series of papers Parham and co-workers[95–97] described a novel approach to the synthesis of [*n*](1,3)naphthalenophanes **39** in which bridged indenes are converted to the corresponding dihalocyclopropanes, which are treated with base to form the desired cyclophanes (Eq. 29):

(CH$_2$)$_n$ → X X (CH$_2$)$_n$ —HX→ X (CH$_2$)$_n$ (29)

39 n = 6, 8, 10

The advantageous energetics in releasing cyclopropyl ring strain as well as in forming the aromatic ring in the final step probably contributed to the good yields (73–89% for final step) and generality of this approach for [*n*](1,3)naphthalenophanes down to n = 6. Indirect evidence suggests the formation of **39** (n = 5)[98], which was later prepared by Grice and Reese.[98a,98b] Loss of HX could not be effected for n = 4. The halogen-substituted cyclophanes **39** were reduced via the corresponding Grignard or lithium compound to the all-carbon members of this series. These workers extended this approach to the formation of bridged quinolines **40**

N (CH$_2$)$_n$ → N Cl Cl (CH$_2$)$_n$ → N Cl (CH$_2$)$_n$ —NH$_2$NH$_2$, Pd-C→ N (CH$_2$)$_n$ (30)

n = 6, 8, 10 **40** n = 8, 10

(Eq. 30) but encountered difficulty in the reduction of the chlorine-substituted cyclophane to the parent system, where $n = 6$; there were indications that the nitrogen-containing ring had been reduced.[99,100] Although Grignard reagents are not generally useful in quinoline systems, these workers succeeded in using the Grignard reagents formed from halogen-substituted bridged quinolines in effecting the required reduction.[101] Application of this approach to benzoquinolines including **41,** found to be a

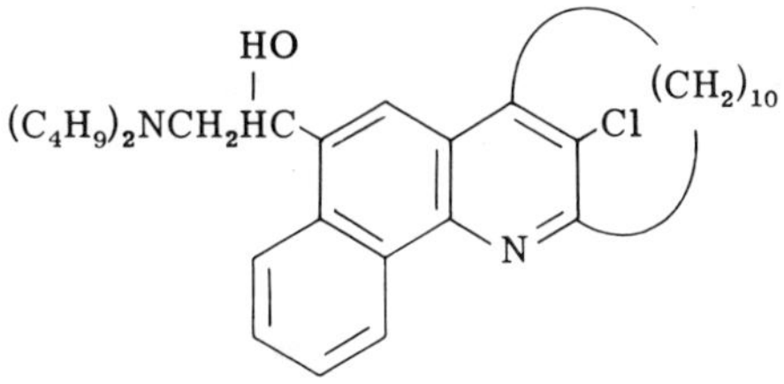

HO
$(C_4H_9)_2NCH_2HC$
$(CH_2)_{10}$
Cl
N

41

curative agent in murine malaria, was also reported. A Parham-type approach was used by Fujita[102] to prepare [7]metacyclophane (Eq. 31):

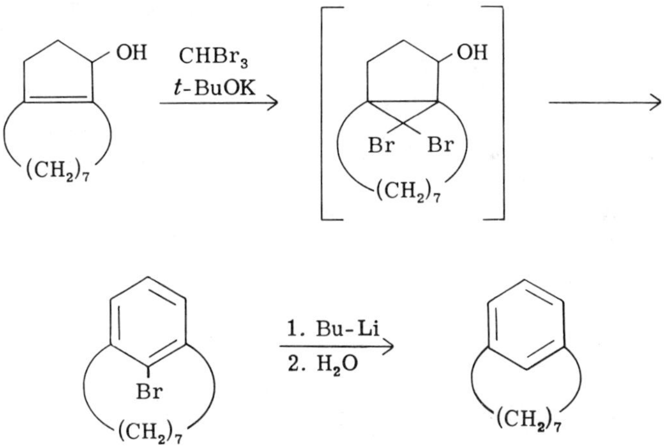

OH
$CHBr_3$
t-BuOK
$(CH_2)_7$
OH
Br Br
$(CH_2)_7$

Br
$(CH_2)_7$
1. Bu-Li
2. H_2O
$(CH_2)_7$
(31)

A successful extension to [6]- and [10]metacyclophanes by Hirano, Fujita, and co-workers[103] suggests the generality of this route in the benzene series, making this an attractive approach (especially because starting materials are readily available and the halogen-substituted cyclophanes can be easily derivatized via lithiation). Bickelhaupt and co-workers[104,105] reported the synthesis of [5]metacyclophane, the smallest known member of the series, in an analogous fashion (Eq. 32).

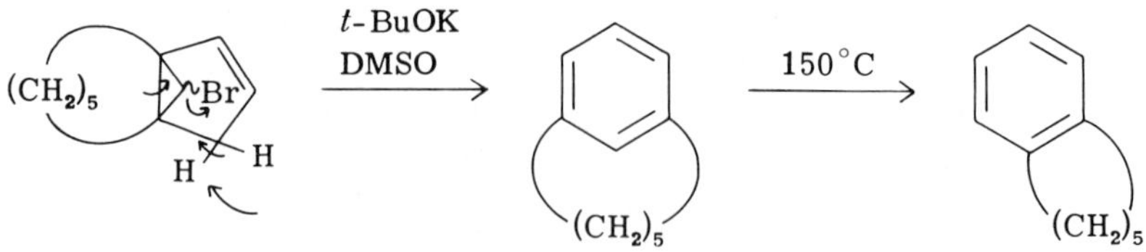

$$\xrightarrow{t\text{-BuOK, DMSO}} \xrightarrow{150^\circ C} \quad (32)$$

[5]Metacyclophane undergoes facile thermal rearrangement, perhaps via a bridged Dewar benzene intermediate. The possible intermediacy of [4]metacyclophane in a route analogous to that shown in Eq. (32), in which the corresponding Dewar isomer (**42,** a *double Bredt olefin*) was

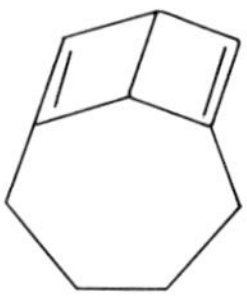

42

isolated, has been suggested.[106] Other attempts to prepare [4]metacyclophane have been unsuccessful.[107] Parham prepared [10](3,5)pyrazolophane[108] as illustrated in Eq. (33) but isolated a mixture, in low yield, when preparing [6](3,5)pyrazolophane.[109]

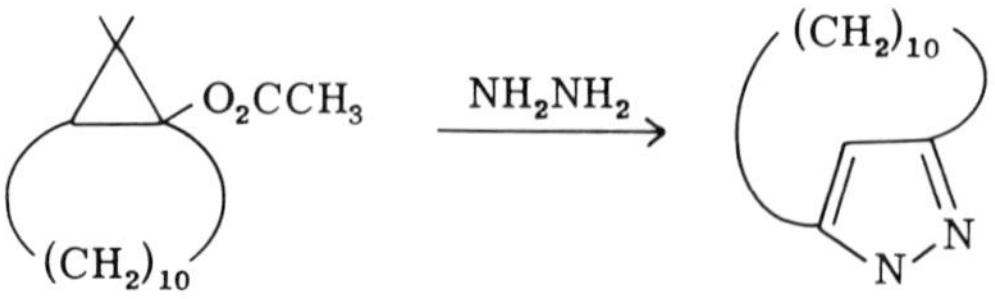

$$\xrightarrow{NH_2NH_2} \quad (33)$$

E. Formation of [*n*]Metacyclophanes by Bridge Rearrangement

Smith[1] described several examples of rearrangement of [*n*]cyclophane bridges generally in the presence of Friedel–Crafts catalysts and typically via carbonium ion intermediates. Such rearrangements have, on occasion, provided entry to the [*n*]metacyclophane series. For example, Märkl and Fuchs[110] observed bridge migration in ortho-bridged benzenes (Eq. 34):

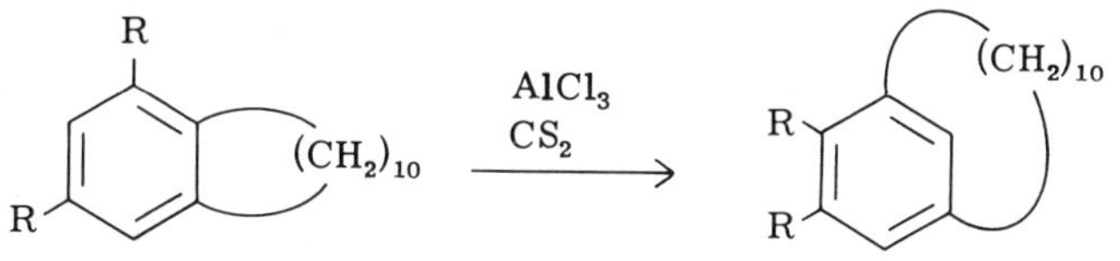

$$\xrightarrow[CS_2]{AlCl_3} \qquad (34)$$

R = H (79%)
R = CH_3 (72%)

[7]Metacyclophane forms upon treatment of [7]paracyclophane with fluorosulfonic and p-toluenesulfonic acids in benzene.[111] An unusual thermal bridge rearrangement yielded the smallest known 1,3-bridged cyclophane (**43,** Eq. 35) as well as its next higher homolog.[112]

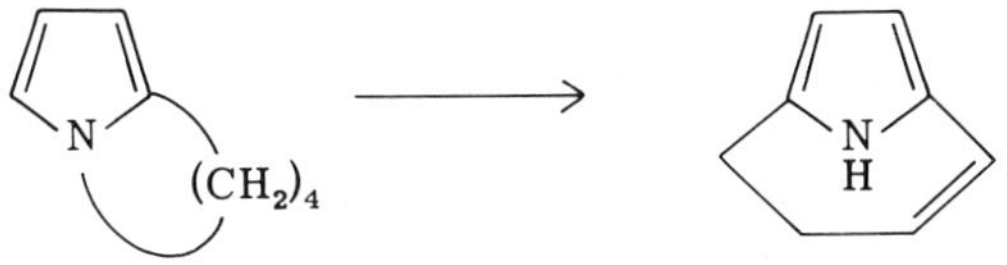

(35)

43

F. Conclusion

In the [n]metacyclophane series the saturated all-carbon bridge compounds with n = 5–10, 12, and 13 have been reported, and n = 5 appears to be the lower limit in this series. The [n](1,3)naphthalenophanes with n = 6, 8, and 10 have been prepared, and although an attempt to prepare the n = 5 compound failed it may yet be obtained through other synthetic approaches. The [n](2,5)thiophenophanes with n = 8–12 have been prepared, whereas only the n = 8 compounds in the corresponding furanophane and n = 5 and 8 compounds in the pyrrolophane series have been reported. The analogous 2,4-bridged pyrroles, furans, and thiophenes have been synthesized for n = 6 and 9, with the exception of [6](2,4)furanophane. The relatively high reactivity of furan and pyrrole may limit the number of attractive routes to these compounds.

IV. SPECTROSCOPY

Spectroscopic data on the [n]cyclophanes have demonstrated the nonplanarity of the aromatic ring for smaller values of n, and yet expectations of a dramatic loss of aromaticity, as evaluated by both chemical and spectroscopic criteria, have not been realized. For example, although initial MNDO calculations of the splittings observed in the photoelectron spectra of [6]- and [7]paracyclophanes[113] suggested bending of the aromatic ring and substantial loss of aromatic character for [6]paracyclophane, reexamination of these results and study of the spectra of [8]paracyclophane and dialkylbenzene analogs[114] showed that the observed splittings were due mainly to the electron-releasing effects of the alkyl chain rather than to the loss of aromatic character. In this regard Wynberg[115] and Allinger and co-workers[116] questioned the rigid planar model of benzene and polynuclear aromatic hydrocarbons on the basis of experimental data and *ab initio* calculations.

Ultraviolet spectra of [*n*]cyclophanes are generally characterized by increasing bathochromic shifts and loss of fine structure as the length of the methylene bridge decreases, an effect ascribed to the bending of the aromatic ring. The epr spectrum of the metastable triplet of [7]paracyclophane has been explained by π-molecular orbital energy changes that occur as ring carbons 1 and 4 are raised from the plane of the aromatic ring.[117] Figure 1 illustrates the trend observed in the uv spectra of [*n*]paracyclophanes for $n = 12$, 10, 9, and 8 ($n = 7$, λ_{max} 216, 245, 283 nm[55]; $n = 6$, λ_{max} 212, 253, 290 nm[54]). The known [*n*](1,4)naphthalenophanes[12] appear to deviate from this trend but the [*n*]metacyclophanes,[102–104] [*n*](1,3)naphthalenophanes,[95,96,98] [*n*](2,4)quinolinophanes,[99] [*n*](2,6)-pyridinophanes,[72,118] [*n*](3,5)pyrazolophanes,[85,92] and [*n*](2,4)heterophanes[83–85,92] all exhibit behavior qualitatively similar to that of the [*n*]paracyclophanes. [9](3,6)-*s*-Tetrazinophane[66] appears exceptional in exhibiting a hypsochromic shift in comparison with its dialkyl analogs.

Ultraviolet absorption spectra have been studied extensively as a measure of benzene ring deformation in the [*n*]paracyclophanes.[6,116] Early (molecular orbital) calculations by Allinger and co-workers,[6] by the method of Pariser and Parr, provided spectra that were fitted to experimental data to determine the strain energy and the angle of ring bending ϕ for $n = 8$, 9, 10, and 12. These workers also predicted the uv spectrum and strain energy (35 kcal/mol) of [7]paracyclophane using a value of $\phi = 25°$ obtained from molecular models. Subsequent X-ray crystallographic analyses of carboxy derivatives of [7]-[60] and [8]paracyclophane[58] showed the values of ϕ to be 17° and 9.1°, respectively, indicating that the calculated values of ϕ and the strain energy were too large. Force field calculations,[116] and more recent molecular mechanics calculations,[118a] yielded values of ϕ in better agreement with experimental data (Table I). Calculated electronic spectra based on the latter values of ϕ were found to be in very good agreement with the experimental data. Analysis of the calculated strain energies indicated that for the smaller values of *n* the major component of the strain energy is due to bending of the aromatic ring,

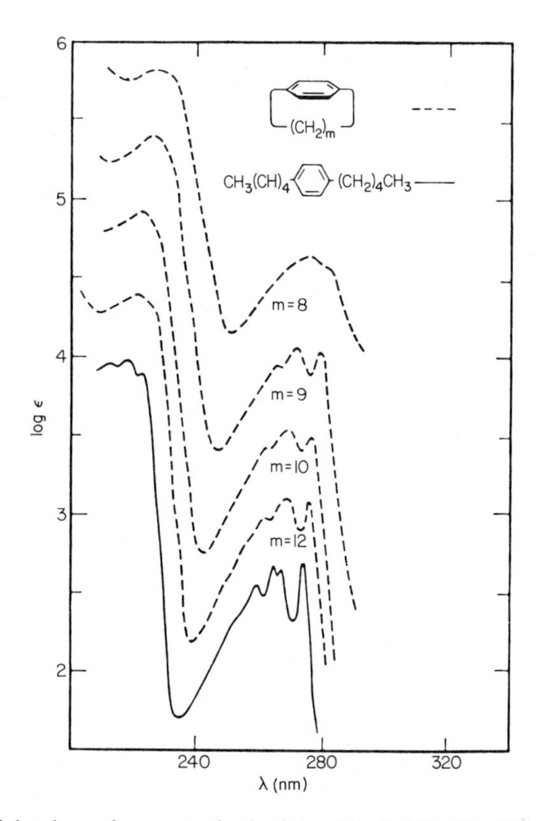

Fig. 1. Ultraviolet absorption spectra in absolute ethanol of [8]- [9]-, [10]-, and [12]paracyclophanes and of open-chain model. The curves are displaced upward successively by 0.5 log unit from the curve immediately below. Reproduced from Cram *et al.*[32] with permission of the publisher.

whereas for $n = 9$ and 10 the strain energy is due mostly to torsional strain caused by poor conformations of the bridge.[116] Stretching of bridge C—C bonds did not appear to be a factor in the strain energy, in agreement with the X-ray crystallographic analyses of 3-carboxy[7]-[60] and 4-carboxy-[8]paracyclophane.[58] These X-ray structures also exhibited C—C—C bond angle compression at the benzylic carbons and C—C—C bond angle expansion of some bridge methylenes (117° being the largest aliphatic bond angle, observed in 3-carboxy[7]paracyclophane[60]).*

In light of the substantial deviation from planarity of the aromatic ring in [8]- and [10]paracyclophanes, Nakazaki and co-workers[33] noted that [8][8]- and [8][10]paracyclophanes would be expected to have twisted boat

* A recently reported crystal structure of a ring-substituted [6]paracyclophane exhibited bond angles as large as 126.5° in the aliphatic bridge.[118b]

TABLE I
Calculated Values of ϕ and Strain Energy of [*n*]Paracyclophanes

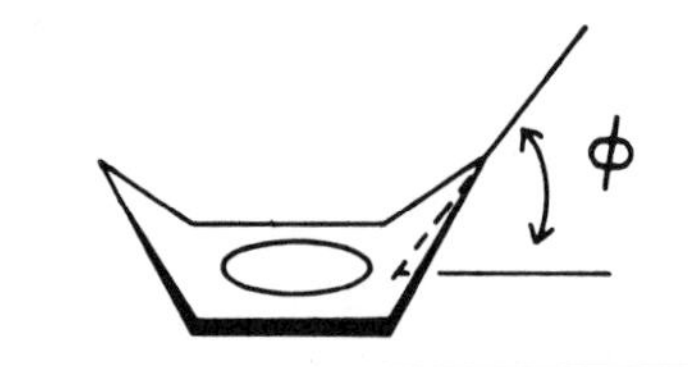

n	ϕ (deg)	Strain energy (kcal/mol)
10	8.4	15.48
9	8.5	12.70
8	12.5	16.83
7	18.2	20.91
6	22.4	28.74
5	26.5	39.05

geometries. Not surprisingly, these compounds exhibit bathochromic shifts and decreased intensities relative to the analogous dimethyl[8]paracyclophane.

The nuclear magnetic resonance spectra of the [*n*]cyclophanes are also treated in Chapters 4 and 6, and the following discussion is intended to highlight the striking features of these spectra. Due to the structural requirements in these systems, some of the bridge protons are forced to lie above the face of the aromatic nucleus and therefore are affected by the diamagnetic ring current.[116] This effect results in large upfield shifts in certain aliphatic resonances in the ^{1}H-nmr spectra of these compounds. The highest-field proton resonances for the [*n*]paracyclophanes are given in Table II. It is interesting that the corresponding ^{1}H-nmr resonances of [10](1,4)naphthalenophane[12] and [10](9,10)anthracenophane[111] occur at δ −0.40 and 0.00 ppm, respectively, farther upfield than for [10]paracyclophane (δ 0.48).[119] These upfield resonances have been used to study ring current theories. For example, in 1957 Waugh and Fessenden[120] studied [10]- and [12]paracyclophanes to examine their "free electron model," and more recently McGlinchy *et al.*[121] studied [10]paracyclophane to examine and compare the Johnson–Bovey and Haigh–Mallion theories.

The [*n*]paracyclophanes and the [*n*](9,10)anthracenophanes have been studied by ^{13}C-nmr spectroscopy to examine the effect of aromatic ring currents on the ^{13}C chemical shifts of the aliphatic bridge.[119,122–125] Misumi

TABLE II

Highest-Field Resonances in the ^{1}H-nmr Spectra of the [*n*]Paracyclophanes

n	δ (ppm)	Reference
12	0.78[a]	119
11	0.68[a]	119
10	0.48[a]	119
9	0.33[a]	119
8	0.19[a]	119
7	−0.3 to −0.9 (m,2H)[b]	55
6	−0.6 (m,2H)[b]	54

[a] Decoupled spectra.
[b] Center of multiplet.

and co-workers[119] suggested that the ring current effect in the ^{13}C-nmr is apparent in [9]- and [10]paracyclophanes and [10](9,10)anthracenophane and is concealed in [8]paracyclophane by the effect of steric compression in the strained bridge.

The ^{1}H-nmr of the [*n*]heterophanes also exhibit upfield resonances, but the bridge protons experience both a ring current effect and an anisotropic effect due to the heteroatom, and these effects cannot be easily separated.[90,118] Marchesini, Bradamante and co-workers[94] noted the greater shielding of bridge protons in the five-membered heteroaromatics than in analogous six-membered heteroaromatics, citing the absence of strong shielding effects in [9](2,4)pyridinophane and [9]metacyclophane.

In 1974 Allinger and co-workers[116] proposed that the ring current of the smaller [*n*]paracyclophanes would be diminished due to the nonplanarity of the benzene moiety, causing an upfield shift in the aromatic ^{1}H resonances. Although this phenomenon has not been observed in the [*n*]paracyclophanes,[60,116] Hirano *et al.*[103] noted this trend in the [*n*]metacyclophane series.

Proton nmr studies have afforded information on the conformational mobility of the methylene chain in [*n*]cyclophanes. Vögtle and Förster[126] discussed the conformational mobility of some substituted [*n*]cyclophanes, and Lai[127] reviewed the conformational behavior of dithia-[*n*]metacyclophanes. Wiberg and O'Donnell[12] found that the methylene bridges in [8]-, [9]-, and [10](1,4)naphthalenophanes are fixed on one side of the aromatic ring or the other at room temperature, whereas [14](1,4)naphthalenophane is conformationally mobile on the nmr time scale. Recently, chain rotation in a ring-substituted [6]paracyclophane has been observed by nmr.[127a,127b] In the [*n*]metacyclophanes, only the

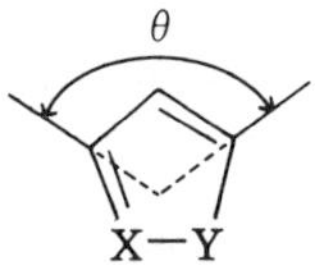

θ

X—Y

Fig. 2

spectra of [5]metacyclophane[104] is not temperature dependent, whereas the bridge of [6]metacyclophane[103] undergoes pseudorotation at room temperature and flips above and below the plane of the benzene ring at 76.5°C. As expected, the movement of the bridge is hindered by the presence of a large substituent on the aromatic ring. For example, whereas [7]- and [10]metacyclophanes exhibit room temperature spectra averaged by flipping of the methylene bridge, 2-bromo[7]- and 2-bromo-[10]metacyclophanes exhibit the nonequivalence of benzylic protons.[103] The conformational behavior of similarly substituted [*n*](1,3)naphthalenophanes,[96,97] [8](2,5)pyrrolophanes,[83] diaza- and dioxa[*n*]metacyclophanes,[81] dithia[*n*]metacyclophanes,[79,128–130] [*n*]paracyclophanes,[131] diaza- and dioxa[16]paracyclophanes,[132] and dioxadioxo[*n*]paracyclophanes[37,38] has also been studied. The conformational motion of the bridge is also affected by the incorporation of heteroatoms in the bridge. For example, the dithia[*n*]metacyclophanes have been found to have greater conformational mobility than the corresponding [*n*]metacyclophanes due to the longer C—S bond and greater flexibility in the C—S—C linkage.[127] The flipping of the aromatic ring and the pseudorotation of the methylene bridge of the smaller [*n*]heterophanes are dependent on the structure of the heteroaromatic ring. Hirano and co-workers[85] found that the energy of flipping, $\Delta G^{\ddagger}$, for the [7]heterophanes increases in an order parallel to the increase in the angle θ (Fig. 2), yielding the order [7](2,4)thiophenophane < [7](2,4)pyrrolophane < [7](3,5)pyrazolophane. Comparison with earlier calculated values of $\Delta G^{\ddagger}$ for [7](2,6)pyridinophane (9.0 kcal/mol)[118] and [7]metacyclophane (11.5 kcal/mol)[103] indicated the greater rigidity of the above five-membered heteroaromatics.

V. CHEMISTRY OF THE BRIDGED NUCLEUS

In this and the following section we focus on the aberrant chemistry exhibited by the aromatic nucleus and by the bridge rather than cataloging the more or less routine transformations reported for these compounds. A sharp division between the chemistry of the aryl ring and that of the bridge is inappropriate in certain instances (most obviously in discussions of transannular reactions), and therefore some ambiguity in the placement of sections may occur. Although several organizational approaches are possible, these sections are structured according to general reaction type. In general, emphasis is placed on the qualitative divergence from "normal" behavior, although structural effects on reaction rate are also discussed.

A. Electrophilic Substitution

1. Transannular Reactions

Cram and Goldstein[8] reported that acetolysis of optically active [9]paracyclophane-4-tosylate leads to the corresponding acetate with 98 ± 2% retention of configuration, suggesting the intermediacy of bridged ion **44.**

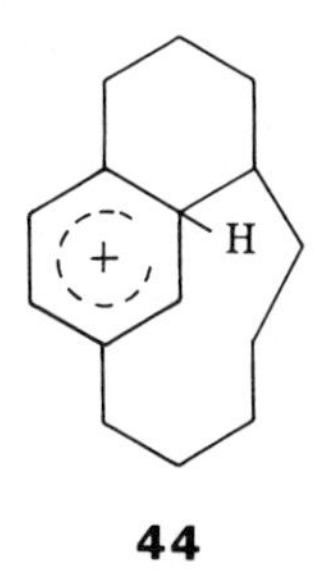

44

Relative rates of acetolysis (Table III) provide additional support for phenyl participation, because moving the tosyl group toward a position over the phenyl ring leads to rate acceleration. The dramatic rate acceleration for [9]paracyclophane-4-tosylate was attributed to strain release

TABLE III
Relative Rates of Acetolysis of Tosylates at 50°C

Substrate	Relative rate
Cyclohexyl tosylate	1
[9]Paracyclophane-3-tosylate	60
[9]Paracyclophane-4-tosylate	1800
[9]Paracyclophane-5-tosylate	170

upon ionization, a suggestion supported by examination of molecular models. Acetolysis of [8]paracyclophane-3-tosylate (Eq. 36a) yields hydrocarbon products, reflecting the importance of neighboring hydrogen and transannular phenyl participation in ionization.[32]

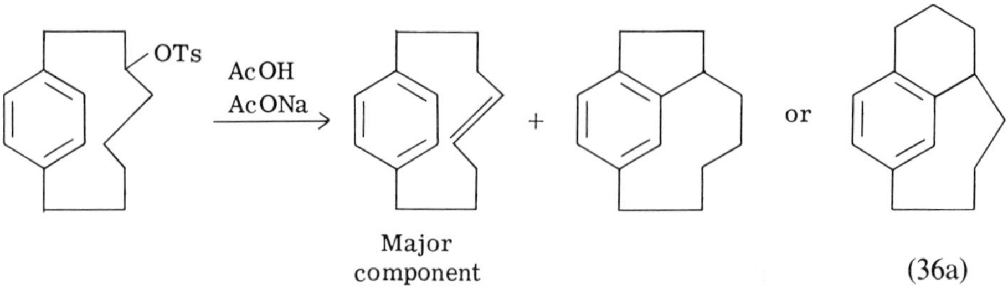

OTs AcOH AcONa → + or

Major component (36a)

Factors cited as of probable importance in determining rates are the replacement of hydrogen π repulsions with carbonium ion π attractions, the inductive effect of the benzene ring, changes in torsional and other nonbonded interactions, and changes in steric inhibition of solvation. Most such solvolysis studies preceded the appearance of the Smith book and are reviewed exhaustively therein.[1] More recently, Hopf and Noble[133] have observed a double transannular substitution upon acetolysis of **45** and a thermally induced transannular reaction yielding **46** (Eq. 36b).

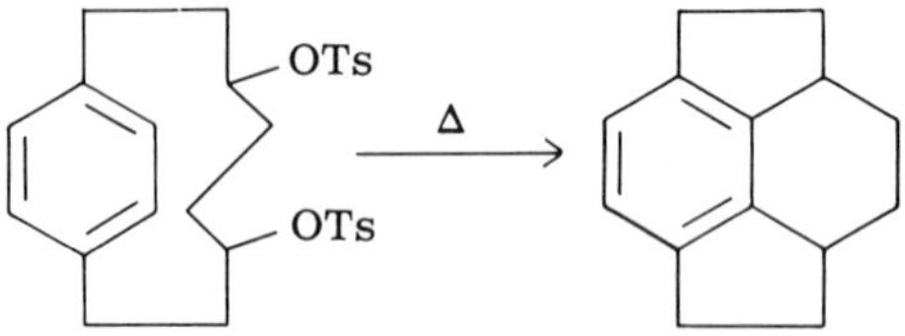

OTs OTs Δ → (36b)

45 **46**

2. Bridge Rearrangement and Cleavage

Naphthalenophane **47** underwent bridge cleavage in the presence of HBr, perhaps as depicted in Eq. (37)[96]. Most such chemistry, however, has involved Lewis acids and generally affords bridge rearrangement products in addition to, or in preference to, cleavage products (see Section III,E). For example, Hopf and Noble[133] attempted bromination of [8]paracyclophane and isolated only products of rearrangement or cleavage (Eq. 38).

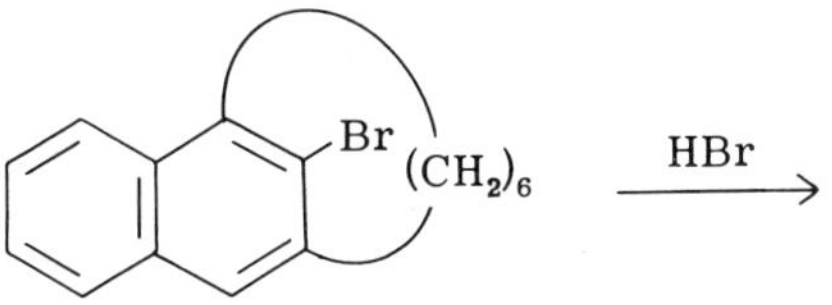

(37)

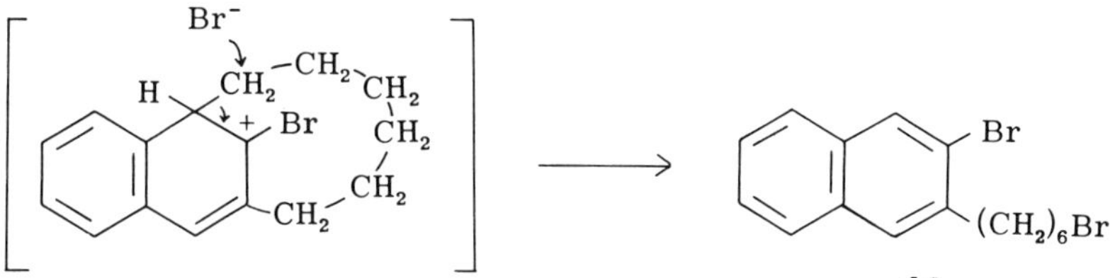

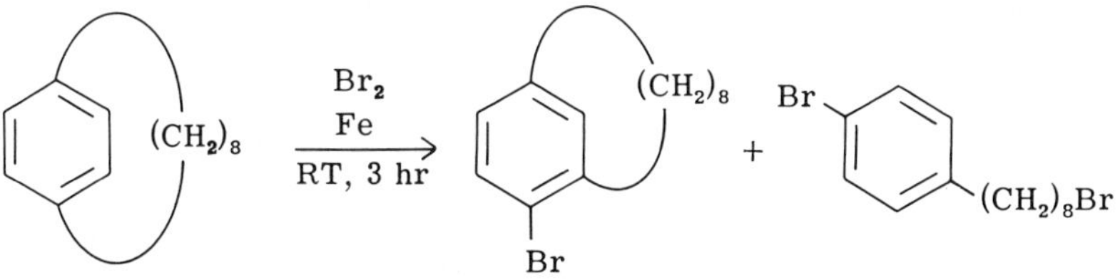

(38)

More vigorous conditions yielded analogous polybromo rearrangement and cleavage products. These workers isolated [8]metacyclophane as the major product (38%) upon treatment of [8]paracyclophane with $AlCl_3$–HCl in methylene chloride at −10°C.

3. Metallated Rings

[*n*]Cyclophanes with halogen (Br or Cl) substituents on the aryl ring are easily converted to the corresponding Grignard reagent or lithium compound. Reaction of these metallated cyclophanes with H_2O or D_2O proceeds in a normal fashion, and this sequence has been used to remove halogen in the synthesis developed by Parham for [*n*](1,3)naphthalenophanes[96,97] and in an analogous fashion for [*n*]metacyclophanes.[102] Grignard reagents in the [*n*](1,3)naphthalenophane series (e.g., **49**) exhibit a preference for deuterium abstraction over carbonyl addition in reaction with hexadeuterioacetone.[97] Reaction of Grignard reagent **49** with O_2 (Eq. 39) leads to abnormal products, reflecting the effets of strain and impinging bridge hydrogens on the course of the reaction:[97]

49 → 1. O_2; 2. aqueous work-up → 42% + 16% + 34% (mixture, $x + y = 9$) (39)

Lithiated [*n*]metacyclophanes **50** have been used to form a variety of derivatives (**51,** Eq. 40):[103]

50 → **51** X = H, D, I, Me, CO_2H, CH(OH)Ph, C(OH)Me_2 (40)

Reaction of **50** with benzaldehyde afforded a mixture of the expected adduct and the product of further Cannizzaro reaction (**51:** X = COPh).

B. Diels–Alder Reaction

Hopf, Jones, Jr., and co-workers[111,133] described the thermal [2 + 4] cycloaddition reactions of [7]- and [8]paracyclophanes with dicyanoacetylene and hexafluoro-2-butyne (Eq. 41):

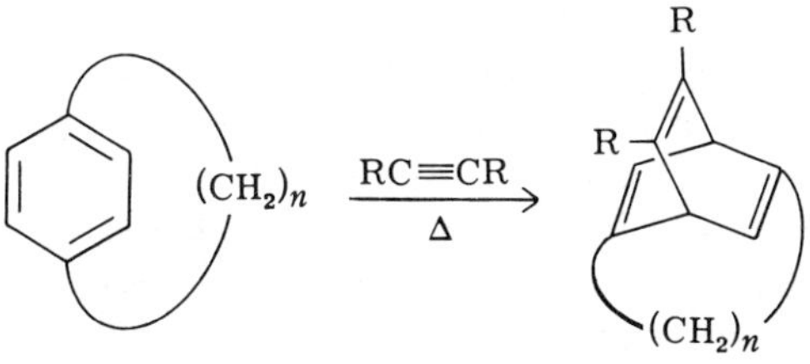

$$\text{(41)}$$

$R = CF_3$, $n = 7, 8$
$R = CN$, $n = 8$

Fair yields of 1 : 1 adducts were obtained at elevated temperatures, except that [7]paracyclophane was unreactive toward dicyanoacetylene. These workers rearranged [7]paracyclophane to [7]metacyclophane in the presence of acid and added hexafluoro-2-butyne to the rearranged cyclophane, yielding the bridged barrelene **52.** Similarly, quantitative yields

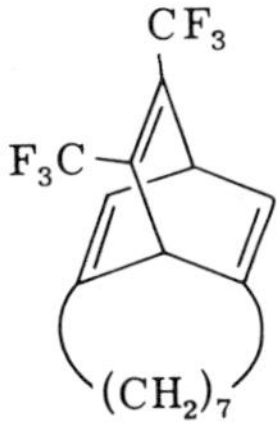

52

for the addition of a variety of dienophiles to [5]metacyclophane have been reported.[133a] In developing a synthetic approach to [2.2.2.*n*]-paddlanes, Wiberg and O'Donnell[12] added dicyanoacetylene to [8]-, [9]-, [10]-, and [14](1,4)naphthalenophanes (Eq. 42):

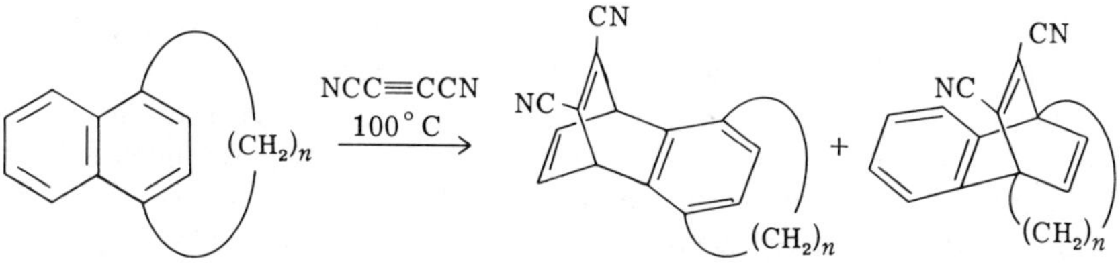

$n = 8, 9, 10, 14$ **53** **54**

(42)

The [9]- and [10](1,4)naphthalenophanes gave only the product of addition at the unsubstituted ring **(53),** but [14](1,4)naphthalenophane gave both adducts, although the desired adduct **(54)** was isolated in only 7% yield. The major product (17% yield) in the addition of dicyanoacetylene to

[8](1,4)naphthalenophane was **53**, but a 2% yield of a product appearing to be derived from bridge rearrangement to [8](1,3)naphthalenophane followed by Diels–Alder addition across the 1,4 positions was isolated in 2% yield. Similarly, Vögtle and Koo Tze Mew[22] added the dienophiles benzyne, maleic anhydride, and *N*-phenylmaleimide to [10]- and [14](9,10)anthracenophanes and to their dithia and bis(sulfone) precursors in a successful, although low-yield, synthesis of paddlane derivatives.

Another successful paddlane synthesis was effected by Helder[13] through the addition of dicyanoacetylene to [8](2,5)furanophane, affording an 80% yield of paddlane **55.** The addition of dicyanoacetylene to

O NC NC $(CH_2)_8$

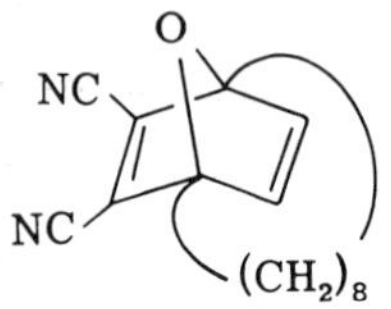

55

[8](2,5)thiophenophane apparently occurred as well, although thermal extrusion of sulfur *in situ* left only an [8]paracyclophane derivative upon work-up.

C. Complexation

The π-base strength of the aryl ring in [n]cyclophanes has been of interest for many years. For example, [9]-, [10]-, and [12]paracyclophanes form 1 : 1 complexes with tetracyanoethylene (TCNE), and the position of the long-wavelength uv absorption band was found to decrease as the complex dissociation constant decreased and is therefore related to π-base strength.[134,135] The energy of this longest-wavelength transition was found to follow the order $n = 9 > 12 > 10$ and was interpreted to result from a combination of transannular delocalization (hyperconjugation) of charge, which increases the capacity of the ring to stabilize positive charge, and inhibition of the transannular delocalization due to nonplanarity of the benzene rings.

D. Photochemistry

Photolysis of [6]paracyclophane (unfiltered, medium-pressure mercury arc lamp) in cyclohexane-d_{12} effects an apparently quantitative conversion to the corresponding bridged Dewar benzene,[64] whereas both [7]paracyclophane[64] and [8]paracyclophane[133] undergo slow polymerization under these conditions. Irradiation of 2-chloro[6](1,3)naphthalenophane **(56)** leads to products of transannular reaction (perhaps via a naphthyl radical) with loss of HCl (Eq. 43)[32]:

Cl $(CH_2)_6$ $\xrightarrow{h\nu}$ +

56 (43)

Similar chemistry is observed in the photolysis of 2-bromo[6]metacyclophane and 2-bromo[7]metacyclophane.[103]

VI. CHEMISTRY OF THE BRIDGE

Studies devoted to the systematic examination of the chemistry of the bridge in [*n*]cyclophanes have been surprisingly limited. These compounds are especially suitable as substrates for analyzing the effects of proximate aryl rings on aliphatic chains. Still, no concerted effort, with the exception of Cram's work on [*n*]paracyclophane tosylate acetolysis,[8,32] has been made to correlate structure (e.g., distance between bridge functionality and aryl ring or rigidity and strain in the bridge) and reactivity. In this section we have gathered the isolated and generally more recent reports of such chemistry, concentrating on reactions in which the bridge exhibits chemical behavior qualitatively different from that of noncyclophane analogs. The more routine transformations appearing in the earlier literature have been reviewed exhaustively by Smith[1] and/or are covered in Sections II and III of this chapter.

A. Nucleophilic Substitution of Bridge-Functionalized Derivatives

In addition to the previously discussed (Section V,A) transannular chemistry observed in cyclophane tosylate solvolysis, attempts at bimolecular nucleophilic substitution of bridge substituents have led to unexpected results. Hopf and Noble found that reduction of **45** with lithium aluminum hydride, diisobutylaluminum hydride, or lithium triethylborohydride led to [8]paracyclophane in only 8 to 19% yield along with various amounts of the corresponding diol of **45,** [8]paracyclophan-3-ene and [8]paracyclophan-4-ene.[133]

B. Keto–Enol Equilibria

Keto–enol equilibria in [*n*]cyclophanes appear to be dependent on the value of *n,* most likely due to variations in strain energy. For example, Cram and Antar[7] reported that diketone **57** shows no tendency to enolize, in contrast to its next higher homolog **58,** which exists as a 1 : 1 keto–enol mixture at equilibrium.[136]

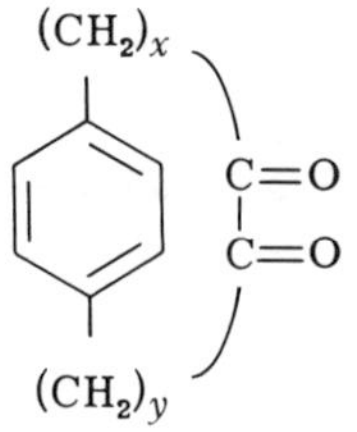

$(CH_2)_x$ C=O C=O $(CH_2)_y$

57 $x = 4,\ y = 3$

58 $x = y = 4$

C. Addition to Multiple Bonds

Hopf and Noble[133,137] examined the chemistry of the rigidly held[42] bridge double bond in *trans*-[8]paracyclophan-4-ene **(59).** Olefin **59** undergoes osmium tetroxide oxidation, yielding the corresponding *cis*-diol, and *m*-chloroperbenzoic acid oxidation, affording the expected epoxide. The hindered epoxide was found to be surprisingly stable but could be opened by treatment with lithium aluminum hydride. Unexpectedly, when lithium aluminum deuteride was used, deuterium incorporation occurred only at the hydroxyl-bearing carbon in the product alcohol. These authors proposed isomerization of epoxide to ketone followed by reduction to explain this unusual result. Addition of molecular bromine in chloroform to **59** proceeded slowly, affording the *cis*-dibromide in low yield along with polymeric material. Carbenoid reagents reacted with **59** in the expected fashion (Eq. 44), and dichlorocyclopropane **60** (X = Cl) was reduced by treatment with sodium in liquid ammonia to **61:**

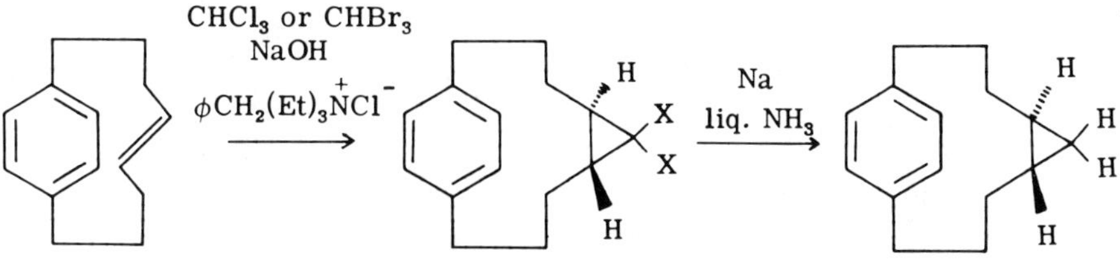

$CHCl_3$ or $CHBr_3$, NaOH, $\phi CH_2(Et)_3\overset{+}{N}Cl^-$ → ; Na, liq. NH_3 →

59 **60** X = Cl (40-45%), X = Br (60%) **61** (42%)

(44)

A low yield of **61** was obtained directly by treatment of **59** with diazomethane in the presence of cuprous chloride, a reaction that also led to unstable adducts formed by attack on the benzene ring. An orthogonal orientation of the cyclopropane and benzene rings in **60** and **61** appears likely from examination of Dreiding models and the cyclopropane proton chemical shifts. The unusual allene **62** was isolated in 65% yield by reaction of **60** (X = Br) with methyllithium.

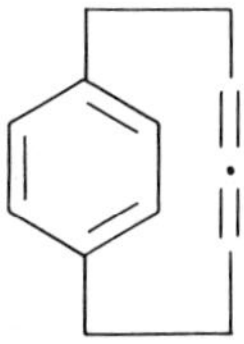

62

In attempting to examine the expected charge-transfer complex of **63** with TCNE, Misumi and co-workers[138] isolated the 1 : 1 TCNE adduct **64,** formed at room temperature (Eq. 45):

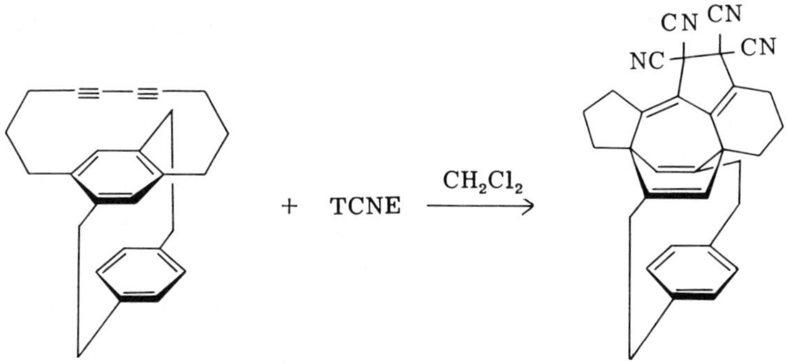

63 + TCNE $\xrightarrow{CH_2Cl_2}$ **64** (45)

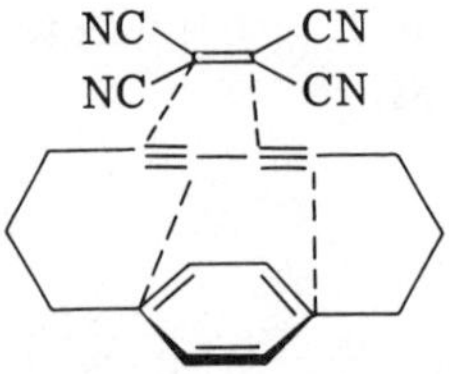

Fig. 3. Suggested mode of cycloaddition of [10]paracyclophane-4,6-diyne with TCNE.

An analogous adduct was isolated from reaction of the simpler [10]paracyclopha-4,6-diyne with TCNE at elevated temperature, and both reactions occur in the absence of light. This unusual thermal cycloaddition apparently proceeds as depicted in Fig. 3.

D. Photochemistry

Benzene solutions of [10](9,10)anthracenopha-4,6-diyne **(65)** afford quantitative yields of the photodimer **66** upon standing for 10 min in sunlight (Eq. 46)[19,20]:

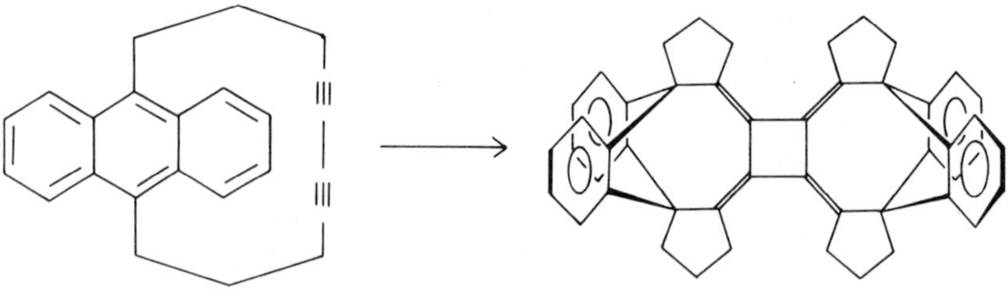

65 → **66** (46)

Analogous adducts form from photoreaction of **65** with both furan and cyclopentadiene. These reactions appear to proceed via an intermediate, observable by uv spectroscopy at 77 K, with proposed structure **67** (Eq. 47):

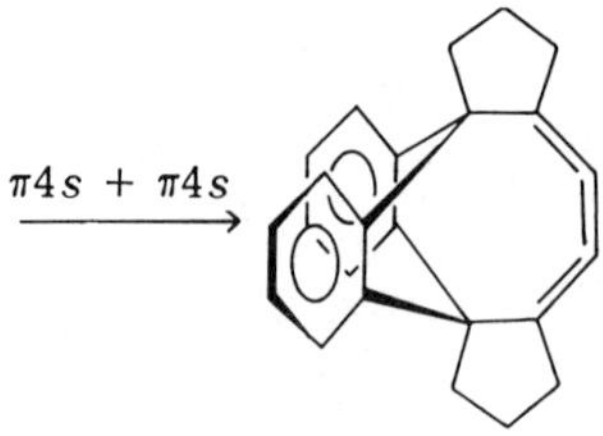

$$\mathbf{65} \xrightarrow{\pi 4s\ +\ \pi 4s} \mathbf{67} \qquad (47)$$

Dimerization of highly strained **67** to form **66** was suggested to occur via a thermally forbidden [$\pi 2s + \pi 2s$] process.[19]

VII. NATURALLY OCCURRING [*n*]CYCLOPHANES

A number of [*n*]cyclophanes that occur in nature have generated interest in recent years, primarily because of their utility as biologically active compounds. The coverage in this section is intended to be suggestive of the structural variation in such systems and is not exhaustive.

Pubescene A **(68)**,[139] one of a number of cyclopeptide alkaloids isolated from plant material, has been found to be a diastereomer of melonovine A. Dihydrozizyphine G **(69)**[140] has the [10]paracyclophane framework in common with **68.**

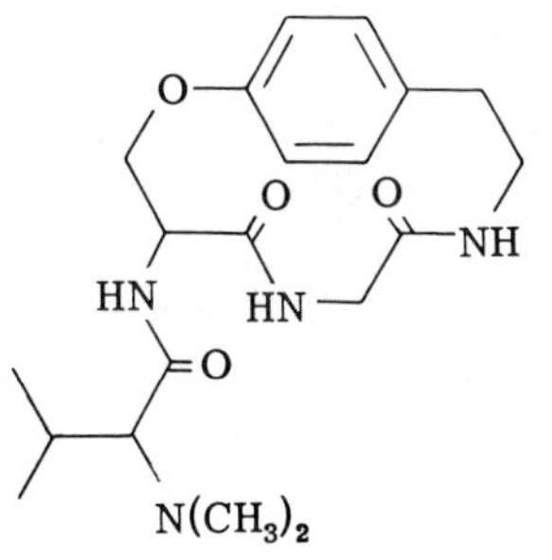

68 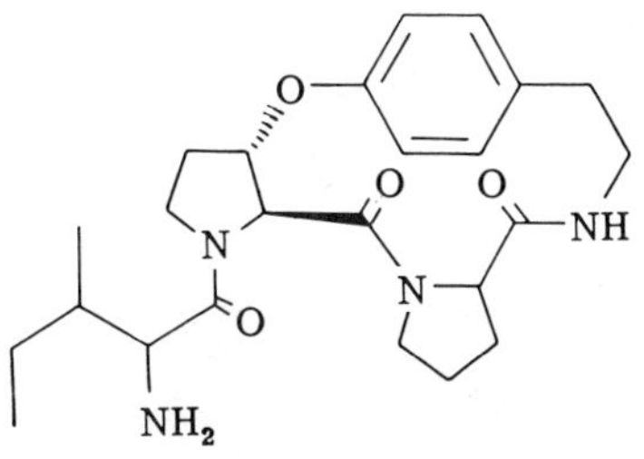 **69**

The rifamycins, used extensively to combat bacterial infections, are members of the class of compounds referred to as ansamycins, which have in common the structural feature of an aromatic ring with an aliphatic bridge. Rifamycins have been extensively characterized spectroscopically,[141–143] and a total synthesis of rifamycin S **(70)** has been accomplished by Kishi.[144,145]

The tumor-inhibiting [16]metacyclophane maytansine **(71)** has generated much interest, and progress toward a total synthesis has been reviewed.[146,147] Other maytansanoid tumor inhibitors have also been isolated.[148]

The sea whip neuromuscular toxin lophotoxin **(72)**[149] has functionality and a multiple ring structure that make it unusual among marine natural

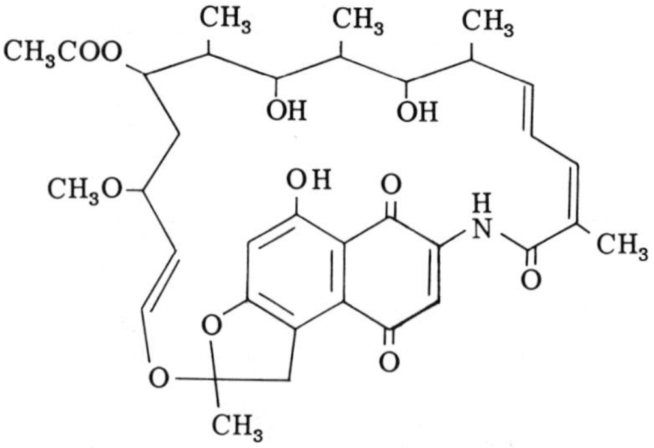

CH_3COO CH_3 CH_3 CH_3
OH OH
CH_3O
OH O H N
CH_3
O
O
O O
CH_3

70

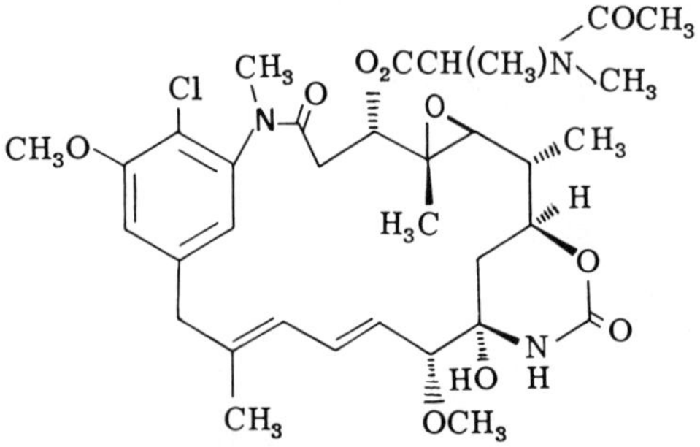

$COCH_3$
CH_3 $O_2CCH(CH_3)N$ CH_3
Cl N O O
CH_3O CH_3
H
H_3C O
O
N
HO H
CH_3 OCH_3

71

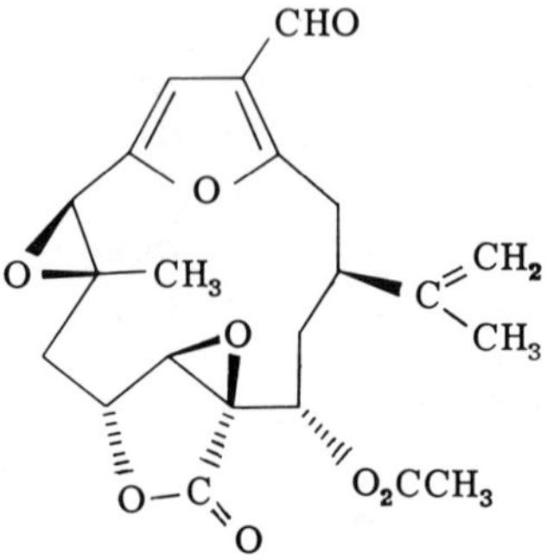

CHO
O
O CH_3 CH_2
C
O CH_3
O_2CCH_3
O C
O

72

products. More surprisingly, a [6]furanophane was recently isolated from the same organism.[150]

Acknowledgments

The authors gratefully acknowledge partial support from the donors of the Petroleum Research Fund, administered by the American Chemical Society, and assistance from the Wellesley College Committee on Faculty Awards. We also thank Julia Johnson for assistance in the preparation of the illustrations and Dr. H. Hopf for generously supplying preprints of his work.

REFERENCES

1. B. H. Smith, "Bridged Aromatic Compounds." Academic Press, New York, 1964.
2. For example, see E. W. Abel, M. Booth, C. A. Brown, K. G. Orrell, and R. L. Woodford, *J. Organomet. Chem.* **214,** 93 (1981).
3. For example, see E. Logemann, K. Rissler, G. Schill, and H. Fritz, *Chem. Ber.* **114,** 2245 (1981).
4. F. Vögtle and P. Neumann, *Tetrahedron Lett.* p. 5329 (1969).
5. F. Vögtle and P. Neumann, *Tetrahedron* **26,** 5847 (1970).
6. N. L. Allinger, L. A. Freiberg, R. B. Hermann, and M. A. Miller, *J. Am. Chem. Soc.* **85,** 1171 (1963).
7. D. J. Cram and M. F. Antar, *J. Am. Chem. Soc.* **80,** 3109 (1958).
8. D. J. Cram and M. Goldstein, *J. Am. Chem. Soc.* **85,** 1063 (1963).
9. K. Wiesner, D. M. MacDonald, R. B. Ingraham, and R. B. Kelly, *Can. J. Res., Sect. B* **28,** 561 (1950).
10. D. J. Cram, N. L. Allinger, and H. Steinberg, *J. Am. Chem. Soc.* **76,** 6132 (1954).
11. K. J. Clark, *J. Chem. Soc.* p. 2202 (1957).
12. K. Wiberg and M. O'Donnell, *J. Am. Chem. Soc.* **101,** 6660 (1979).
13. R. Helder, Ph.D. Thesis, Rijksuniversiteit te Groningen (1974).
14. W. M. Schubert, W. A. Sweeney, and H. K. Latourette, *J. Am. Chem. Soc.* **76,** 5462 (1954).
15. R. Huisgen, W. Rapp, I. Ugi, H. Walz, and I. Glogger, *Justus Liebigs Ann. Chem.* **586,** 52 (1954).
16. R. Huisgen and I. Ugi, *Chem. Ber.* **93,** 2693 (1960).
17. T. Matsuoka, Y. Sakata, and S. Misumi, *Tetrahedron Lett.* p. 2549 (1970).
18. T. Matsuoka, T. Negi, T. Otsubo, Y. Sakata, and S. Misumi, *Bull. Chem. Soc. Jpn.* **45,** 1825 (1972).
19. S. Misumi, *Mem. Inst. Sci. Ind. Res., Osaka Univ.* **36,** 37 (1979).
20. T. Inoue, T. Kaneda, and S. Misumi, *Tetrahedron Lett.* p. 2969 (1974).
21. T. Otsubo and S. Misumi, *Synth. Commun.* **8,** 285 (1978).
22. F. Vögtle and P. Koo Tze Mew, *Angew. Chem., Int. Ed. Engl.* **17,** 60 (1978).

23. For a review of sulfone pyrolysis, see F. Vögtle and L. Rossa, *Angew. Chem., Int. Ed. Engl.* **18,** 515 (1979).
24. J. Chung and S. M. Rosenfeld, *J. Org. Chem.* **48,** 387 (1983).
25. P. N. Swepston, S. T. Lin, A. Hawkins, S. Humphrey, S. Siegel, and A. W. Cordes, *J. Org. Chem.* **46,** 3754 (1981).
26. H. Saito, Y. Fujise, and S. Ito, *Koen Yoshishu—Hibenzenkei Hokozoku Kagaku Toronkai* [*oyobi*] *Kozu Yuki Kagaku Toronkai, 12th, 1979* p. 149 (1979); *Chem. Abstr.* **92,** 180716u.
27. M. Iwata, H. Kuzuhara, and S. Emoto, *Chem. Lett.* p. 983 (1976).
28. H. Kuzuhara, M. Iwata, and S. Emoto, *J. Am. Chem. Soc.* **99,** 4173 (1977).
29. M. Iwata and H. Kuzuhara, *Chem. Lett.* p. 5 (1981).
30. T. Komatsu, M. Ando, F. Suguwara, and H. Kuzuhara, *Koen Yoshishu—Tennen Yuki Kagobutsu Toronkai, 22nd, 1979* p. 627 (1979); *Chem. Abstr.* **93,** 26407h.
31. D. J. Cram and G. R. Knox, *J. Am. Chem. Soc.* **83,** 2204 (1961).
32. D. J. Cram, C. S. Montgomery, and G. R. Knox, *J. Am. Chem. Soc.* **88,** 515 (1966).
33. M. Nakazaki, K. Yamamoto, and S. Tanaka, *J. Org. Chem.* **41,** 4081, (1976).
34. M. Nakazaki, K. Yamamoto, M. Ito, and S. Tanaka, *J. Org. Chem.* **42,** 3468 (1977).
35. E. Langer, H. Lehner, and K. Schlögl, *Tetrahedron* **29,** 2473 (1973).
36. K. Sakamoto and M. Oki, *Bull. Chem. Soc. Jpn.* **46,** 270 (1973).
37. K. Sakamoto and M. Oki, *Bull. Chem. Soc. Jpn.* **50,** 3388 (1977).
38. K. Sakamoto and M. Oki, *Bull. Chem. Soc. Jpn.* **47,** 2739 (1974).
39. K. Sakamoto and M. Oki, *Bull. Chem. Soc. Jpn.* **49,** 3159 (1976).
40. T. Tsuji, S. Nishida, and H. Tsubomura, *Chem. Commun.* p. 284 (1972).
41. T. Tsuji and S. Nishida, *J. Am. Chem. Soc.* **95,** 7519 (1973).
42. T. Tsuji, T. Shibata, Y. Hienuki, and S. Nishida, *J. Am. Chem. Soc.* **100,** 1806 (1978).
43. T. Tsuji and S. Nishida, *Chem. Lett.* p. 1335 (1973).
44. T. Tsuji and S. Nishida, *J. Am. Chem. Soc.* **96,** 3649 (1974).
45. T. Tsuji and S. Nishida, *Chem. Lett.* p. 631 (1977).
46. G. L. Closs and M. S. Czeropski, *Chem. Phys. Lett.* **45,** 115 (1977).
47. Y. Hienuki, T. Tsuji, and S. Nishida, *Tetrahedron Lett.* **22,** 867 (1981).
48. Y. Hienuki, T. Tsuji, and S. Nishida, *Tetrahedron Lett.* **22,** 863 (1981).
49. T. Shibata, T. Tsuji, and S. Nishida, *Bull. Chem. Soc. Jpn.* **53,** 709 (1980).
50. F. Kataoka and S. Nishida, *Chem. Commun.* p. 864 (1978).
51. F. Katoaka, S. Nishida, T. Tsuji, and M. Murakami, *J. Am. Chem. Soc.* **103,** 6878 (1981).
52. M. F. Bartlett, S. K. Figdor, and K. Wiesner, *Can. J. Chem.* **30,** 291 (1952).
53. P. G. Gassman, T. F. Bailey, and R. C. Hoye, *J. Org. Chem.* **45,** 2923 (1980).
54. V. V. Kane, A. D. Wolf, and M. Jones, Jr., *J. Am. Chem. Soc.* **96,** 2643 (1974).
55. A. D. Wolf, V. V. Kane, R. H. Levin, and M. Jones, Jr., *J. Am. Chem. Soc.* **95,** 1680 (1973).
56. M. F. Pero, C. M. Cotell, K. A. Choe, and S. M. Rosenfeld, *Synth. Commun.* **12,** 299 (1982).
57. N. L. Allinger and L. A. Freiberg, *J. Org. Chem.* **27,** 1490 (1962).
58. M. G. Newton, T. J. Walter, and N. L. Allinger, *J. Am. Chem. Soc.* **95,** 5652 (1973).
59. N. L. Allinger and T. J. Walter, *J. Am. Chem. Soc.* **94,** 9267 (1972).
60. N. L. Allinger, T. J. Walter, and M. G. Newton, *J. Am. Chem. Soc.* **96,** 4588 (1974).
61. I. J. Landheer, W. H. de Wolf, and F. Bickelhaupt, *Tetrahedron Lett.* p. 2813 (1974).
62. I. J. Landheer, W. H. de Wolf, and F. Bickelhaupt, *Tetrahedron Lett.* p. 349 (1975).
63. J. W. van Straten, I. J. Landheer, W. H. de Wolf, and F. Bickelhaupt, *Tetrahedron Lett.* p. 4499 (1975).

64. S. L. Kammula, L. D. Iroff, M. Jones, Jr., J. W. van Straten, W. H. de Wolf, and F. Bickelhaupt, *J. Am. Chem. Soc.* **99,** 5815 (1977).
65. J. W. van Straten, W. H. de Wolf, and F. Bickelhaupt, *Recl. Trav. Chim. Pays-Bas* **96,** 88 (1977).
66. E. M. Beccalli, M. Del Puppo, E. Licandro, and A. Marchesini, *J. Heterocycl. Chem.* **18,** 685 (1981).
67. S. A. Sherrod, R. L. da Costa, R. A. Barnes, and V. Boekelheide, *J. Am. Chem. Soc.* **96,** 1565 (1974).
68. R. W. Griffin, *Chem. Rev.* **63,** 45 (1963).
69. S. Fujita and H. Nozaki, *Soc. Synth. Org. Chem. Jpn.* **30,** 679 (1972).
70. Y. L. Goldfarb, S. Z. Taits, and L. I. Belenkii, *Izv. Akad. Nauk SSR, Otd. Khim. Nauk* p. 1262 (1957).
71. S. Z. Taits and Y. L. Goldfarb, *Izv. Akad. Nauk SSR, Otd. Khim. Nauk* p. 1289 (1963).
72. K. Tamao, S. Kodama, T. Nakatsuka, Y. Kiso, and M. Kumada, *J. Am. Chem. Soc.* **97,** 4405 (1975).
73. K. Biemann, G. Büchi, and B. H. Walker, *J. Am. Chem. Soc.* **79,** 5558 (1957).
74. A. W. Lee, P. M. Keehn, S. M. Ramos, and S. M. Rosenfeld, *Heterocycles* **7,** 81 (1977).
75. R. H. Mitchell and V. Boekelheide, *Tetrahedron Lett.* p. 2013 (1969).
76. F. Vögtle, *Tetrahedron Lett.* p. 5221 (1968).
77. F. Vögtle, *Tetrahedron* **25,** 3231 (1969).
78. F. Vögtle, *Tetrahedron Lett.* p. 3193 (1969).
79. F. Vögtle and P. Neumann, *Tetrahedron* **26,** 5299 (1970).
80. F. Vögtle and E. Weber, *Angew. Chem.* **86,** 126 (1974).
81. F. Vögtle and P. Neumann, *Tetrahedron Lett.* p. 115 (1970).
82. H. Nozaki, T. Koyama, T. Mori, and R. Noyori, *Tetrahedron Lett.* p. 2181 (1968).
83. H. Nozaki, T. Koyama, and T. Mori, *Tetrahedron* **25,** 5357 (1969).
84. S. Fujita, T. Kawaguti, and H. Nozaki, *Tetrahedron Lett.* p. 1119 (1971).
85. S. Hirano, T. Hiyama, S. Fujita, T. Kawaguti, Y. Hayashi, and H. Nozaki, *Tetrahedron* **30,** 2633 (1974).
86. H. H. Wasserman, G. C. Rogers, and D. D. Keith, *J. Am. Chem. Soc.* **91,** 1263 (1969).
87. H. H. Wasserman, D. D. Keith, and J. Nadelson, *J. Am. Chem. Soc.* **91,** 1264 (1969).
88. A. T. Balaban, M. Gavăt, and C. D. Nenitzescu, *Tetrahedron* **18,** 1079 (1962).
88a. A. T. Balaban, *Tetrahedron Lett.* p. 4643 (1968).
89. U. K. Georgi and J. Rétey, *Chem. Commun.* p. 32 (1971).
90. S. Bradamante, R. Fusco, A. Marchesini, and G. Pagani, *Tetrahedron Lett.* p. 11 (1970).
91. S. Fujita, Y. Hayashi, and H. Nozaki, *Tetrahedron Lett.* p. 1645 (1972).
92. S. Fujita, T. Kawaguti, and H. Nozaki, *Bull. Chem. Soc. Jpn.* **43,** 2596 (1970).
93. E. M. Beccalli, L. Majori, A. Marchesini, and C. Torricelli, *Chem. Lett.* p. 659 (1980).
94. A. Marchesini, S. Bradamante, R. Fusco, and G. Pagani, *Tetrahedron Lett.* p. 671 (1971).
95. W. E. Parham and J. K. Rinehart, *J. Am. Chem. Soc.* **89,** 5668 (1967).
96. W. E. Parham, D. R. Johnson, C. T. Hughes, M. K. Meilahn, and J. K. Rinehart, *J. Org. Chem.* **35,** 1048 (1970).
97. W. E. Parham, R. W. Davenport, and J. K. Rinehart, *J. Org. Chem.* **35,** 2662 (1970).
98. W. E. Parham, D. C. Egberg, and W. C. Montgomery, *J. Org. Chem.* **38,** 1207 (1973).
98a. P. Grice and C. B. Reese, *Tetrahedron Lett.* p. 2563 (1979).
98b. P. Grice and C. B. Reese, *J. Chem. Soc., Chem. Commun.,* p. 424 (1980).
99. W. E. Parham, R. W. Davenport, and J. B. Biasotti, *Tetrahedron Lett.* p. 557 (1969).

100. W. E. Parham, R. W. Davenport, and J. B. Biasotti, *J. Org. Chem.* **35,** 3775 (1970).
101. W. E. Parham, D. C. Egberg, and S. Salgar, *J. Org. Chem.* **37,** 3248 (1972).
102. S. Fujita, S. Hirano, and H. Nozaki, *Tetrahedron Lett.* p. 403 (1972).
103. S. Hirano, H. Hara, T. Hiyama, S. Fujita, and H. Nozaki, *Tetrahedron* **31,** 2219 (1975).
104. J. W. van Straten, W. H. de Wolf, and F. Bickelhaupt, *Tetrahedron Lett.* p. 4670 (1977).
105. L. A. M. Turkenburg, P. M. L. Blok, W. H. de Wolf, and F. Bickelhaupt, *Tetrahedron Lett.* **22,** 3317 (1981).
106. L. A. M. Turkenburg, J. W. van Straten, W. H. de Wolf, and F. Bickelhaupt, *J. Am. Chem. Soc.* **102,** 3256 (1980).
107. L. A. Pacala, Ph.D. Thesis, Princeton University, Princeton, New Jersey (1980).
108. W. E. Parham and J. F. Dooley, *J. Am. Chem. Soc.* **89,** 985 (1967).
109. W. E. Parham and J. F. Dooley, *J. Org. Chem.* **33,** 1476 (1968).
110. G. Märkl and R. Fuchs, *Tetrahedron Lett.* p. 4695 (1972).
111. K.-L. Noble, H. Hopf, M. Jones, Jr., and S. L. Kammula, *Angew. Chem.* **90,** 629 (1978).
112. J. M. Patterson, J. Brasch, and P. Drenchko, *J. Org. Chem.* **27,** 1652 (1962).
113. H. Schmidt, A. Schweig, W. Thiel, and M. Jones, Jr., *Chem. Ber.* **111,** 1958 (1978).
114. R. Gleiter, H. Hopf, M. Eckert-Maksic, and K.-L. Noble, *Chem. Ber.* **113,** 3401 (1980).
115. H. Wynberg, W. C. Nieuwpoort, and H. T. Jonkman, *Tetrahedron Lett.* p. 4623 (1973).
116. N. L. Allinger, J. T. Sprague, and T. Liljefors, *J. Am. Chem. Soc.* **96,** 5100 (1974).
117. E. Wasserman, R. S. Hutton, and F. B. Bramwell, *J. Am. Chem. Soc.* **98,** 7429 (1976).
118. S. Fujita and H. Nozaki, *Bull. Chem. Soc. Jpn.* **44,** 2827 (1971).
118a. L. Carballeira, J. Casado, E. Gonzalez and M. A. Rios, *J. Chem. Phys.* **77,** 5655 (1982).
118b. Y. Tobe, K. Kakiuchi, Y. Odaira, T. Hosaki, Y. Kai and N. Kasai, *J. Am. Chem. Soc.* **105,** 1376 (1983).
119. T. Kaneda, T. Otsubo, H. Horita, and S. Misumi, *Bull. Chem. Soc. Jpn.* **53,** 1015 (1980).
120. J. S. Waugh and R. W. Fessenden, *J. Am. Chem. Soc.* **79,** 846 (1957).
121. A. Agarwal, J. A. Barnes, J. L. Fletcher, M. J. McGlinchey, and B. G. Sayer, *Can. J. Chem.* **55,** 2575 (1977).
122. R. H. Levin and J. D. Roberts, *Tetrahedron Lett.* p. 135 (1973).
123. T. Kaneda, T. Inoue, Y. Yasufuku, and S. Misumi, *Tetrahedron Lett.* p. 1543 (1975).
124. N. Mori and T. Takemura, *J. Chem. Soc., Perkin Trans. 2* p. 1259 (1978).
125. K. Sakamoto and M. Oki, *Chem. Lett.* p. 257 (1976).
126. H. Förster and F. Vögtle, *Angew. Chem., Int. Ed. Engl.* **16,** 429 (1977).
127. Y. H. Lai, *Heterocycles* **16,** 1739 (1981).
127a. C. Wolff, J. Liebe, and W. Tochtermann, *Tetrahedron Lett.* **23,** 1143 (1982).
127b. J. Liebe, C. Wolff, and W. Tochtermann, *Tetrahedron Lett.* **23,** 2439 (1982).
128. F. Vögtle, J. Grütze, R. Nätscher, W. Wieder, E. Weber, and R. Grün, *Chem. Ber.* **108,** 1694 (1975).
129. E. Weber, W. Wieder, and F. Vögtle, *Chem. Ber.* **109,** 1002 (1976).
130. F. Vögtle, *Chem. Ber.* **102,** 1784 (1969).
131. M. Nakazaki, K. Yamamoto, and S. Okamoto, *Tetrahedron Lett.* p. 4597 (1969).
132. K. Sakamoto and M. Oki, *Tetrahedron Lett.* p. 3989 (1973).
133. K.-L. Noble, Ph.D. Thesis, Universität Wurzburg (1980).
133a. L. A. M. Turkenburg, P. M. L. Blok, W. H. De Wolf, and F. Bickelhaupt, *Angew. Chem.* **94,** 291 (1982).
134. R. E. Merrifield and W. D. Phillips, *J. Am. Chem. Soc.* **80,** 2778 (1958).

135. D. J. Cram and R. H. Bauer, *J. Am. Chem. Soc.* **81,** 5971 (1959).
136. D. J. Cram and H. U. Daeniker, *J. Am. Chem. Soc.* **76,** 2743 (1954).
137. K.-L. Noble and H. Hopf, *Chem. Ber.,* in press (1983).
138. T. Kaneda, T. Ogawa, and S. Misumi, *Tetrahedron Lett.* p. 3373 (1973).
139. R. Tschesche, D. Hillebrand, and R. C. Bick, *Phytochemistry* **19,** 1000 (1980).
140. U. Schmidt, A. Lieberknecht, H. Griesser, and J. Haeusler, *Angew. Chem.* **93,** 272 (1981).
141. G. Lancini and W. Zanichelli, *in* "Structural Activity Relationships Among the Semisynthetic Antibiotics" (F. Perlman, ed.), pp. 531–560. Academic Press, New York, 1977.
142. P. Salvadori, C. Bertucci, C. Rosini, M. Zandomeneghi, G. G. Gallo, E. Martinelli, and P. Ferrari, *J. Am. Chem. Soc.* **103,** 5553 (1981).
143. E. Martinelli, P. Gironi, and G. G. Gallo, *Farmaco, Ed. Sci.* **36,** 671 (1981).
144. H. Nagaoka, G. Schmid, I. Hideo, and Y. Kishi, *Tetrahedron Lett.* **22,** 899 (1981).
145. Y. Kishi, *Pure Appl. Chem.* **53,** 1163 (1981).
146. B. Ganem, *Proc. Asian Symp. Med. Plants Spices, 4th, 1980* Vol. 1, p. 235 (1981).
147. D.-L. Bai, *Yu Chi Hua Hsueh* **1,** 1 (1981); *Chem. Abstr.* **95,** 80766d.
148. R. G. Powell, D. Weisleder, and C. R. Smith, Jr., *J. Org. Chem.* **46,** 4398 (1981).
149. W. Fenical, R. K. Okuda, M. M. Bandurraga, P. Culver, and R. S. Jacobs, *Science* **212,** 1512 (1981).
150. M. M. Bandurraga, W. Fenical, S. F. Donovan, and J. Clardy, *J. Am. Chem. Soc.* **104,** 6463 (1982).

Index

A

B

C

D

E

F

G

H

I

K

L

M

N

O

P

Q

R

S

T

U

V

W

X

Z

ORGANIC CHEMISTRY

A SERIES OF MONOGRAPHS

EDITOR

HARRY H. WASSERMAN

Department of Chemistry
Yale University
New Haven, Connecticut

1. Wolfgang Kirmse. CARBENE CHEMISTRY, 1964; 2nd Edition, 1971
2. Brandes H. Smith. BRIDGED AROMATIC COMPOUNDS, 1964
3. Michael Hanack. CONFORMATION THEORY, 1965
4. Donald J. Cram. FUNDAMENTALS OF CARBANION CHEMISTRY, 1965
5. Kenneth B. Wiberg (Editor). OXIDATION IN ORGANIC CHEMISTRY, PART A, 1965; Walter S. Trahanovsky (Editor). OXIDATION IN ORGANIC CHEMISTRY, PART B, 1973; PART C, 1978; PART D, 1982
6. R. F. Hudson. STRUCTURE AND MECHANISM IN ORGANO-PHOSPHORUS CHEMISTRY, 1965
7. A. William Johnson. YLID CHEMISTRY, 1966
8. Jan Hamer (Editor). 1,4-CYCLOADDITION REACTIONS, 1967
9. Henri Ulrich. CYCLOADDITION REACTIONS OF HETEROCUMULENES, 1967
10. M. P. Cava and M. J. Mitchell. CYCLOBUTADIENE AND RELATED COMPOUNDS, 1967
11. Reinhard W. Hoffmann. DEHYDROBENZENE AND CYCLOALKYNES, 1967
12. Stanley R. Sandler and Wolf Karo. ORGANIC FUNCTIONAL GROUP PREPARATIONS, VOLUME I, 1968; VOLUME II, 1971; VOLUME III, 1972
13. Robert J. Cotter and Markus Matzner. RING-FORMING POLYMERIZATIONS, PART A, 1969; PART B, 1; B, 2, 1972
14. R. H. DeWolfe, CARBOXYLIC ORTHO ACID DERIVATIVES, 1970
15. R. Foster. ORGANIC CHARGE-TRANSFER COMPLEXES, 1969
16. James P. Snyder (Editor). NONBENZENOID AROMATICS, VOLUME I, 1969; VOLUME II, 1971
17. C. H. Rochester. ACIDITY FUNCTIONS, 1970
18. Richard J. Sundberg. THE CHEMISTRY OF INDOLES, 1970
19. A. R. Katritzky and J. M. Lagowski. CHEMISTRY OF THE HETEROCYCLIC *N*-OXIDES, 1970
20. Ivar Ugi (Editor). ISONITRILE CHEMISTRY, 1971
21. G. Chiurdoglu (Editor). CONFORMATIONAL ANALYSIS, 1971
22. Gottfried Schill. CATENANES, ROTAXANES, AND KNOTS, 1971
23. M. Liler. REACTION MECHANISMS IN SULPHURIC ACID AND OTHER STRONG ACID SOLUTIONS, 1971
24. J. B. Stothers. CARBON-13 NMR SPECTROSCOPY, 1972

ORGANIC CHEMISTRY

25. Maurice Shamma. THE ISOQUINOLINE ALKALOIDS: CHEMISTRY AND PHARMACOLOGY, 1972

26. Samuel P. McManus (Editor). ORGANIC REACTIVE INTERMEDIATES, 1973

27. H. C. Van der Plas. RING TRANSFORMATIONS OF HETEROCYCLES, VOLUMES 1 AND 2, 1973

28. Paul N. Rylander. ORGANIC SYNTHESES WITH NOBLE CATALYSTS, 1973

29. Stanley R. Sandler and Wolf Karo. POLYMER SYNTHESES, VOLUME I, 1974; VOLUME II, 1977; VOLUME III, 1980

30. Robert T. Blickenstaff, Anil C. Ghosh, and Gordon C. Wolf. TOTAL SYNTHESIS OF STEROIDS, 1974

31. Barry M. Trost and Lawrence S. Melvin, Jr. SULFUR YLIDES: EMERGING SYNTHETIC INTERMEDIATES, 1975

32. Sidney D. Ross, Manuel Finkelstein, and Eric J. Rudd. ANODIC OXIDATION, 1975

33. Howard Alper (Editor). TRANSITION METAL ORGANOMETALLICS IN ORGANIC SYNTHESIS, VOLUME I, 1976; VOLUME II, 1978

34. R. A. Jones and G. P. Bean. THE CHEMISTRY OF PYRROLES, 1976

35. Alan P. Marchand and Roland E. Lehr (Editors). PERICYCLIC REACTIONS, VOLUME I, 1977; VOLUME II, 1977

36. Pierre Crabbé (Editor). PROSTAGLANDIN RESEARCH, 1977

37. Eric Block. REACTIONS OF ORGANOSULFUR COMPOUNDS, 1978

38. Arthur Greenberg and Joel F. Liebman, STRAINED ORGANIC MOLECULES, 1978

39. Philip S. Bailey. OZONATION IN ORGANIC CHEMISTRY, VOLUME I, 1978; VOLUME II, 1982

40. Harry H. Wasserman and Robert W. Murray (Editors). SINGLET OXYGEN, 1979

41. Roger F. C. Brown. PYROLYTIC METHODS IN ORGANIC CHEMISTRY: APPLICATIONS OF FLOW AND FLASH VACUUM PYROLYTIC TECHNIQUES, 1980

42. Paul de Mayo (Editor). REARRANGEMENTS IN GROUND AND EXCITED STATES, VOLUME I, 1980; VOLUME II, 1980; VOLUME III, 1980

43. Elliot N. Marvell. THERMAL ELECTROCYCLIC REACTIONS, 1980

44. Joseph J. Gajewski. HYDROCARBON THERMAL ISOMERIZATIONS, 1981

45. Philip M. Keehn and Stuart M. Rosenfeld (Editors). CYCLOPHANES VOLUME I, 1983; VOLUME II, 1983

In preparation

12. Stanley R. Sandler and Wolf Karo. ORGANIC FUNCTIONAL GROUP PREPARATIONS, 2ND EDITION, VOLUME I, 1983